P9-DCI-263

Practical Design Using Programmable Logic

DAVID PELLERIN

MICHAEL HOLLEY

Data I/O Corporation

PRENTICE HALL
Englewood Cliffs, NJ 07632

Library of Congress Cataloging-in-Publication Data

Pellerin, David.
Practical design using programmable logic / David Pellerin and Michael Holley.
p. cm.
Includes bibliographical references and index.
ISBN 0-13-723834-7
1. Programmable logic devices. 2. Digital electronics.
I. Holley, Michael. II. Title.
TK7872.L64P44 1991
621.39'5—dc20 90-47805
CIP

Editorial/production supervision and
interior design: Fred Dahl and Rose Kernan
Cover design: Wanda Lubelska
Pre-press Manufacturing buyer: Kelly Behr
P&B Buyer: Susan Brunke
Acquisitions Editor: Karen Gettman

Printed in the United States of America
10 9 8 7 6 5 4 3 2

ISBN 0-13-723834-7

Prentice-Hall International (UK) Limited, *London*
Prentice-Hall of Australia Pty. Limited, *Sydney*
Prentice-Hall Canada Inc., *Toronto*
Prentice-Hall Hispanoamericana, S.A., *Mexico*
Prentice-Hall of India Private Limited, *New Delhi*
Prentice-Hall of Japan, Inc., *Tokyo*
Simon & Schuster Asia Pte. Ltd., *Singapore*
Editora Prentice-Hall do Brasil, Ltda., *Rio de Janeiro*

Contents

Preface

A new era of digital circuit design is upon us! The development and recent explosion in popularity of user-programmable integrated circuits has been as dramatic a development as the introduction of the microprocessor two decades ago. Programmable logic devices, like microprocessors, provide new alternatives to circuit designers, but also require some changes to traditional design methods. The purpose of this book is to help users of programmable logic get the most benefit from this powerful new technology.

Practical Design Using Programmable Logic is intended for both the first-time user of programmable logic devices and for the experienced designer who already has some programmable logic experience. While previous knowledge about programmable logic devices (PLDs) isn't required, it is assumed that the reader has a basic understanding of electronics and digital logic, and has some experience developing digital circuits. In this book we present a wide variety of PLDs, examine the history of these devices, and see how they are used in modern digital systems. We explore in detail how CAE tools can be used to speed the PLD design process, and we provide concrete examples in the form of real circuits that have been developed using these devices and tools.

In Chapters 1 through 5, we provide an overview of the PLD industry and survey the programmable logic devices currently available. We also investigate some of the many design tools available for users of these devices. Design tools aid in the conversion of design concepts into programmed devices. The importance of PLD design tools is comparable to the importance of development tools for microprocessor applications.

In Chapter 1, we begin with a general description of PLDs, comparing them to other integrated circuit types. In Chapter 2, we trace the history of programmable devices beginning with the earliest devices produced by Harris in the late sixties. The number of different PLDs now available is overwhelming; technological and architectural differences account for many thousands of distinct devices available from over two dozen different manufacturers. Because there are so many different devices (and new devices are being introduced in a seemingly endless stream), it is not our intention to catalog them all here. Rather, we have concentrated on those features which distinguish different PLD types.

In Chapter 3, we describe a wide variety of commonly used PLDs and show how a simple circuit is implemented in one of these devices. In this chapter, we also describe many of the recent device enhancements that have given greater flexibility to the basic PLD. In Chapter 4, we examine a few of the new devices that have been collectively named *field programmable gate arrays*, or FPGAs. These types of devices offer many of the advantages of mask programmed gate arrays, but are user-programmable.

CAE tools such as assemblers, compilers, and debuggers are considered indispensable for microprocessor design, and these tools have their counterparts for PLD design. Like microprocessors, PLDs have, over time, increased in complexity and in their ability to solve more general design problems. In addition, computer-based logic synthesis tools have been developed to aid in the conversion of high-level design concepts into actual implementations. In Chapter 5, we provide an overview of the design tools available to PLD users. Design tools for PLDs have been an important factor in their acceptance by mainstream designers. Covered in this chapter are tools that assist in design entry, logic optimization and synthesis, simulation, and test vector generation.

The goal of Chapters 6 through 11 is to present techniques that can be used to produce PLD-based designs quickly, efficiently, and reliably. We start with the basics of logic design and quickly progress to advanced PLD-specific design techniques. Programmable logic devices have rapidly gained acceptance by the designers of digital circuits and sytems. Some designers have been reluctant to use these devices, however, because of the perception that PLDs are complex and difficult to use and because of the belief that an unjustifiable investment in specialized tools and equipment will be required.

Chapter 6 is intended for the first-time PLD user, and in it we suggest some low-cost ways in which to begin using PLDs. We also provide useful information in this chapter to guide the user in putting together a complete PLD-design system. Such a system will allow experimentation with the design techniques that are presented in later chapters. Chapter 7 provides a review of those digital design techniques that are useful to designers using PLDs, and demonstrates how some relatively simple techniques can dramatically improve the efficiency of PLD-based circuits.

During the first few years of their existence, PLDs were used primarily as a replacement for fixed-function TTL parts. One such use was to provide interface (commonly referred to as "glue") logic between major components of a microprocessor-based system. In this capacity, PLDs are realtively easy to use, requiring only rudimentary tools and little specialized knowledge on the part of the user.

Recently, however, PLDs are being used for more complex applications, requiring higher-level design techniques for implementation. Circuits are now being developed that rely almost exclusively on PLDs to perform complex tasks such as complete ALUs or large controllers. In designs such as these, often involving a large number of PLDs on a single board, it becomes increasingly difficult to produce efficient PLD implementations without the use of high-level design techniques and computer-aided design tools. These design methods are explored in detail in Chaper 8, where we present advanced techniques that vastly simplify the expression and verification of complex circuits. High-level design methods for circuits such as counters, decoders, and state machines are presented. We also explore various design options and compromises can affect the efficiency and feasibility of a PLD implementation.

In Chapter 9, we provide an overview of how design methods for FPGAs differ from those for PLDs. To do this, we examine how design tools are used to convert a simple design concept into a working circuit using one particular FPGA device—the *logic cell array* (LCA).

Because of the ease with which a design can be created and modified, products which utilize PLDs can be brought to market quickly. In addition, future design changes can often be made simply by redesigning the logic programmed into the PLDs, with no need to redesign the circuit board. The use of PLDs in production designs does, however, create a number of new problems that must be taken into account before a product can be reliably produced.

The reluctance of many engineers to use PLDs in new designs stems not only from the perceived difficulty of designing for these devices, but also from the real problem of testing them once they have been programmed. Designs involving one or more PLDs require sophisticated testing methods to ensure that each programmed device functions exactly as it should.

First-time users of PLDs are often unpleasantly surprised by poor reliability or testability when their designs get to production, so we devote Chapter 10 to these subjects. We examine the problems of circuit hazards and metastability, and present design-for-test and testability analysis techniques. We also describe some computer-assisted test tools for PLDs that can help with the device testing phase. In Chapter 11, the final chapter, we present a PLD-based design that provides a solid example of how PLDs and high-level design tools can be fully utilized.

Acknowledgments

This book is written from the perspective of "insiders" in the PLD industry. The authors have each spent the past decade at Data I/O Corporation working on PLD-related projects (including being members of the ABEL development team since that product's inception). We would therefore like to thank our co-workers, management and customers for the valuable support, experience and insights we have received over the years.

The information presented in this book would not have been complete without the help of others, both within and outside of Data I/O. The historical information presented in chapter 2 was compiled from discussions over the years with many individuals, including: David Greer, formerly with General Electric, Max

Pillie and George James of Data I/O, John Birkner and H. T. Chua, both formerly with MMI, Napoleone Cavlan, formerly with Signetics, Robin Jigour of International CMOS Technology, Dean Suhr and Dave Rutledge of Lattice Semiconductor, and Paul Franklin, formerly with MMI and now with Actel Corporation. We would also like to thank the reviewers, who included John Wakerley of Stanford University, Khaled A. El-Ayat of Actel, and Kim-Fu Lim, Bob Hamilton, Adam Zilinskas, Mike McClure and Michael Bradley of Data I/O.

Finally, we would like to extend special thanks to cartoonist Don Blackstone, whose drawings are worth at least eleven thousand words!

TRADEMARK ACKNOWLEDGMENTS

Foreword

A PERSONAL TECHNOLOGY

This is a personal book. It comes not only from long experience with programmable logic devices, but from deep feelings about their significance. Through what have come to be called PLDs, we are seeing the configuration of logic devices being taken out of the semiconductor factory and put in the hands of system designers, and there is a real need for all electronics engineers to understand this rapid change.

I refer to the technology of programmable logic as personal because it enables engineers to translate their ideas more exactly into hardware, in effect, putting much more of themselves into their creations. PLDs remove barriers and compromises that traditional standard logic devices have imposed on engineers for decades. And that trend will continue even more forcefully into the 1990s.

Engineers like PLDs and, in my opinion, have responded more strongly to them than to the advent of any other circuit technology, including integrated circuits themselves. While ICs were at first perceived as expensive and even a threat to the jobs of circuit designers, PLDs were quickly recognized as the engineer's friend, helping cut through difficult design and production problems. They simplify, save time and allow exploration of more ideas.

Following in that spirit, Dave Pellerin and Mike Holley write in the first person, in a clear declarative style and keep away from unessential elaborations. This makes more effective their endeavor to open PLD technology to a broader community—of students, practicing engineers and both engineering and marketing managers.

This treatise comes at a time of transition in the ways digital systems are designed. PLDs are now well established, and necessary. But their complexities are suddenly growing rapidly and so are the design methods to make this complexity useable.

PLDs were first a way to integrate glue logic into fewer packages. But they are now a means of getting performance and function at the right cost that standard devices cannot match. And they are becoming a methodology, across the engineering spectrum, from small designs with short production runs, to high-volume cost-sensitive systems. PLD devices and design methodologies are showing increasing abilities to support higher and higher levels of systems performance as well as complexity.

In the persistent effort by the semiconductor and CAD industries to provide system designers with the capabilities to form their systems to their exact wishes, rapidly and economically, PLDs and PLD design tools have become central elements.

This is an industry-wide trend requiring that the skills and knowledge of engineers improve more and more rapidly. It's making development times shorter and engineering more efficient. It's a syndrome tightening the competitive pressures on every engineer. And PLDs are right in the middle of it.

Laying the basis for the newcomer and perspective for the well initiated, this book chronicles a decade of development that established the importance of PLDs. And it's important to understand the technical and philosophical paths that brought us to a time of transition. They point to the future, which is the fundamental concern of engineers.

As this work explores and explains PLDs and their design practices, it hopefully provides what's more important, the ability to address the much greater complexities at our doorstep.

Stan Baker
Senior Editor
Electronics Engineering Times

1

Introduction

Since the birth of the integrated circuit (IC), the majority of small- to medium-size logic circuits have been constructed by engineers who have selected and interconnected off-the-shelf, fixed-function logic ICs. Larger circuits, or circuits intended for large quantity production, have often been implemented with custom ICs.

ICs are powerful building blocks for digital circuits, but their existence and increasing complexity have led to the emergence of a dichotomy of circuit designers. One group of engineers we might call the Makers, while the other group of engineers are the Users.

The Makers are represented by those who have been trained in the art and science of IC design. These engineers and scientists are quickly hired by high technology companies and, once working in the industry, might never again pick up a soldering iron or logic probe. Instead, the job of the IC designer usually entails scribbling with colored pens on enormous sheets of paper or squinting for hours at enormous and colorful CRT displays. The fruits of their labors are seen in the huge number of different ICs available today, from simple fixed-function logic devices to complex microprocessors and other special purpose *very large scale integration* (VLSI) chips.

The Makers are able to use their considerable talents in logic design, physics, and chemistry to come up with ICs that can be used for a wide variety of purposes. Some Makers work for IC manufacturers while others work for companies that develop custom ICs for their own uses.

The Users, on the other hand, are that much larger group of circuit designers who create actual end-products and systems by putting to practical use the ICs developed by the Makers. The task of the User is determining how to make the available chips work in real circuits. This often results in bottom-up design methods since off-the-shelf ICs all have their particular functional peculiarities and circuit interface requirements. The User is often forced to make major design compromises in order to work within the constraints of a particular family of fixed-function ICs.

In the past several years, programmable logic devices have emerged as the technology that has finally given Users what they need—the power to design their own custom ICs. It doesn't take an experienced IC designer to use programmable logic. With this technology, Users can become Makers.

1.1 WHAT IS A PLD?

We'll start by defining what we mean by the term *programmable logic device*. This will help us to better understand the design process for these devices as well as allowing us to quickly recognize whether new devices will require significantly different design techniques.

In recent years, many new programmable devices have appeared that are radical departures from traditional PLDs. This has caused many device manufacturers and industry watchers to create new device categories and coin new acronyms (see appendix A for a tongue twisting sampling of these), greatly confusing the potential new user of PLDs. While some of these new devices do require new design techniques, the majority of them are simply extensions to the original PLD concept, and require similar methods.

Classifying Integrated Circuits

Before diving into the internal construction of typical PLDs, it's important to understand where these devices fit in the larger world of integrated circuits and digital circuit design in general. Integrated circuits fall into two general categories; *fixed-function* (sometimes called *off-the-shelf*) and *application specific*.

Fixed-function ICs are designed by their manufacturers to meet the needs of a wide variety of applications. A fixed-function IC may be as simple as a comparator, or as complex as a microprocessor. Fixed-function ICs are typified by the 7400-series of logic devices. These devices are cheap and readily available and have, therefore, earned the nickname "jelly beans."

To design a circuit using fixed-function ICs, you select those devices that seem most appropriate for your circuit, usually working from a block diagram design concept. As the actual devices are selected, the design may have to be modified to meet the special requirements of those devices.

The advantages of this method of logic design are many, and include low development costs and fast turnaround of designs. In addition, designs implemented in standard logic devices are relatively easy to test, and require few specialized skills on the part of the designer.

Alternatively, designs that are too complex for fixed-function devices have often been implemented using *application-specific ICs* (ASICs). ASICs are those devices that are designed by the end-user to meet the specific requirements of a circuit. ASICs are usually produced by an IC manufacturer as specified by the end-user.

ASICs have many convincing advantages over fixed-function devices including reduced space and power requirements. In addition, designs implemented using ASICs are frequently lower in cost than the equivalent design composed of fixed-function devices, if the design is produced in large volumes.

ASICs are designed by the end-user and are then fabricated for that user by an ASIC manufacturer (often called a *foundry*). Because an ASIC is designed for a specific application, large reductions in circuit size are possible through the use of high levels of integration. The use of ASICs can also create a formidable barrier to would-be competitors since designs implemented in them are difficult or impossible to copy.

The disadvantages of ASICs are many, however. The initial development costs can be enormous and when the ASIC is finally ready for use in the circuit, developing adequate testing methods for the new device can become a nightmare.

Programmable logic devices, or PLDs, first appeared in the mid-1970s (see Chapter 2). PLDs are small scale ASICs that can be configured by the end-user to implement a specific logic function. They offer many of the advantages of fixed-function devices (short design cycles, low development cost, less reliance on specialized skills) and of ASICs (higher densities, lower quantity production costs, design security).

In terms of density, a typical PLD can be used to replace anywhere from a few to a few dozen fixed-function devices depending on the application. This is compared to the hundreds (or even thousands) of fixed-function devices that can be replaced with a single large scale ASIC.

PLDs require a much lower up-front cost than foundry-produced ASICs while still providing substantial advantages over fixed-function devices in terms of circuit densities and design security. And when design changes are required, PLDs have significant advantages over any other implementation method.

PLDs can be used to replace fixed-function devices if a product is being redesigned to decrease board sizes or reduce power requirements. PLDs are also often used when prototyping large ASIC designs since many of the design techniques used are common to all types of ASICs, large and small.

In the past decade, PLDs have grown in popularity to the point where it's becoming difficult to find a digital system that doesn't contain at least one PLD. More and more frequently, designs are implemented all the way from concept to final product through the use of PLDs.

The first Apple Macintosh, for example, contained six PLDs on one PC board. It's now common to find boards containing dozens of PLDs. Using PLDs instead of gate arrays or other foundry produced ASICs allows designers more flexibility to experiment with designs and reduces the possibility of huge cost overruns when designs require major changes. PLDs can be reprogrammed in seconds to correct design deficiencies, while foundry-produced ASICs may require months of costly redevelopment.

What constitutes a PLD? Generally speaking, a PLD is an integrated circuit that is user configurable and capable of implementing digital logic functions. A typical PLD is composed of a programmable array of logic gates and interconnections with array inputs and outputs connected to the device pins through fixed logic elements such as inverters and flip-flops. A typical PLD is diagrammed in Figure 1.1. Most PLDs are composed of a single *programmable logic array* that

Figure 1.1 Block diagram of a PLD

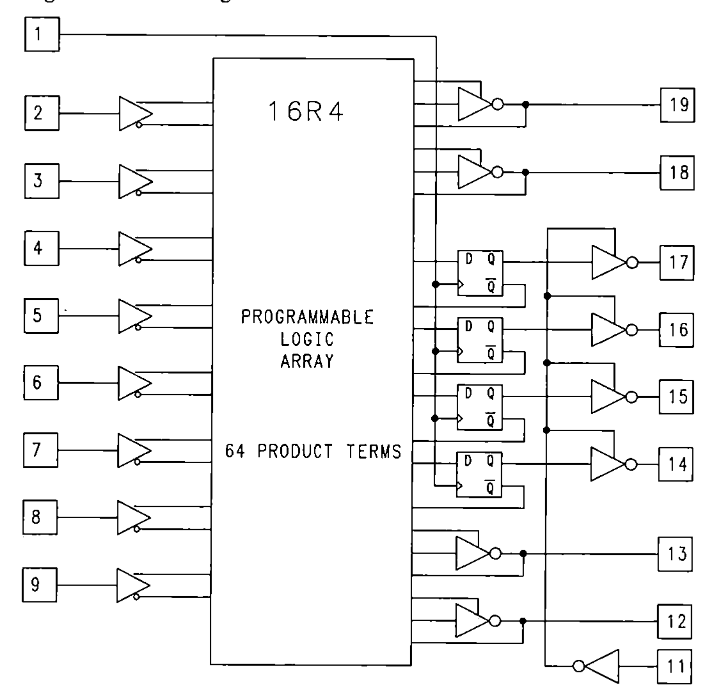

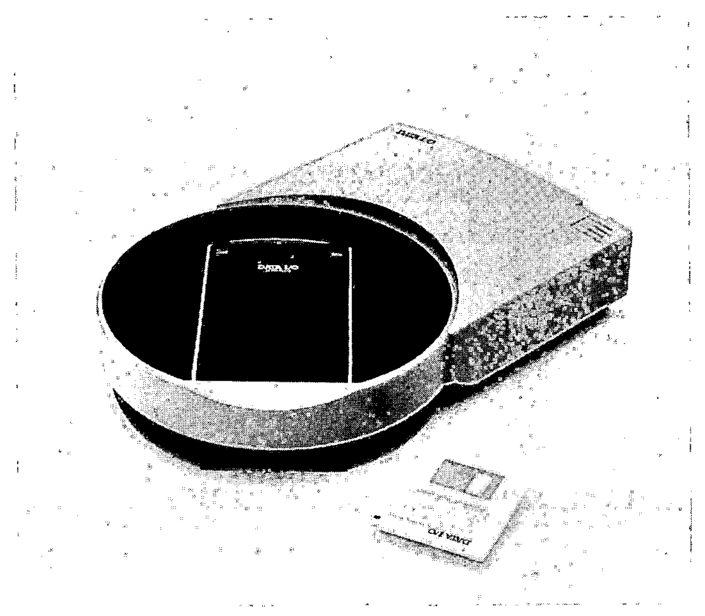

Figure 1.2 Modern PLD programmer (Courtesy of Data I/O Corporation)

is surrounded by input and output circuitry of varying complexity. This device, a PAL 16R4, features four memory elements in the form of D-type flip-flops.

The devices are programmed through the use of a *device programmer*. Device programmers are available from many suppliers and can cost anywhere from a few hundred to many thousands of dollars depending on the features required and number of different devices to be programmed (see Figure 1.2).

1.2 SUMMARY

In this chapter we have introduced the concept of the programmable logic device and have shown a diagram of a simple device. In the next chapter we'll trace the history of these devices. In Chapter 3 we'll examine in detail how these devices are actually constructed and used.

1.3 REFERENCES

Data I/O Corporation, *Programmable Logic—A Basic Guide for the Designer,* 2nd edition, Data I/O Corp., Redmond, WA, 1986.

Robinson, Phillip. "Overview of Programmable Hardware." *Byte* (January 1987): 197–203.

2

History

Most industry watchers point to the introduction, in 1975, of the *field programmable logic array* (FPLA) as the birth of the PLD, but the history of programmable logic actually goes back a few years farther. The PLD we know today is a result of research carried out in the 1960s and early 1970s by Harris Semiconductor, IBM, and General Electric.

2.1 THE HISTORY OF THE PLD

The first commercial programmable logic device, the *fuse configurable diode matrix* (Figure 2.1), was developed in the mid-1960s by Harris Semiconductor (which was then known as Radiation, Inc.).

MONOLITHIC DIODE MATRICES

Features

- FIELD PROGRAMMABLE
- CMOS COMPATIBLE
- ZERO POWER DISSIPATION
- FAST SWITCHING
- FIVE POPULAR ORGANIZATIONS

Description

Designed with the CMOS circuit engineer in mind, these versatile diode matrices allow the application of logically powerful programmable solutions to low power CMOS system applications.

These devices incorporate an advanced dielectric isolation process to eliminate the need for power supply pins and allow parasitic free operation.

Programming is accomplished by cleanly vaporizing a fusible link by application of a brief high voltage pulse to a selected array element. This operation open circuits a row to column orring diode eliminating their former interaction.

Monolithic Structure

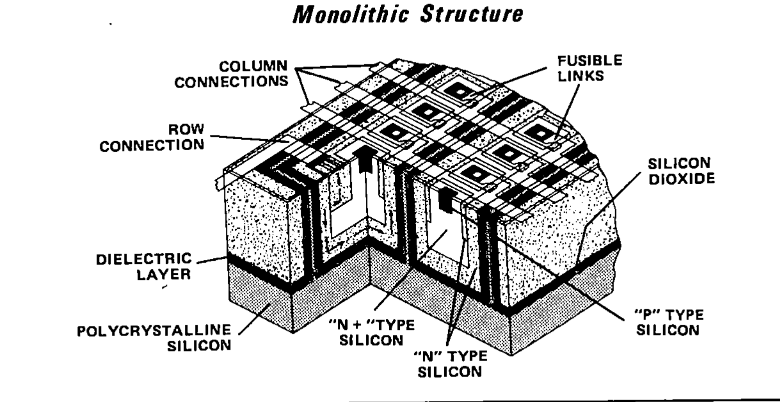

Fusible Link System

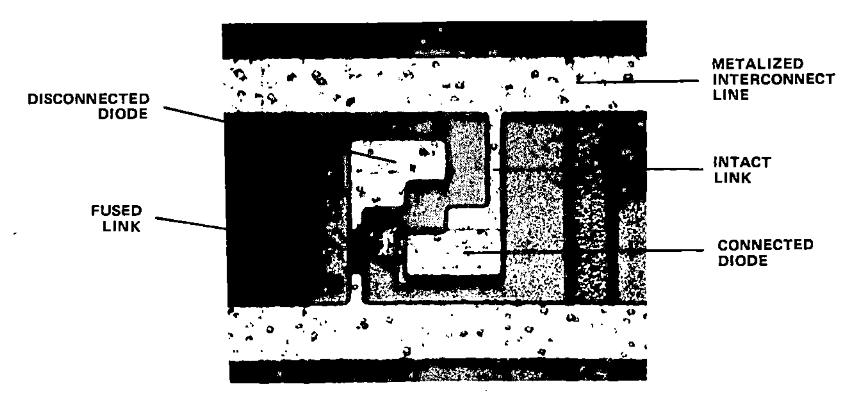

Figure 2.1 Harris diode matrix (Courtesy of Harris Corporation)

Diode matrices were used in many simple logic applications (one example of their use was in television channel selectors). Simple logic functions can be programmed into this matrix by disconnecting specific signal paths. This is done by disabling (through the process of melting fuses) the diodes associated with undesired paths.

Compared to today's PLDs, the diode matrix was quite primitive. The matrix was small and programming it was tricky since a high current was required to blow the fuses. There were no commercially available device programmers, so users of these devices either had to have Harris do the programming for them or construct a device programmer of their own from instructions provided in the Harris data book. The programming circuit allowed programming of only one fuse at a time and required nearly a full amp of current to blow each fuse.

The Programmable Logic Array (PLA)

The PLDs in use today are based not on diode matrices but on arrays of logic gates. The most common of these array structures are based on the *programmable logic array*, or PLA. The first on-chip PLAs were developed at IBM and were described by IBM researchers in 1969 as *read-only associative memory*, or ROAM. The term *associative* referred to the fact that the array outputs responded with a predictable output value whenever a pattern was applied to the inputs that corresponded to a pattern of connections programmed into the array.

The PLA approach was found to be a highly effective technique for integrated circuit development, and large sections of many large scale custom ICs and microprocessors have since been based on the PLA concept. The structure of a PLA allows arbitrary logic functions to be easily implemented by specifying interconnections in a matrix of logic gates and intersecting signals.

The Mask-programmable PLA

In the early 1970s, Texas Instruments developed a mask-programmable IC based entirely on IBM's ROAM structure. This device, the TMS 2000, was programmed by altering the mask prior to production as specified by the end-user. The TMS 2000 could handle up to seventeen inputs and eighteen outputs and featured eight JK-type flip-flops as memory elements. TI actually coined the now-familiar term *programmable logic array*.

The TMS 2000 wasn't well accepted in the market due to its complexity and the relatively high nonrecurring costs of using it. A similar fate was in store for another masked-programmed PLA device (the CRC 3506/7) offered by Collins Radio in 1971.

Then, in 1973, National Semiconductor came out with their own mask-programmable PLA device. National's device was similar in many respects to the TI device, but featured only fourteen inputs and eight outputs with no on-chip memory elements. The data sheet for this device, the DM7575/DM8575, appears in Figure 2.2. The National device was less complex and somewhat more popular than the earlier devices, but circuit designers were still reluctant to use the new technology.

The DM7575/DM8575 device is important because its design was the basis for later PLDs produced by Intersil and Signetics. Before discussing the first modern PLDs, however, let's look at how modern user-programmable devices evolved. The first such device was the programmable read only memory (PROM).

The Birth of the PROM

The first PROM devices were developed at Harris in 1970. The new PROMs combined Harris' nichrome fusible link technology with the simplified array structure that had become popular in the form of mask-programmed ROMs (read-only memory). These devices were useful for a variety of program and data storage functions and were quickly accepted by circuit designers. Within a few years, PROMs had overtaken masked ROMs in popularity, and many IC manufacturers were producing the new devices in a variety of shapes and sizes.

Like the earlier diode matrices, these first PROMs featured a fuse matrix. A *fuse matrix* consists of actual metal bridges that connect intersecting signals in the matrix. Programming this matrix involves raising certain I/O pins to specific (higher than normal) voltages while applying address and data information to other I/O pins. When a specified sequence of inputs and voltages is applied to the device pins, a high current results across the target fuse and it melts.

These specific patterns of voltage waveforms, referred to as programming algorithms, are unique to each device. Since each programmable device has unique programming requirements, users of early PROMs had to construct and maintain a different programming circuit for each device they wanted to use.

The data books for these first PROMs included descriptions of how to build programming circuitry, since there were no commercial device programmers available. Use of these customer-built programmers often resulted in poor programming yields and unreliable programmed devices.

By the early 1970s, there were a growing number of PROM device manufacturers including Intel, Intersil, Harris, MMI, and TI. Each of these manufacturers had a large list of devices to offer. Intel introduced an *ultra-violet* (UV) *erasable PROM* in 1971, further accelerating the growth of the industry.

Commercial Programmers Appear

The rising popularity of PROMs was the reason companies like Prolog and Data I/O were formed. These companies specialized in programming hardware that met the exact specifications as provided by PROM manufacturers for their devices.

With the availability of reliable programming hardware, PROMs gained quick acceptance and an increasing number of manufacturers threw their hats into the ring. Soon there were a bewildering number of different devices available.

DM7575/DM8575, DM7576/DM8576 programmable logic array (PLA)

general description

The DM7575/DM8575 and DM7576/DM8576 are mask-programmable logic arrays designed for use in applications where random logic is required. The devices have fourteen data inputs and eight outputs. Each output provides a sum of product terms where each product term can contain any combination of 14 variables or their complements. The total number of product terms which can be provided is 96. Any product term which is repeated is counted only once. Since some functions are more easily represented in their inverted form, an option is provided to allow for either the true or complement of the function on each output. The products are particularly useful in providing control logic for digital systems. The DM7575/DM8575 has a conventional totem-pole output whereas the DM7576/DM8576 is provided with a passive pullup output. This latter configuration is useful in expanding functions by connection of outputs of different packages.

features

- A 2^{14}-by-8 (128k) bit memory would be needed to provide equivalent function
- Typical delay 90 ns
- Typical power dissipation 550 mW
- Series 54/74 compatible

logic and connection diagrams

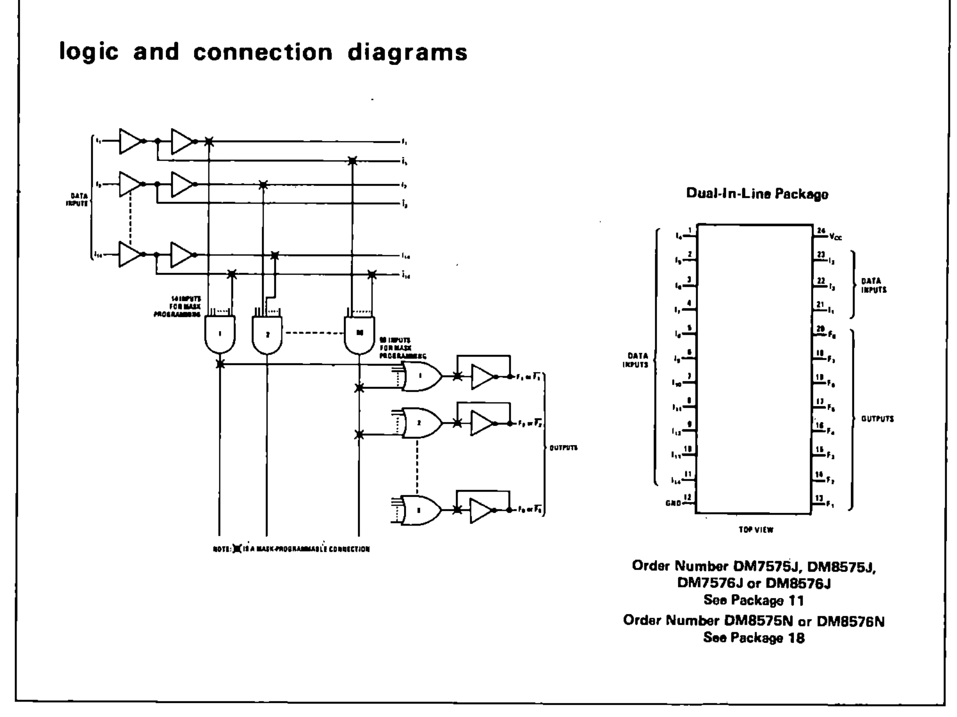

Order Number DM7575J, DM8575J, DM7576J or DM8576J
See Package 11
Order Number DM8575N or DM8576N
See Package 18

Figure 2.2 The National DM7575/DM8575 (Courtesy of National Semiconductor Corporation)

Because of the large number of new devices and manufacturers, it was clear that some package and pinout standardization was needed, so a committee was formed as a part of JEDEC (*Joint Electron Device Engineering Council,* a part of the Electronics Industry Association) to create standards. The JC-42 committee helped to create a more sane environment for users of PROMs.

While the PROM was capable of being used for simple logic functions in addition to its primary use as a data storage device, the IC manufacturers recognized that there was a need for a device more appropriate for logic applications. PROM users had, from the start, used these devices to implement simple logic functions such as address decoding and state machine applications. It was apparent, however, that PROMs were not the ideal devices when applied to logic functions with more than four or five inputs.

The First User-Configurable PLAs

Intersil Corporation and Signetics Corporation, both manufacturers of PROMs, realized that the structure of National's DM7575/DM8575 mask-programmed PLA device was ideal for a field programmable device. It became a race between Intersil and Signetics to bring the first such devices to market. Both of these companies called their proposed new devices FPLAs for *field programmable logic array*.

In the June 2, 1975 issue of *EE Times,* Intersil announced the IM5200 Field Programmable Logic Array (FPLA). Soon after, Signetics introduced its 82S100. These devices were ahead of their time in many ways. The data sheet for Intersil's IM5200 FPLA is shown in Figure 2.3.

These first FPLAs were powerful devices (the 82S100 is still available, under the new name of PLS100). They provided a reasonable number of inputs and outputs, and were flexible enough in their design to be used for a wide variety of logic applications.

While on the surface the IM5200 and 82S100 looked very similar (the 82S100 featured two more inputs than the IM5200), they were in fact quite different in terms of their programming technology. Rather than use the well understood fusible link technology for their new device, Intersil chose to use a new type of programmable element that had been developed for their PROM devices. This programming element was designed to improve the programming yields over earlier programmable devices. The technology, called *avalanche induced migration,* or AIM, utilizes an open base NPN transistor as the programming element. To program an AIM element, a high current is forced through the transistor from the emitter to the collector. This causes a short from the emitter to the base, which leaves the transistor to operate as a diode.

Unfortunately, the actual devices that rolled out of the Intersil's foundry turned out to be unreliable, with low programming yields, and the Intersil devices were not successful in the market.

The Signetics 82S100, which utilized the fusible link programming technol-

IM5200 Field Programmable Logic Array (FPLA)

FEATURES

- Avalanche Induced Migration (AIM) Programmability
- 48 Product Terms, 14 Inputs, 8 Outputs
- Output Active Level – High or Low
- Product Term Expandability
- Edit Flexibility
- DTL/TTL Compatible Inputs and Outputs
- tpd – typically 65 ns
- 5 Volt ± 5% Power Supply
- Passive Pullup Outputs

APPLICATIONS

- Random Combinatorial Logic
- Code Conversion
- Microprogramming
- Look-up Tables
- Control of Sequential Circuits, Counters, Registers, RAMs, etc.
- Character Generators
- Decoders or Encoders

GENERAL DESCRIPTION

The IM5200, field programmable logic array (FPLA), is useful in a wide variety of logic applications. The device has 14 inputs and 8 outputs. The FPLA may have up to 48 product terms. Each product term may have up to 14 variables and each one of the outputs provides a sum of the product terms. The FPLA is functionally equivalent to a collection of AND gates which may be OR'ed at any of its outputs. Since some functions are more easily represented in their inverted form, the output level is also programmable to either a high or low active level. The IM5200 is provided with passive pullup outputs. This output configuration is useful for product term expansion by wire-ANDing the outputs of different IM5200's.

LOGIC DIAGRAM

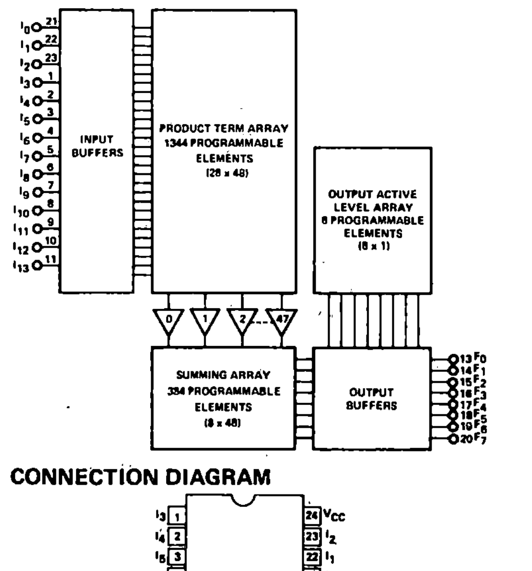

PACKAGE DIMENSIONS

CONNECTION DIAGRAM

Signal	Pin	Pin	Signal
I_3	1	24	V_{CC}
I_4	2	23	I_2
I_5	3	22	I_1
I_6	4	21	I_0
I_7	5	20	F_7
I_8	6	19	F_6
I_9	7	18	F_5
I_{10}	8	17	F_4
I_{11}	9	16	F_3
I_{12}	10	15	F_2
I_{13}	11	14	F_1
GND	12	13	F_0

TOP VIEW

ORDERING INFORMATION

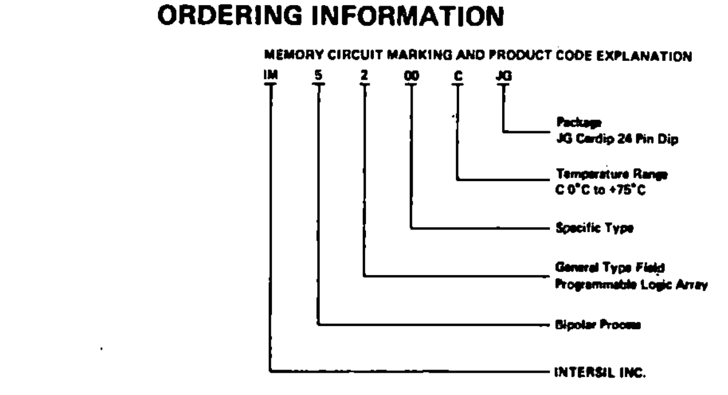

Figure 2.3 Intersil IM5200 FPLA (Courtesy of Harris Corporation)

ogy, was more reliable and more successful in the market. Another factor in the success of the Signetics devices was the efforts of Napoleone Cavlan who was, at that time, Signetics' Manager of Advanced Products. The new devices were completely unfamiliar to most circuit designers and required a much higher level of user education and promotion than the earlier programmable devices.

Even with a high level of promotion, documentation, and applications support, most digital designers chose not to use these first PLDs because the devices were still perceived to be too difficult to use and because of the risk of relying on the new technology for critical circuit elements. Even with the reasonable reliability of the Signetics devices, the FPLA had a relatively slow maximum operating speed (due to the two programmable arrays), was expensive, and had a poor reputation for testability.

Another factor limiting the acceptance of the FPLA was the large package. The Signetics part, for example, was contained in a 28-pin DIP (*dual inline package*) that was over a half an inch wide. Today, 28 pins is standard for PLCC (*plastic leaded chip carrier*) type packages but, in the 1970s, smaller packages were the norm.

Physical and operating limitations aside, other difficulties were experienced by designers who used the devices. The challenge in using the FPLA devices stemmed from a conceptual difference between how the designers had created circuits in the past and how they were forced to create them using FPLAs.

Designs that were to be implemented in FPLAs had to be described in an unfamiliar format that bore little resemblance to the familiar schematics or Boolean equations that logic designers had used in the past. To design a PLA circuit and program a device, the user would first convert the design into a tabular form called H&L. A sample of the H&L format is shown in Figure 2.4.

The data from the form was then entered into a device programmer, such the Data I/O Model 10 shown in Figure 2.5. This programmer was designed specifically for the Intersil and Signetics devices and had a CRT and built-in H&L editor. A major advance that this programmer had over the previous PROM programmers was its ability to exercise the programmed part functionally and compare its operation to the results calculated in the programmer. This meant that faulty or misprogrammed parts could be quickly identified and rejected.

GE's Associative Logic

Although Intersil and Signetics were the first companies to successfully market field programmable logic devices, they were by no means the first to develop them. As early as 1971, General Electric Company was developing a programmable logic device based on the new PROM technology. The device was developed by David Greer, then with GE in Syracuse, NY. GE's experimental device improved on IBM's ROAM structure by providing an internal path for OR-plane signals to directly reenter the AND plane. This allowed the use of multilevel logic with no waste of I/O pins.

```
STX                                   *A  LLLLLLLL

*P  ØØ  *I  LL---L-------H   *F  -----AA-
*P  Ø1  *I  LLLLLH--------   *F  A----AAA
*P  Ø2  *I  HLLLLH--------   *F  AA---AA-
*P  Ø3  *I  HHLLLH--------   *F  -A---A--
*P  Ø4  *I  LHLLLH------H-   *F  --A--AA-
*P  Ø5  *I  LHLLLH-----HL-   *F  ---A-AA-
*P  Ø6  *I  LHLLLH----HLL-   *F  --AA-AA-
*P  Ø7  *I  LHLLLH---HLLL-   *F  ----AAA-
*P  Ø8  *I  LHLLLH--HLLLL-   *F  --A-AAA-
*P  Ø9  *I  LHLLLH-HLLLLL-   *F  ---AAAA-
*P  1Ø  *I  LHLLLHHLLLLLL-   *F  --AAAAA-
*P  11  *I  LHLLLHLLLLLLL-   *F  --------
*P  12  *I  LL---L------HL   *F  --A--AA-
*P  13  *I  LLHLLH--------   *F  A-A--AAA
*P  14  *I  HLHLLH--------   *F  AAA--AA-
*P  15  *I  HHHLLH--------   *F  -AA--A--
*P  16  *I  LHHLLH-----H--   *F  ---A-AA-
*P  17  *I  LHHLLH----HL--   *F  --AA-AA-
*P  18  *I  LHHLLH---HLL--   *F  ----AAA-
*P  19  *I  LHHLLH--HLLL--   *F  --A-AAA-
*P  2Ø  *I  LHHLLH-HLLLL--   *F  ---AAAA-
*P  21  *I  LHHLLHHLLLLL--   *F  --AAAAA-
*P  22  *I  LHHLLHLLLLLL-H   *F  -----AA-
*P  23  *I  LHHLLHLLLLLL-L   *F  --------
*P  24  *I  LL---L-----HLL   *F  ---A-AA-
*P  25  *I  LLLHLH--------   *F  A--A-AAA
*P  26  *I  HLLHLH--------   *F  AA-A-AA-
*P  27  *I  HHLHLH--------   *F  -A-A-A--
*P  28  *I  LHLHLH----H---   *F  --AA-AA-
*P  29  *I  LHLHLH---HL---   *F  ----AAA-
*P  3Ø  *I  LHLHLH--HLL---   *F  --A-AAA-
*P  31  *I  LHLHLH-HLLL---   *F  ---AAAA-
*P  32  *I  LHLHLHHLLLL---   *F  --AAAAA-
*P  33  *I  LHLHLHLLLLL--H   *F  -----AA-
*P  34  *I  LHLHLHLLLLL-HL   *F  --A--AA-
*P  35  *I  LHLHLHLLLLL-LL   *F  --------
*P  36  *I  LL---L----HLLL   *F  --AA-AA-
*P  37  *I  LLHHLH--------   *F  A-AA-AAA
*P  38  *I  HLHHLH--------   *F  AAAA-AA-
*P  39  *I  HHHHLH--------   *F  -AAA-A--
*P  4Ø  *I  LHHHLH---H----   *F  ----AAA-
*P  41  *I  LHHHLH--HL----   *F  --A-AAA-
*P  42  *I  LHHHLH-HLL----   *F  ---AAAA-
*P  43  *I  LHHHLHHLLL----   *F  --AAAAA-
*P  44  *I  LHHHLHLLLL---H   *F  -----AA-
*P  45  *I  LHHHLHLLLL--HL   *F  --A--AA-
*P  46  *I  LHHHLHLLLL--LL   *F  ---A-AA-
*P  47  *I  LHHHLHLLLL-LLL   *F  --------

                                              ETX
```

Figure 2.4 H&L data entry format

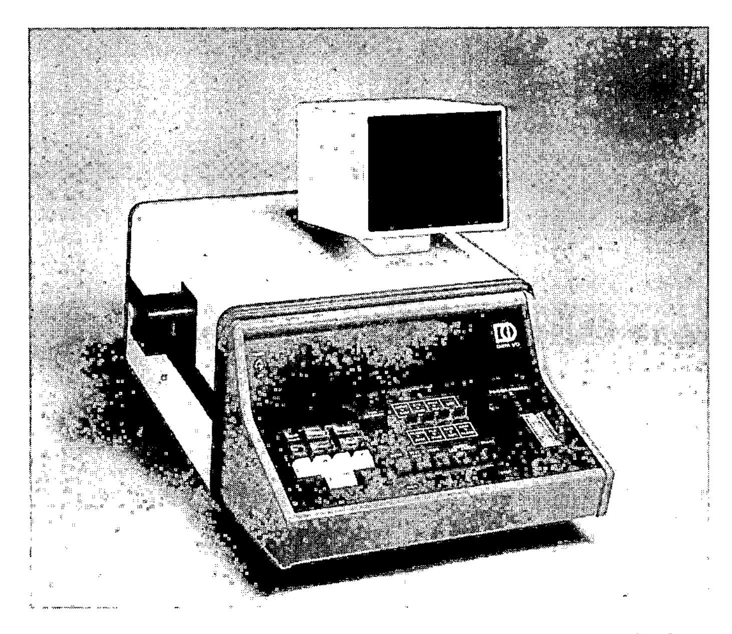

Figure 2.5 Data I/O Model 10 programmer (Courtesy of Data I/O Corporation)

Late in 1971, an experimental MOS programmable logic device was completed at the General Electric Research and Development Center in Schenectady, NY by Gerry Michon and Hugh Burke. Not only did this device feature the improved PLA-type logic array, but also used the floating gate UV-erasable technology announced earlier that year by Intel. The GE device (shown in Figures 2.6 and 2.7) was actually the first erasable PLD ever developed, predating commercially available EPLDs by over a decade. Researchers at General Electric not only developed the first working FPLAs and EPLDs, but also described and patented a folded array structure remarkably similar to the folded arrays that began appearing in complex PLDs nearly fifteen years later.

In 1974, under the terms of a patent and trade secrets agreement with GE, Monolithic Memories began the development of a mask-programmable logic device incorporating the GE innovations. The device was named the *programmable associative logic array*, or *PALA*. The device (MMI part number 5760/6760) was completed in 1976, and could implement multilevel or sequential circuits of well over 100 equivalent gates. In addition to its advanced structure, the device was supported by a highly automated design environment developed by GE. Designs were entered using Boolean equations and converted automatically to a mask pattern. The system even included a facility for test vector generation and simulation. While the MMI/GE device was never marketed, it did serve as a model for PLDs produced by a number of manufacturers in later years.

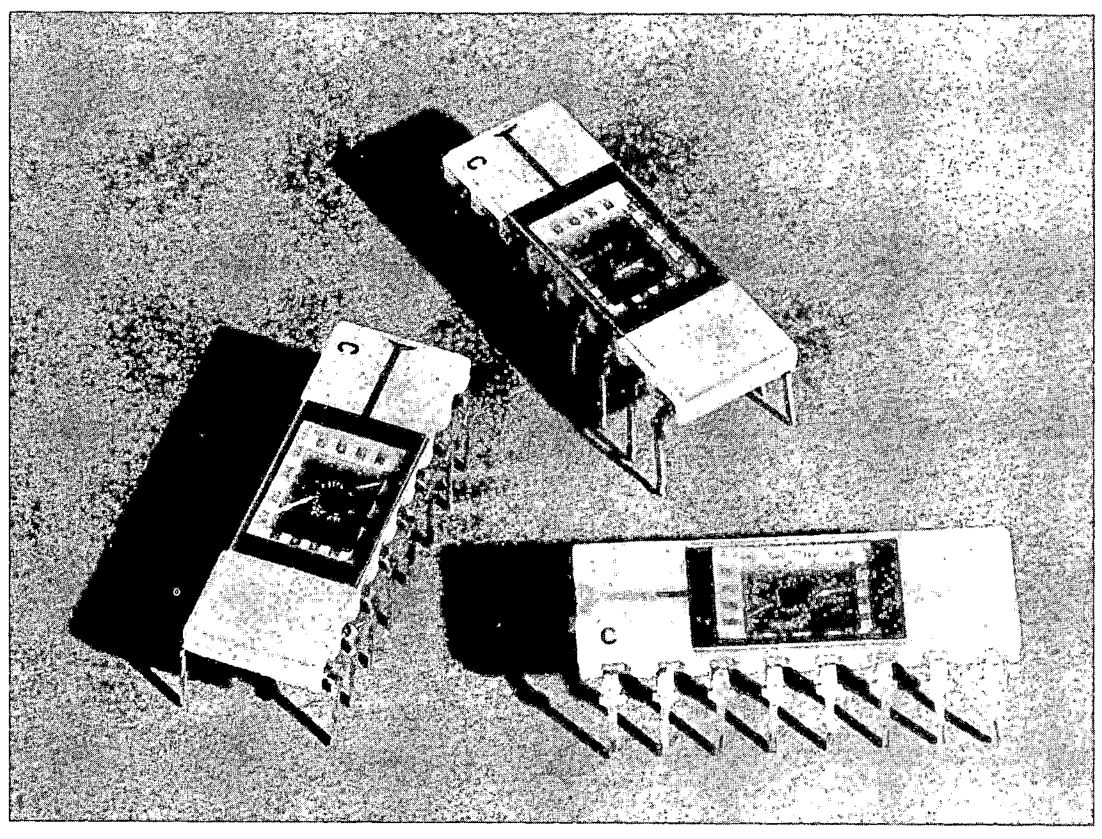

Figure 2.6 Prototype UV-erasable PLDs produced by General Electric in 1971 (Courtesy of General Electric Company)

MMI's PAL

Although the Signetics devices enjoyed some success, PLDs didn't really gain widespread acceptance until the late 1970s, when MMI introduced the PAL device. After working with GE on the PALA device, MMI's first effort in the design of their devices was to reproduce the Intersil and Signetics FPLAs. A few hundred copies of the Signetics 82S100 were produced for internal evaluation, but these devices were never released. From this experience, and MMI's earlier experience with GE, the PAL device was born. The new family of devices was announced in the summer of 1978.

The project to create the PAL device was managed by John Birkner and the actual PAL circuit was designed by H. T. Chua. Birkner had come from Computer Automation, Inc., where he had developed a 16-bit processor using 80 standard logic devices. His experience with standard logic led him to believe that user programmable devices would be more attractive to users if the devices were de-

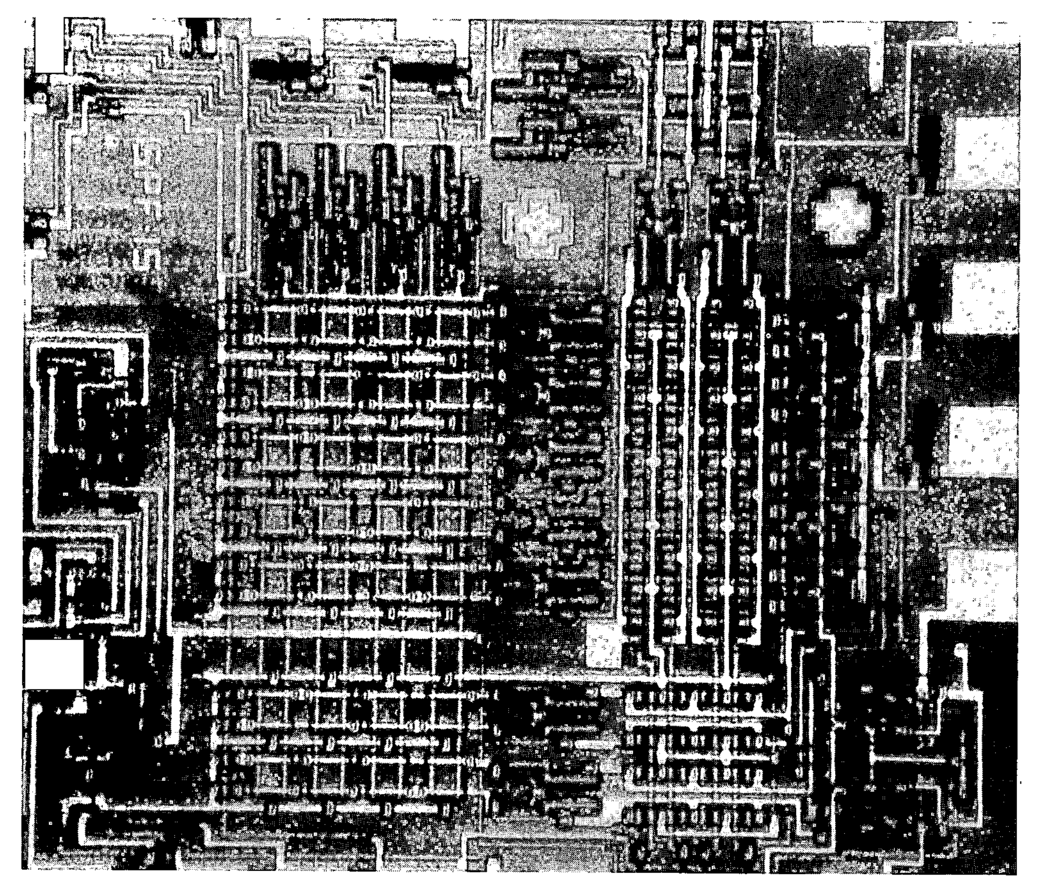

Figure 2.7 Photomicrograph of GE's early EPLD (Courtesy of General Electric Company)

signed to replace standard logic. This meant that the package sizes had to be more typical of the existing devices, and the speeds had to be improved. The new devices that resulted from this thinking were a breakthrough and a huge success in the market.

The PAL devices utilized the now mature and reasonably reliable PROM fuse technology and featured only one programmable array. This combination resulted in a device with much faster operation than the earlier FPLAs. Programming of the devices was simple since they were implemented in industry standard packages and used well understood PROM fuse technology.

One factor in the success of the PAL devices was the high level of customer support offered by MMI in the form of applications and user documentation that served to demystify the design process. The PAL Handbook, written by Birkner himself, provided a conceptual bridge between the discrete logic methods of the past and the high-level design methods of the future.

Another factor in the success of the PAL devices was PALASM, which stands for PAL Assembler. PALASM was a computer program written by John Birkner that converted design descriptions composed of Boolean equations directly into programming data for a specified PAL device.

Boolean equations have been used since the early days of logic design (long before the integrated circuit was conceived of) to express logic functions. Boolean algebra is at the core of any university course in digital logic, so it's a form of representation that is more familiar to engineers than tabular forms.

PALASM was a simple computer program written in FORTRAN. The entire program required only six pages of FORTRAN source code and was published in the PAL data book. A typical PALASM design file is shown in Figure 2.8.

The simplicity of the PALASM language made it possible for programmer manufacturers to implement the language directly in the programming hardware, as was done in the Data I/O LogicPak, the Structured Design programmer, and other similar PAL programmers.

Figure 2.8 PALASM design file

```
PAL16R4 PAL             PAL DESIGN SPECIFICATION
CNT4SC
4 bit counter with synchronous clear
Michael Holley and Dave Pellerin
Clk  Clear  NC  NC  NC  NC  NC  NC  NC  GND
 OE  NC     NC /Q3 /Q2 /Q1 /Q0  NC  NC  VCC

  Q3 :=  Clear
      + /Q3 * /Q2 * /Q1 * /Q0
      +  Q3 *  Q0
      +  Q3 *  Q1
      +  Q3 *  Q2

  Q2 :=  Clear
      + /Q2 * /Q1 * /Q0
      +  Q2 *  Q0
      +  Q2 *  Q1

  Q1 :=  Clear
      + /Q1 * /Q0
      +  Q1 *  Q0

  Q0 :=  Clear
      + /Q0

FUNCTION TABLE
OE Clear Clk   /Q0 /Q1 /Q2 /Q3
------------------------------
 L   H    C     L   L   L   L
 L   L    C     H   L   L   L
 L   L    C     L   H   L   L
 L   L    C     H   H   L   L
 L   L    C     L   L   H   L
 L   H    C     L   L   L   L
------------------------------
```

One indication of the success of the PAL is that, ten years after the introduction of the first 16L8 and 16R8 PALs, these devices still made up the majority of all PLDs used even though there were over 200 unique PLD device types available at that time. In recent years, more flexible devices such as the 22V10 have begun eroding the 16L8/16R8 monopoly, but even these devices are based in large part on the original Birkner/Chua design. Similarly, the PALASM language remains a widely used method of PLD design description. Most, if not all, of the current PLD design tools can trace their heritage back to PALASM.

Another factor in the success of the PAL was the level of programming support. The earlier FPLAs had suffered from the same programming problems that plagued early PROMs; unreliable customer-built programmers and little understood programming algorithms.

John Birkner and his crew were determined to make the PAL a success, so they worked very closely with programmer companies to ensure that reliable programming would be accessible to the device users. The first PAL programmer was developed as a joint effort between MMI and Data I/O and actually utilized two PALs as part of its construction. This meant that the first prototype PAL programmer had to be bootstrapped by emulating the function of its own PALs with PROMs and some additional TTL devices.

The first few years of PAL production weren't without their problems, however. Production couldn't keep up with demand, and production yields suffered. In addition, the programming algorithms (the specific programming voltage and waveform specifications) hadn't been finalized, leading to often unacceptable programming yields.

The PAL shortage had a serious impact on end-users of the time. Tracy Kidder, in his book *Soul of a New Machine,* chronicles the development of the Data General MV8000 computer. The designers of the MV8000 had gambled on the new PAL devices, and were soon feeling the effects of limited supply. Kidder describes a plaque awarded to Data General engineers working on the redesigned Eclipse computer, another large design that used a number of PALs in its construction. The plaque was named the PAL award, but where a commemorative device should have been, there was instead an empty socket.

Digital Equipment Corporation also felt the effects of limited PAL supplies when designing their new VAX 730 computer. To overcome the supply problems, DEC's semiconductor liaison suggested to MMI that they should produce a mask programmable version of the PAL for large quantity production. This was done, and the resulting product was named the HAL for *hard array logic*. HAL devices corresponding to all of the PAL devices were produced. These devices were mask programmed as specified by DEC and other customers.

The HAL devices weren't without their own problems; in one early production run, MMI neglected to properly label the many HAL silicon wafers that were being produced for the VAX. According to Paul Franklin, then with MMI, it became necessary to carefully examine every wafer with a microscope in order to determine exactly which HAL devices they were composed of. The PAL survived its early teething pains and exploded in the market, earning MMI millions of dollars. The HAL turned out to be a marketable product as well and is still available.

For their efforts, MMI rewarded both John Birkner and H. T. Chua with unique bonuses—new cars every year. The October, 1982 issue of *National Geographic,* in an article about silicon valley's glory years, pictured Birkner and Chua with their fourth such annual bonuses—a Mercedes and a Porsche.

The JEDEC Programming Data Standard

The programming support for PLDs was improved dramatically by the introduction of JEDEC standard 3. This standard provides PLD users with a common data interchange format for device programming data. The standard format was first proposed in 1980, and defines a computer file format that makes it possible to transfer design and testing data between devices, and device programmers from different manufacturers. A example of the JEDEC format is shown in Figure 2.9. The idea of a common data format wasn't new; Intel and Motorola had already created defacto standards for PROM device data. The JEDEC standard was important because it addressed the specific needs of the PLD, including the requirement that testing data be included.

Figure 2.9 JEDEC standard programming data file

```
ABEL(tm) 3.20 Data I/O Corp.  JEDEC file for: P16R4 V8.0
4 bit counter with synchronous clear
Michael Holley and Dave Pellerin*
QP20* QF2048* QV6* F0*
NOTE Table of pin names and numbers*
NOTE PINS Q3:14 Q2:15 Q1:16 Q0:17 Clk:1 Clear:2 OE:11*
L0512 01111111111111111111111111111111*
L0544 11111111101111111111111111111111*
L0768 01111111111111111111111111111111*
L0800 11111111101110111111111111111111*
L0832 11111111110111011111111111111111*
L1024 01111111111111111111111111111111*
L1056 11111111101110111011111111111111*
L1088 11111111110111111101111111111111*
L1120 11111111111111011101111111111111*
L1280 01111111111111111111111111111111*
L1312 11111111101110111011101111111111*
L1344 11111111110111111111110111111111*
L1376 11111111111111011111110111111111*
L1408 11111111111111111101110111111111*
V0001 C1XXXXXXXN0XXLLLLXXN*
V0002 C0XXXXXXXN0XXLLLHXXN*
V0003 C0XXXXXXXN0XXLLHLXXN*
V0004 C0XXXXXXXN0XXLLHHXXN*
V0005 C0XXXXXXXN0XXLHLLXXN*
V0006 C1XXXXXXXN0XXLLLLXXN*
C337C*
```

The JEDEC format makes it possible to write a design in any PLD design language and have it automatically processed into a form understandable to any JEDEC compatible device programmer.

Other Companies Leap into the PAL Business

After MMI pioneered and patented the PAL, other device manufacturers began turning out PAL type devices either under license from MMI, or with just enough changes to allow them to claim ownership of the designs. In this way, the PAL became the basis for scores of new devices and literally dozens of patent infringement lawsuits and countersuits (it has been suggested that, in some years, silicon valley lawyers made as much money from the PAL patents as the companies that originated them). The largest of these competitors was Advanced Micro Devices. AMD eventually acquired MMI and the two PLD product lines were merged.

While the PAL was trademarked and patented, PALASM was placed in the public domain and flourished on many different computer systems. PALASM was certainly a boon to the PAL user, but it did have many weaknesses. The most obvious problem with PALASM was that it only supported PAL devices (in particular, PAL devices made by MMI) unless the language was modified for the peculiarities of other devices.

In addition to the lack of universal PLD device support, PALASM suffered from a complete lack of flexibility in design entry methods. The PALASM user had to convert whatever representation of the design he had into the exact sum-of-product equations that were required to be programmed into the device. In many cases, this meant hours of Boolean algebra calculations and manual logic minimizations to create a set of equations that would fit in the selected PAL. It was obvious that most of these hand calculations could be automated, and in fact many universities had already developed computer-based algorithms to perform these tasks.

FPLA Users Demand More Support

Meanwhile, the FPLA devices being produced by Signetics, having for the most part overcome their initial problems of speed and package sizes, were now becoming more popular and were gaining a devoted following of users. Signetics had followed the 82S100 device with a number of other devices of varying sizes. Some of these devices featured highly advanced register elements and were oriented toward complex state machine applications.

FPLA users were demanding a better design environment—one that would support all of the currently available PLDs and would allow more flexibility in the specification and processing of the large and complex sequential designs typical of the newer FPLA devices. Research into such a design tool began at Data I/O in the spring of 1981. Initially, the only results of this research were some sketchy product requirements, and the choice of a product name, *Advanced Boolean Expression Language,* or ABEL.

An important technological advance that was recognized during this research was in the area of state machine design. State machines were a common application for PLDs, particularly the Signetics registered FPLAs. At the Design Automation Conference in 1981, two significant papers were published. The first was by Douglas Brown of Tektronix which described of a system called SMS, for *state machine synthesizer.* This system converted a high-level description of a state machine into Boolean equations. These Equations were then fed into a logic minimization algorithm called *PRESTO,* which had been developed by Antonin Svoboda. SMS was used extensively within Tektronix for PAL development. Another paper, written by S. Kang and W. M. vanCleemput of Stanford University, described a PLA synthesis system that converted a circuit description into minimized and partitioned PLA implementations. The work on this system was also funded by Tektronix.

These papers, and others to follow, were important influences in the design of ABEL, and were also important pieces of work in the area of more general logic minimization algorithms.

Bill Wiley Smith—Signetics Missionary

In the summer of 1982, Bill Wiley Smith of Signetics made a road trip searching for support for a better design environment for the Signetics devices. He carried with him a straw man specification for a PLD design tool called BEE+ (*Boolean Equation Entry*). The specification called for a variety of design entry options, including Boolean equations, state notation, truth tables, and graphics that would allow a designer to draw schematics at the level of standard logic parts. The specification also included the requirement for automatic logic minimization and logic simulation.

Smith was a strong proponent of third-party development software for PLDs. His wasn't the only opinion, however. Within Signetics (as within other PLD manufacturers) there was pressure to develop proprietary design tools that would give the device manufacturers an advantage over their competition.

After Smith's visit, the decision was made to step up efforts within Data I/O to create a new design tool and a small group of engineers began work on the product under the leadership of Dr. Kyu Lee. The ABEL project was announced to a number of PLD manufacturers in December of 1982, and was well received. On that same trip to California, Data I/O project leaders met with Bob Osann of Assisted Technology. Bob Osann had earlier founded Assisted Technology and, with the help of Tracy Kahl and others, had begun work on a software product he called CUPL, for *common universal tool for programmable logic.* Osann had suggested that Data I/O should distribute his software through an OEM arrangement. Instead, Data I/O decided that their own development efforts should continue.

The First Universal PLD Tools

In March of 1983, Assisted Technology released the Beta version of CUPL. It supported nearly all the PALs supported by PALASM and a limited number of Signetics combinatorial FPLAs. CUPL provided users with a Boolean equation design language and limited automatic logic minimization. A subsequent release included a PLD simulator.

In late 1983, Data I/O released the Beta version of ABEL which provided support for virtually all of the PLDs available including the complex registered devices from Signetics. The first version of ABEL also featured a powerful state diagram syntax, giving designers even more flexibility in design description for sequential designs. In the year immediately following its release, ABEL proved to be a huge success, taking its creators by surprise.

Most device manufacturers immediately understood the benefits of high-level tools such as ABEL and CUPL and provided Data I/O and Assisted Technology with advanced information about PLDs under development in order to ensure that both of these products supported their new devices.

Device Manufacturers Develop Their Own Tools

At the same time, many device manufacturers began working on PLD support software of their own. This was done to protect their interests, since their could be no guarantee that Data I/O, Assisted Technology, or any other third-party would provide software support for every device in time for its introduction. As PLD users became more dependent on software tools, the timely support of devices became an increasingly important factor in the market success of a new device.

Signetics developed their BEE+ proposal into one of the first of these post-PALASM software packages and called it AMAZE (which stands for *automated map and zap of equations*). AMAZE is still the design tool of choice for many users of Signetics' devices.

National Semiconductor released a design tool for PALs called PLAN (*programmable logic analysis by National*). In creating PLAN, National's Bob Nelson enhanced the popular PALASM software, adding interactive design entry, automatic device selection and pin assignment, and other advanced features. These packages were followed by other development tools developed and distributed by virtually every PLD manufacturer.

PAL Devices Become More Complex

After the tremendous success of the PAL, MMI's new device efforts were primarily aimed at enhancing and scaling up the basic PAL device. The most striking example of this approach was the 64R32 MegaPAL device. This huge (84-pin) device was the equivalent of four standard PALs and consumed a large amount of power. The MegaPAL was a market failure because there was little advantage to be gained from using a single 64R32 over using four smaller PLDs.

One of MMIs successes of this time was the 20RA10, the first PLD to feature independently clocked flip-flops. (According to John Birkner, the 20RA10 was produced in response to demands from customers who insisted on using PALs to build asynchronous circuits such as ripple counters. Birkner once referred to the 20RA10 as "ripple wine for the masses" since such designs were considered difficult to test and, perhaps, less elegant than comparable synchronous circuits.)

Meanwhile, AMD produced a series of PAL devices that extended the basic PAL structure by providing additional user configurable features, such as the programmable register bypass and feedback of the 22V10. MMI responded with their own configurable devices such as the 32VX10. This trend continued with new devices generally being distinguished not by pin counts or array sizes, but by the configuration and programmability of the I/O circuitry. For a number of years, the predominant device technology in use remained the bipolar PROM technology, until the CMOS revolution reached PLDs.

Altera Enters the PLD Business

The entry of Altera into the programmable logic business in 1984 brought a new level of sophistication to devices and design tools. Prior to Altera, device manufacturers were generally content to let other companies take care of the design and marketing of programming hardware and advanced PLD design tools. Some device manufacturers, in fact, were considering discontinuing support for their own design tools. Altera, however, was developing a completely new series of devices that were, in Altera's opinion, of sufficient complexity that much more powerful design tools would be required—tools designed specifically for those devices.

Unlike previous PLDs, the Altera devices were based on CMOS EPROM technology. The erasable nature of these devices made them ideal for the design environment and also helped in the production environment because of their inherently greater level of testability. The use of CMOS technology allowed Altera to produce devices of very high density and complexity. This increased complexity made it impractical to use traditional design techniques so, more than for any previous type of PLD, users of the Altera devices would clearly require high-level design tools. The quality of the design tools was seen as critical to the success of the new devices.

Altera announced their first device, the EP300, in the spring of 1984. There is a popular story (related by Clive McCarthy of Altera) that the chief silicon designer of the EP300 placed a bet of a full year's salary with Altera's vice president of engineering, Robert Hartmann, that the first silicon would work. As he had predicted, the first chips produced did function correctly. Unfortunately, this meant that Altera had devices to sell but no design tool support, as the software was many months from completion.

Rather than delay the release of the EP300, Altera produced an interim fuse-level design-entry product called ALTRANS and approached Data I/O and Assisted technology about software support in the ABEL and CUPL products. With information provided by Altera, a version of ABEL was quickly produced that supported the EP300.

Within a few months after the release of the first Altera EPLDs (erasable PLDs), Altera's complete development system was ready. The system, which utilized the IBM PC as a platform, featured a highly integrated design entry system (called A+PLUS) with advanced features targeted specifically at the Altera EPLDs. The system also included a simple device programmer that plugged into one of the PC's expansion slots.

Other Startups Join the CMOS Revolution

Altera, of course, had no monopoly on CMOS technology. In fact, Altera has never actually produced any ICs at all; all of their device production is carried out by other manufacturers under various technology exchanges and second-source agreements. One of the first of the IC manufacturers to work with Altera was Intel. Intel now produces second-source Altera EPLDs, as well as other CMOS PLDs of their own design. The Altera design system was also made available to Intel and is sold by Intel (with Intel's own enhancements) as the iPLDS system.

In addition to Altera, many other small device companies have joined in the fray. These smaller device developers have tended to produce the most state of the art PLDs, usually CMOS based, and often rely heavily on support from PLD tool suppliers to ensure success for their devices.

One of these companies was Cypress Semiconductor. The first devices produced by Cypress were high-speed CMOS versions of the popular PALs, produced under an agreement with MMI. Cypress was a start-up company producing high-speed CMOS memory products. The need for CMOS technology had been recognized by MMI and the technology exchange between the two companies gave Cypress limited rights on the PAL architecture, helping to protect the company from the PAL litigation slugfest that occurred soon after.

Another company specializing in erasable CMOS technology is Lattice Semiconductor, which was founded in 1983. Lattice developed a family of electrically erasable PAL-type devices called *generic array logic*, or GAL, devices. These devices prompted a string of lawsuits from MMI which nearly ruined Lattice. When the dust had settled, MMI had the right to produce GAL devices (although not under the GAL acronym). After surviving the MMI lawsuits, Lattice went on to develop and release more advanced PLDs that go beyond the standard PAL structure with devices such as the 39V18 (now called the Lattice 6001), an electrically erasable CMOS FPLA.

Two of the key players at Lattice were Dave Rutledge and Dean Suhr, both of whom had previously worked for Harris. Harris Semiconductor had worked closely with IBM on early PLD research but hadn't followed through with a PLD product line in the early days of PLDs. If the early Harris/IBM research had been published, much of the Lattice versus MMI litigation grief might have been avoided.

From the beginning of its existence, Lattice worked closely with PLD design tool suppliers to ensure that their devices were fully supported in the software from the moment the devices became available to customers. A tutorial disk produced as a cooperative effort between Lattice and Assisted Technology helped to educate new PLD users about the benefits of the devices and on high-level design techniques.

International CMOS Technology (ICT) was also founded in 1983. ICT developed the PEEL (programmable electrically erasable logic) family of devices. These devices, like those developed by Lattice, use CMOS electrically erasable (EE) technology. The first production PEEL devices shipped in late 1986. ICT also developed PLD design tools for their devices. The APEEL assembler and simulator were written by John Birkner, who worked for ICT after leaving MMI.

Xilinx Introduces the Logic Cell Array

In 1984, Xilinx Corporation announced that they would soon be releasing a completely new concept in user programmable devices. Xilinx's device, the *Logic Cell Array* (LCA), is a radical departure from earlier programmable ICs. The LCA (which is described in detail in Chapter 4) is composed of many small logic cells that are each similar to a PROM in architecture. Each cell is capable of implementing a logic function of four or five (depending on the LCA device) inputs and two outputs. The cells are interconnected by programmable connections of various types.

The unique architecture of the LCA makes the design process for these devices quite different than for PLDs, so Xilinx also created a set of design tools for LCA users. These design tools include automatic placement and routing functions that help to efficiently utilize the device.

Exel Microelectronics' ERASIC

Another small company whose devices have unique software requirements (although not as extensive as those of the LCA device) is Exel Microelectronics, founded in 1983. The Exel XL78C800 ERASIC device utilizes a single folded array that allows multiple levels of logic to be implemented (see Chapter 3). This device requires design tool capabilities not available at the time of the device's development. The product manager for the XL78C800, Erich Goetting, determined that the best solution to the software support dilemma was for Exel to produce a set of software programs that could be added to the ABEL product, giving ABEL the capability to support the new device. Exel worked closely with

Data I/O when producing these add-on programs, resulting in timely support for the new devices.

Third Generation Design Tools Appear

Soon after the release of the second revision of ABEL, Data I/O initiated a project to develop a next generation design tool. This was to be the third generation of PLD tool evolution, the first generation having been PALASM and other PLD assemblers, while the second generation had been tools like ABEL and CUPL. The third generation tools would be distinguished by the fact that logic designs could be entered, optimized and verified before a target device technology was chosen.

The project was code named Zorba and the result was a product called FutureDesigner. FutureDesigner is a combination of the FutureNet DASH schematic editor and the Gates design entry system. The Gates design entry system allows logic designs to be entered and modified in a manner somewhat analogous to the way data can be entered and manipulated in a spreadsheet program. When the design has been finalized, it can be implemented in one or more PLDs, or in other technologies (including gate arrays) through the use of automatically generated schematics.

Another third generation PLD design tool has been produced by MINC, Incorporated of Boulder, Colorado. MINC's PLDesigner product is intended to be a nearly automatic system for converting large designs into multiple-PLD implementations. PLDesigner was the first product available that was capable of automatically partitioning a large design into multiple PLDs based on user-specified constraints. Soon after PLDesigner was introduced, MINC sold the PLDesigner source code to Mentor Graphics, who have since busied themselves integrating the technology into their design environment.

FPGAs Take Off

As this book goes to press, the most dynamic segment in the PLD market is in the area of FPGAs. After Xilinx released its first FPGA devices, other manufacturers (most of them small startup companies), began working on other types of FPGAs.

In 1988, Actel Corporation introduced an FPGA quite different from the Xilinx devices. The ACT 1, as it's called, has a density that is comparable to mask-programmed gate arrays and uses a completely new programming technology. Like the LCA, the Actel device requires placement and routing of logic functions to be used effectively. The software to perform this placement and routing was developed at Actel in parallel with the development of the device.

In 1989, Plessey Semiconductor announced that they would soon be producing FPGAs. While all of the FPGA devices currently available or announced share some similar attributes, all of them have unique architectures that are major departures from earlier PLDs.

2.2 SUMMARY

The development of the programmable logic device has been rapid and its history filled with Silicon Valley innovation and intrigue. As of this writing, the electronics industry as a whole is in somewhat of a decline, when compared to the frantic pace of a decade ago. The PLD business, however, is booming. New devices are being introduced every month, and the design tools are maturing into sophisticated parts of complete CAE systems. In later chapters, we will examine these devices and design tools in more detail.

2.3 REFERENCES

Baker, Stan. "John Birkner: Born To Innovate." *VLSI Systems Design* (July 1988):16.

Birkner, John. "Reduce Random-Logic Complexity by Using Arrays of Fuse-Programmable Circuits." *Electronic Design* (August 16, 1978).

Birkner, John Martin. "The Evolution of PALs." *Byte* (January 1987):208.

Blakeslee, T. R. *Digital Design with Standard MSI and LSI,* John Wiley and Sons, New York, NY, 1975.

Cavlan, Napoleone and Cline, Ron. "Field-PLAs Simplify Logic Design." *Electronic Design* (September 1, 1975):84.

Greer, D. L. "Associative Logic for Logic Network Implementation." *IEEE ISSCC Digital Technical Papers* (February 1976):18–19.

Greer, D. L. "An Associative Logic Matrix." *IEEE Journal of Solid State Circuits* (October 1976):679–91.

Greer, D. L. *Electrically Programmable Logic Circuits,* U. S. Patent 3,818,452 (June 18, 1974).

Greer, D. L. *Multiple Level Associative Logic Circuits,* U. S. Patent 3,816,725 (June 11, 1974).

Greer, D. L. *Segmented Associative Logic Circuits,* U. S. Patent 3,849,638 (November 19, 1974).

Hemel, A. "The PLA: A "Different Kind" of ROM." *Electronic Design* (January 5, 1976):78–84.

Johnston, Moira. "Silicon Valley." *National Geographic* (October 1982):459–76.

Kidder, Tracy. *The Soul of a New Machine,* Avon Books, New York, NY, 1981.

Lancaster, Don. *TTL Cookbook,* Howard W. Sams and Company, Indianapolis, IN, 1974.

Milliman, Jacob. *Micro-Electronics—Digital and Analog Circuits and Systems,* McGraw-Hill, New York, NY, 1979.

Mrazek, D. "PLAs Replace ROMs for Logic Designs." *Electronic Design* (October 25, 1973):66–70.

3

A Tour of PLDs

In this chapter, we'll take a tour through the land of PLDs. The devices we've selected for this tour have been chosen either because they are extremely popular, or because they have particularly useful features that are found in other devices. First we'll discuss the different *architectures* of these devices. The architecture of a PLD affects the logic applications for which the device can be used. As of this writing, there are nearly 300 unique architectures of PLDs available. This number is constantly growing as manufacturers introduce new devices.

We'll conclude this chapter with an overview of current PLD technologies. Technology attributes include such things as manufacturing process (CMOS, bipolar, ECL, and so on), package type, speed grade, eraseability, and so forth. If we factor in all the available technology options, the number of distinct PLDs available rises into the many thousands. Technology differences affect how a programmed device will operate in a larger circuit, how the device is actually programmed, and in what kind of systems it can be utilized.

3.1 PLD ARCHITECTURES

When we refer to the architecture of a PLD, we are referring to those device attributes that affect the logical construction of the device—those attributes significant to the logic of a design to be implemented. Architectural attributes include such things as the configuration of the pins, the arrangement and size of the programmable array(s), and the configuration of the input and output interface logic.

PROMs, PALs, and PLAs

A programmable logic array is normally composed of a specific number of input lines connected through a fixed or programmable array to a set of AND gates, which are in turn connected to a fixed or programmable array of OR gates. The OR gates provide the output signals from the logic array.

A simplified programmable logic array composed of two inputs and one output is shown in Figure 3.1. Notice that, to provide all possible combinations of inputs, each input is routed to the array in both its true and complemented form. These inputs are then connected to AND gates via programmable interconnection points. These *product terms* are ORed together to form a *sum-of-products* logic array.

Since a typical PLD has many more inputs and outputs than the simple array shown above, a special notation, called a *logic diagram,* is used to graphically describe the complex PLA structures associated with these devices. When the array is programmed to implement a particular logic function, the desired inter-

Figure 3.1 Schematic of two-input, one-output PLA

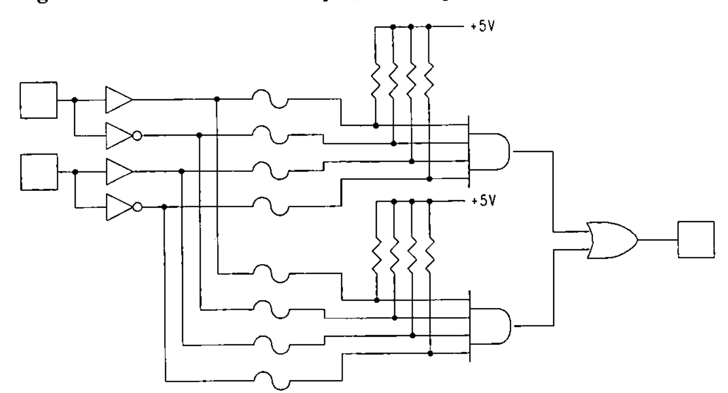

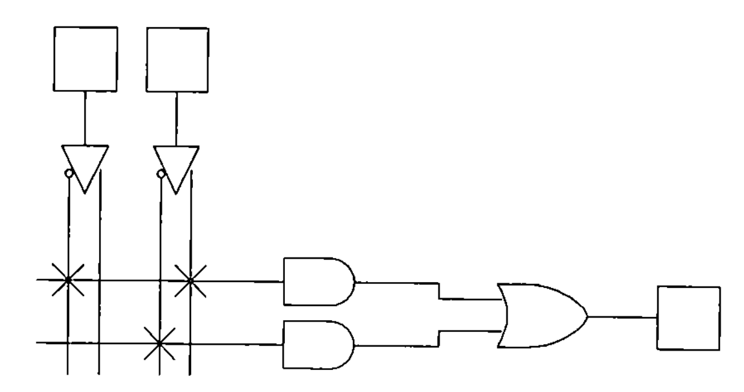

Figure 3.2 Logic diagram of two-input, one-output PLA

connections can be indicated on the logic diagram with Xs. Figure 3.2 shows the two-input, one-output logic array drawn in logic diagram form and programmed with a simple logic function.

In some devices, the AND/OR structure is replaced by either NAND/NAND or NOR/NOR structures (we'll see how this is done later in this chapter), but the result is the same: with a large enough array, any logic function can be implemented.

Programming of the device is accomplished by enabling or disabling interconnections in the device's programmable array. The actual connections can be provided in a variety of ways depending on the device technology.

Programmable read-only memories (PROMs) are the oldest form of programmable logic devices, but were not actually designed for logic applications. PROMs have historically been used to store data such as bootstrap programs and microcode instructions. They have also been used extensively for decoding functions and simple state machines (when provided with external storage registers) and as such can be considered programmable logic devices. The PROM structure can be used for any general purpose logic circuit since it incorporates a sum-of-products logic array.

The basic PROM array is shown in the logic diagram of Figure 3.3. In this figure, we have indicated all programmable fuse locations with Xs. To program this PROM array, you decide which of the interconnect points (each intersection of a horizontal and vertical line) are to be preserved, and disable the remaining connections.

For a PROM with n inputs, there are 2^n possible input values, all of which are provided for in the AND array. The programmable OR array allows each of these input values to be decoded into any value consisting of k outputs, where k equals the number of OR gates (and corresponding outputs) in the array. This makes PROMs particularly well suited for read-only memory applications and this is their most common use.

Since a PROM can map any of its input states into an arbitrary output state, it can be used quite effectively for implementing n-input, k-output combinational logic functions. Each unique input state corresponds to one unique product term

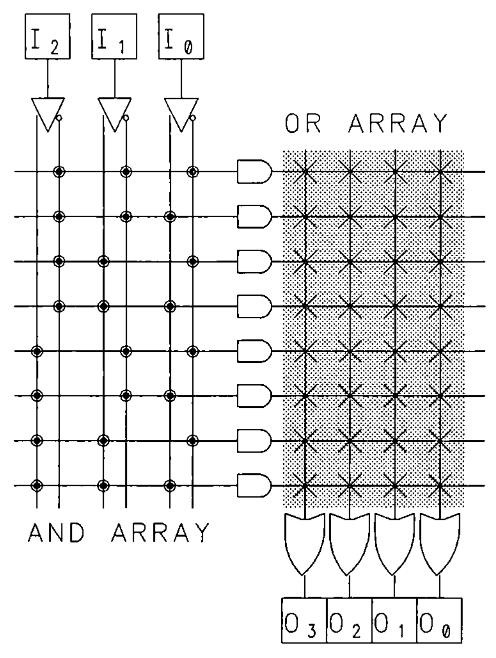

Figure 3.3 PROM-type array

in the array. To use a PROM for logic functions, the designer need only specify the truth table for the set of functions. There is no need for logic minimization, since all possible input combinations are provided in the AND array.

As we pointed out, a PROM provides the complete set of input combinations in the AND array. For most logic functions, however, this is completely unnecessary, and results in a tremendous amount of wasted circuitry on the chip. Particularly, when a large number of inputs are required, the PROM structure becomes impractical. Consider, for example, when a logic function of sixteen input variables and eight output variables is desired. To implement such a function in a PROM, you would have to use a 64K by 8-bit PROM device, regardless of the complexity of the logic function. A PROM of this size would be a highly inefficient vehicle for most logic functions.

Most n-variable logic functions can be implemented with far less than 2^n product terms. To more efficiently map logic functions with a larger number of inputs, the PLA (programmable logic array) and PAL (programmable array logic) devices were developed. The PLA structure is the basis for virtually all PLDs in use today. The complete PLA structure is the basis for a variety of PLDs, and provides the greatest flexibility in how product terms are allocated to the OR gates and associated outputs.

The PLA structure, shown in Figure 3.4, features both a programmable AND array, and a programmable OR array. For a PLA with n inputs, any input variable

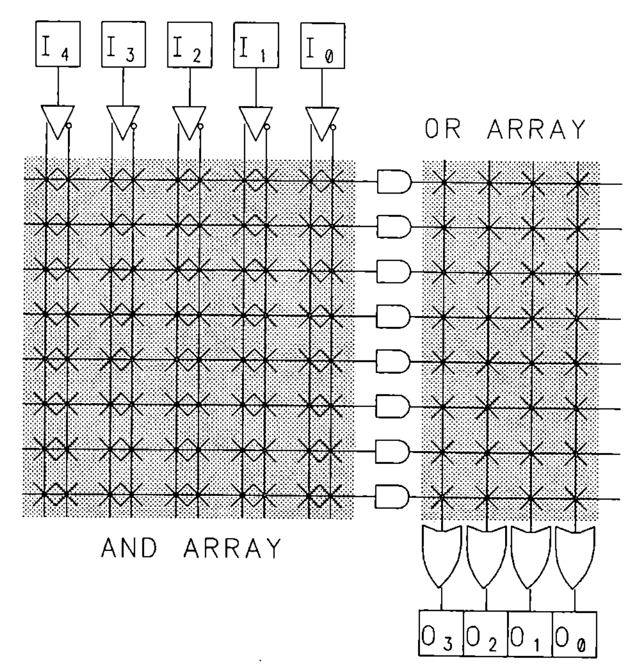

Figure 3.4 PLA-type array

(or its complement) may be an input to any AND gate. Therefore, any AND gate in the AND array can be configured to implement any of 3^n possible product terms (3^n because each input to an AND gate has three possible values—true, complement, and no-connect). The design of the PLA allows any product term in the array to be connected to the OR gate of any output. This feature is sometimes referred to as *product-term sharing.*

The PLA provides the most flexibility for implementing logic circuits, particularly for large designs in which many common logic elements can be shared between circuit outputs. PLA-type devices are generally slower in operation than PROMs and PALs since there are two programmable arrays through which signals must propagate. Unlike a PROM, the total number of product terms available in a PLA is limited, so logic minimization is important when implementing designs. We'll examine the problem of logic minimization in Chapter 7.

The PAL (programmable array logic) structure is similar to the PLA, but has a fixed OR array, as illustrated in Figure 3.5. Every output in a standard PAL-type device has one OR gate that is unique to that output. There is no provision for product term sharing (although, as we'll see later, some PAL-based devices do allow for various forms of product term reallocation).

These product terms are then gated together by fixed OR gates to drive the device outputs. Since a limited number of product terms are provided for each

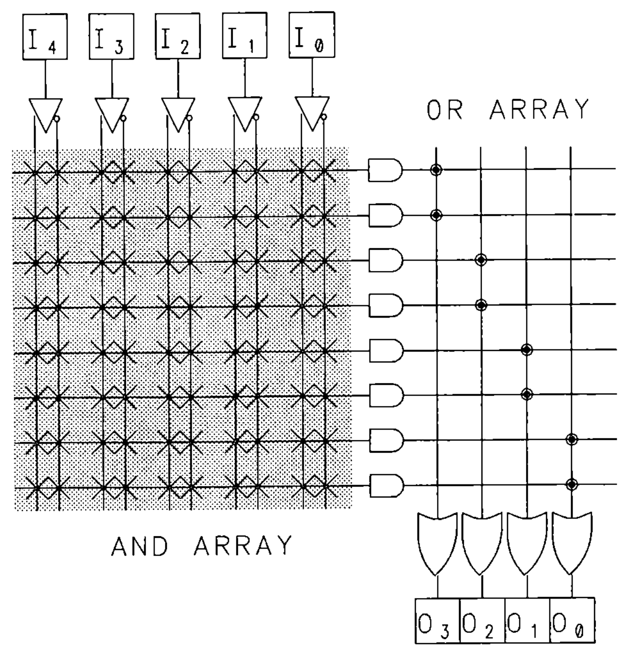

Figure 3.5 PAL-type Array

PAL output, logic minimization techniques become important when logic circuits are implemented in PAL devices.

PAL Devices

The most common PALs in use today are the 16L8 and 16R4/6/8 series of devices. These 20-pin devices originated at MMI (now a part of AMD) are now available from many manufacturers. The devices are intended to replace standard logic parts and are, therefore, designed to operate with TTL (transistor-transistor logic) signal levels.

Combinational PALs

Combinational PALs are those devices that are based on a PAL structure, and do not contain any memory elements. Combinational PALs are useful for a wide variety of random logic functions, including decoders, interface logic, and other applications that require a simple decoding of device inputs. The 16L8 represents the typical combinational PAL and is diagrammed in Figure 3.6.

The diagram for the 16L8 shows that the device has ten dedicated inputs (pins 1 through 9 and pin 11) to the programmable AND array. Each input to the

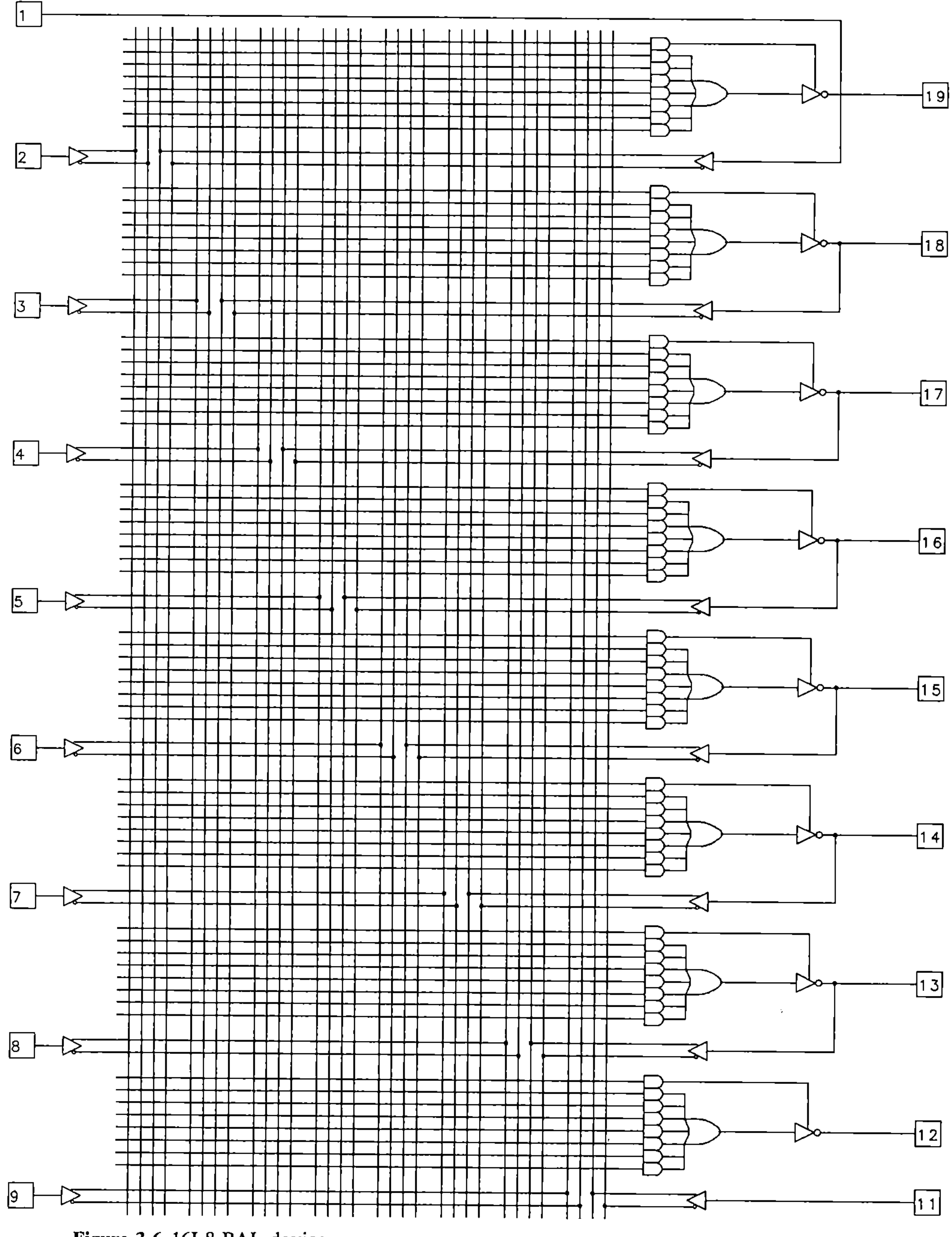

Figure 3.6 16L8 PAL device

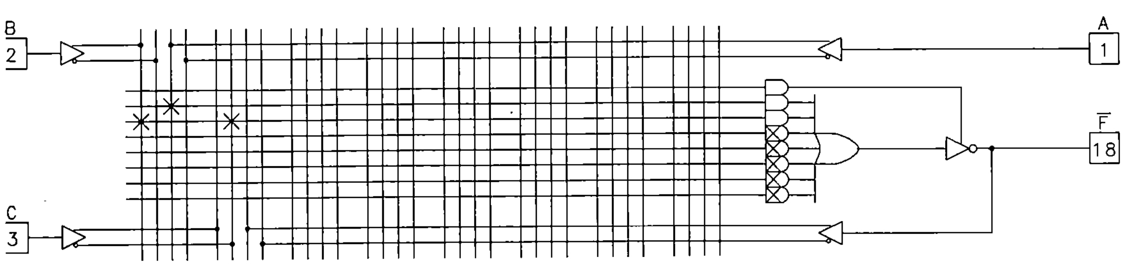

Figure 3.7(a) Implementing a combinational logic function in the 16L8

array is available in its true or complemented form allowing any combination of the inputs to be expressed on any row of the array. Each row of the 16L8 array corresponds to one product term of the device.

The 16L8 has eight outputs, each of which is fed by a seven-input OR gate. Each output of this device is capable of implementing a logic function composed of seven or fewer product terms. The eighth product term is used to control the three-state output buffer, the function of which we will examine in a moment.

A simple logic function is shown implemented in the 16L8 in Figure 3.7(a). The X's shown on the diagram indicate fuse interconnections that have been left intact to implement the function. A large X inside of an unused AND gate indicates that all of the fuses for that row are to be disconnected. When all of the fusible

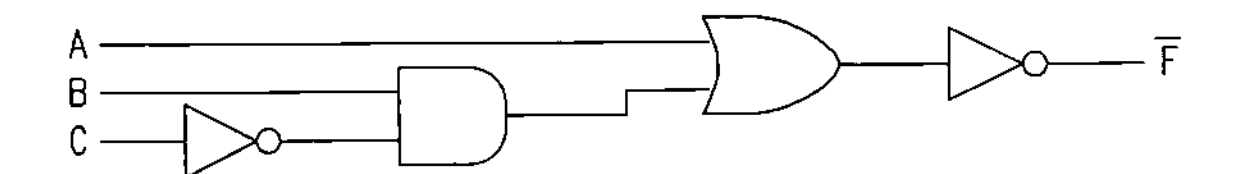

Figure 3.7(b) The same logic function expressed as a schematic

links for a row are disconnected, the associated OR gate input floats high. Figure 3.7(b) shows the same logic function expressed as a traditional schematic.

Notice that there is an inverter associated with the PAL output. There are PAL devices available that don't have the output inverter (often indicated with an H in the part name, such as, 16H8 instead of 16L8) but these devices are less frequently used since active-low logic is prevalent and these logic functions normally fit better in a PLD with inverted outputs.

In addition to the dedicated inputs, there are six I/O pins (pins 13 through 18) on the device that may be used as array inputs as well. These pins can be used in a variety of ways. To use these I/O pins as dedicated inputs, you must disable the three-state output buffer associated with that pin. When a three-state buffer appears on the output of a PLD, it's called an *output enable*. Since the output enable for the 16L8 is controlled from the array, you disable it by leaving intact the fuses for the dedicated product term that controls the enable.

Leaving all of the fuses intact (or, for that matter, any pair of true and

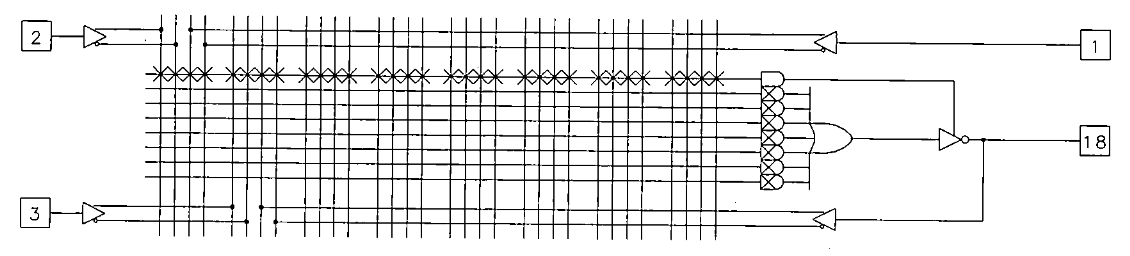

Figure 3.8 Disabling the 16L8's three-state output buffer

complement array inputs for a single input pin) for any product term in the device results in a logic level 0 on that product term. This is shown in Figure 3.8.

Using the I/O pins as inputs, it's possible to use the 16L8 to implement logic functions with as many as sixteen separate inputs, at the expense of usable outputs. The I/O pins can also be used as dedicated outputs, by permanently enabling the output enable. This is done by blowing all of the fuses for the output enable's product term as was done in Figure 3.7(a). When all of the inputs to an AND gate in this device are disconnected, the AND gate floats high, in this case, enabling the output enable.

When the output enable of a 16L8 I/O pin is enabled, the input to the array is still active, and can be used to feed the output back into the array. This is useful for multilevel logic applications (see Chapters 7 and 8). This feature can also be used to create oscillating or asynchronous sequencer circuits, although these applications aren't generally recommended.

The most common use of the output enable is for traditional three-state purposes such as bus interfacing. To use the output enable dynamically, you simply program the output enable product term with the desired logic as shown in Figure 3.9.

Programmable Output Polarity

Quite often, there are situations in which a design can't be implemented in a 16L8 or 16H8 due to the need for some of the outputs to be inverted while others are not. For these applications, a version of the 16L8 is available, called the 16P8,

Figure 3.9 Using the 16L8's output enable dynamically

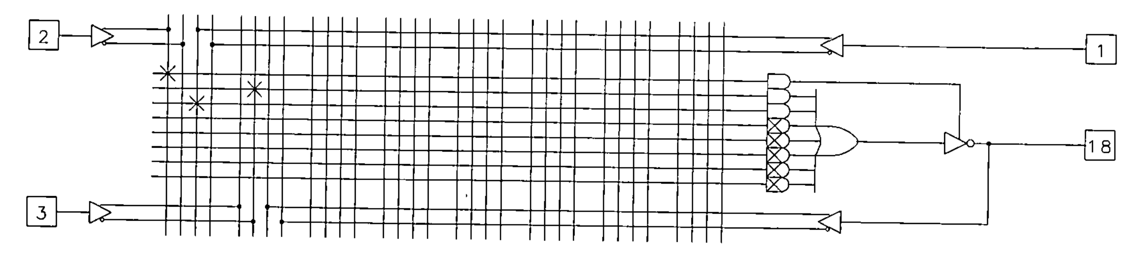

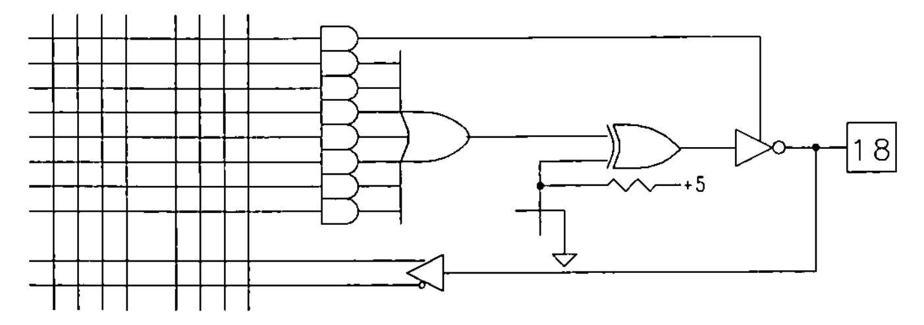

Figure 3.10 Programmable output polarity

that has a feature called *programmable output polarity*. Figure 3.10 illustrates how programmable output polarity is implemented in the 16P8 device.

To provide programmable output polarity, each output of the 16P8 includes an XOR gate. One input to this XOR is the output of one of the PAL's OR gates, while the other can be either connected to ground through a fuse or disconnected and allowed to float high. Some devices utilize other methods such as multiplexers or transmission gates for polarity control, but the result is the same.

Other Related Combinational PALs

There are a wide variety of simple PALs that have architectures similar to the 16L8. Most of these are stripped down versions that are somewhat less expensive. One of the least complex of these PALs is the 10L8, illustrated in Figure 3.11. This device has only dedicated input and output pins, no output enable, and no feedback lines. The small number of product terms (sixteen total) and small number of inputs means that this device has only 320 fuses, and is only slightly more complex than a simple 32 by 8-bit PROM device. These smaller PAL devices are rapidly losing popularity, as most PLD users find that it's most economical (because of inventory costs) to use the 16L8 or 16P8 for all of their strictly combinational applications, rather than attempt to save a few pennies per device by using simpler PAL devices.

Registered PALs

The most commonly used registered PAL devices are represented by the 16R4, the 16R6, and the 16R8 devices. The 16R4 device is shown in Figure 3.12.

The design of the 16R4 device is similar, in most respects, to that of the 16L8. The difference is found on output pins 14 through 17. These outputs feature edge-triggered D-type flip-flops. The Q output of each flip-flop is routed to the PAL output through the output enable and each flip-flop's $\overline{Q}$ output is routed back to the AND array. Like the fed back combinational signals on the other outputs, these signals are provided to the array in both their true and complement form.

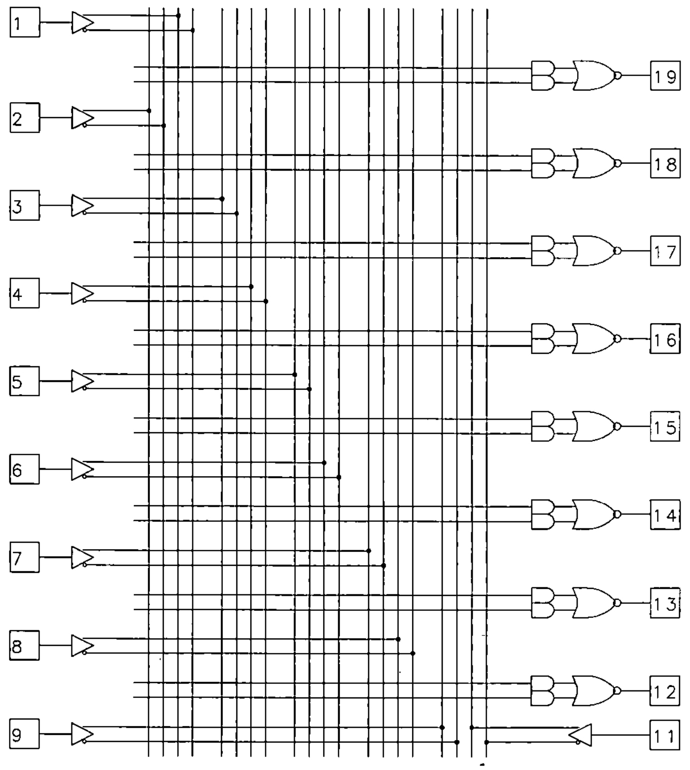

Figure 3.11 10L8 PAL device

Figure 3.12 16R4 registered PAL

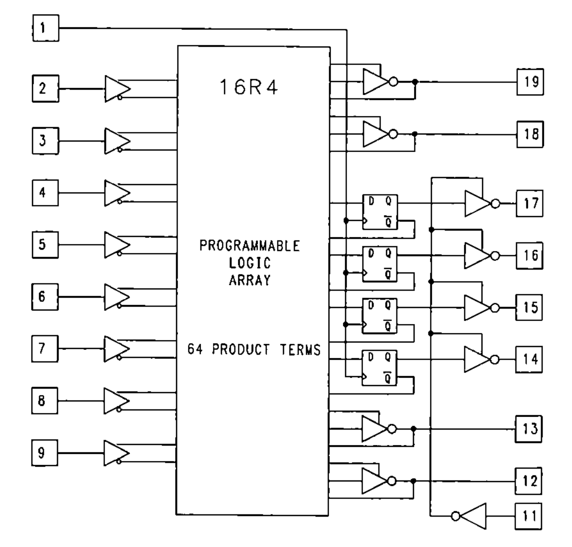

The flip-flops are all controlled by a common clock which is tied directly to pin 1 on the device. This implies, of course, that pin 1 can't be used as an input, as it can in the 16L8. This is also true of pin 11, which is used as a dedicated input for the output enable of the flip-flops. This method of enabling outputs is common to most of the simpler PAL devices. In general, combinational outputs are enabled from a product term, while registered outputs are enabled from a dedicated pin. Since no product term is used for the output enable of the registered outputs, the eighth product term is made available for use as an input to the OR gate for those outputs. For most applications that require output synchronization or state memory, the output enable feature will be used globally, so a complex output enable is not required. Another tradeoff in the design of these devices' output enables is speed; a pin controlled output enable will have a faster pin-to-enable speed.

The architecture of the 16R4 is well suited for simple state machine applications. The registered outputs can be dedicated for use as state memory registers, while the combinational outputs can be used for either state machine outputs or additional control inputs. Note that it isn't possible to use the registered output pins of the 16R4 as inputs under any circumstances.

Implementing a Sequential Design in the 16R4

To show how devices like the 16R4 are used for sequential designs, we will implement a simple circuit. This design is a simple 4-bit counter, the Boolean equations for which are shown in Figure 3.13. The design file shown is written in the ABEL language and uses simple Boolean equations that represent the logic of the counter. In ABEL, the *&* symbol represents an AND operation, the *#* symbol represents an OR operation, and the *!* symbol represents a NOT operation (refer to Chapters 7 and 8 for more information about the ABEL language).

How these equations were derived is the subject of another chapter; we won't go into the specifics of counter design here. For the time being, simply accept the fact that this counter increments a 4-bit number by one every time the device is clocked. The design utilizes active-low logic, so the values observed on the device's outputs will be the complement of the actual number stored in the registers.

Figure 3.14 shows how the equations for the 4-bit counter are implemented in the device. The least significant bit of the counter (represented by the *Q0* variable in the counter equations) is mapped to pin 14 of the device and requires the least logic of any of the counter outputs. The remaining bits of the counter, *Q1* through *Q3*, require correspondingly more logic, since they must each decode the previous counter bits to determine whether a change in state is required.

The JEDEC standard format file that corresponds to the counter design is shown in Figure 3.15. This file is created by ABEL and is used to download programming data to a device programmer. The rows of ones and zeroes represent fuse locations and corresponding fuse values where a one indicates a blown fuse

```
module Counter
title '4-Bit Counter with clear'

        counter device 'P16R4';

        Q3,Q2,Q1,Q0       pin 14,15,16,17;
        Clk,Clear,OE      pin  1, 2,11;

Equations
        !Q3 :=   Clear
               # Q3 & Q2 & Q1 & Q0
               # !Q3 & !Q0
               # !Q3 & !Q1
               # !Q3 & !Q2;

        !Q2 :=   Clear
               # Q2 & Q1 & Q0
               # !Q2 & !Q0
               # !Q2 & !Q1;

        !Q1 :=   Clear
               # Q1 & Q0
               # !Q1 & !Q0;

        !Q0 :=   Clear
               # Q0;

Test_Vectors
       ([Clk,Clear,OE] -> [Q3..Q0])
        [.C.,  1  , 0] ->  0;
        [.C.,  0  , 0] ->  1;
        [.C.,  0  , 0] ->  2;
        [.C.,  0  , 0] ->  3;
        [.C.,  0  , 0] ->  4;
        [.C.,  1  , 0] ->  0;
End
```

Figure 3.13 4-bit counter described in ABEL

and a zero indicates an intact fuse. (The format of the JEDEC file is described in more detail in Chapter 5.)

A timing diagram of the counter circuit is shown in Figure 3.16. The diagram shows the relationship between the 16R4's clock input and the resulting counter values appearing on the counter's four outputs. To ensure that the counter can be initialized to a known value immediately after power-up, we have provided this counter with a synchronous clear input. When the clock signal (pin 1) goes from low to high, the registers change state to reflect the new counter value, which then appears on the device outputs as logic level 1 for all outputs (corresponding to a counter value of 0).

Since there is a delay between the time the new register values appear on the flip-flop outputs and the time these values propagate back through the programmable array, the counter's flip-flop inputs will require a certain amount of time to stabilize before being clocked again. This limits the speed at which this

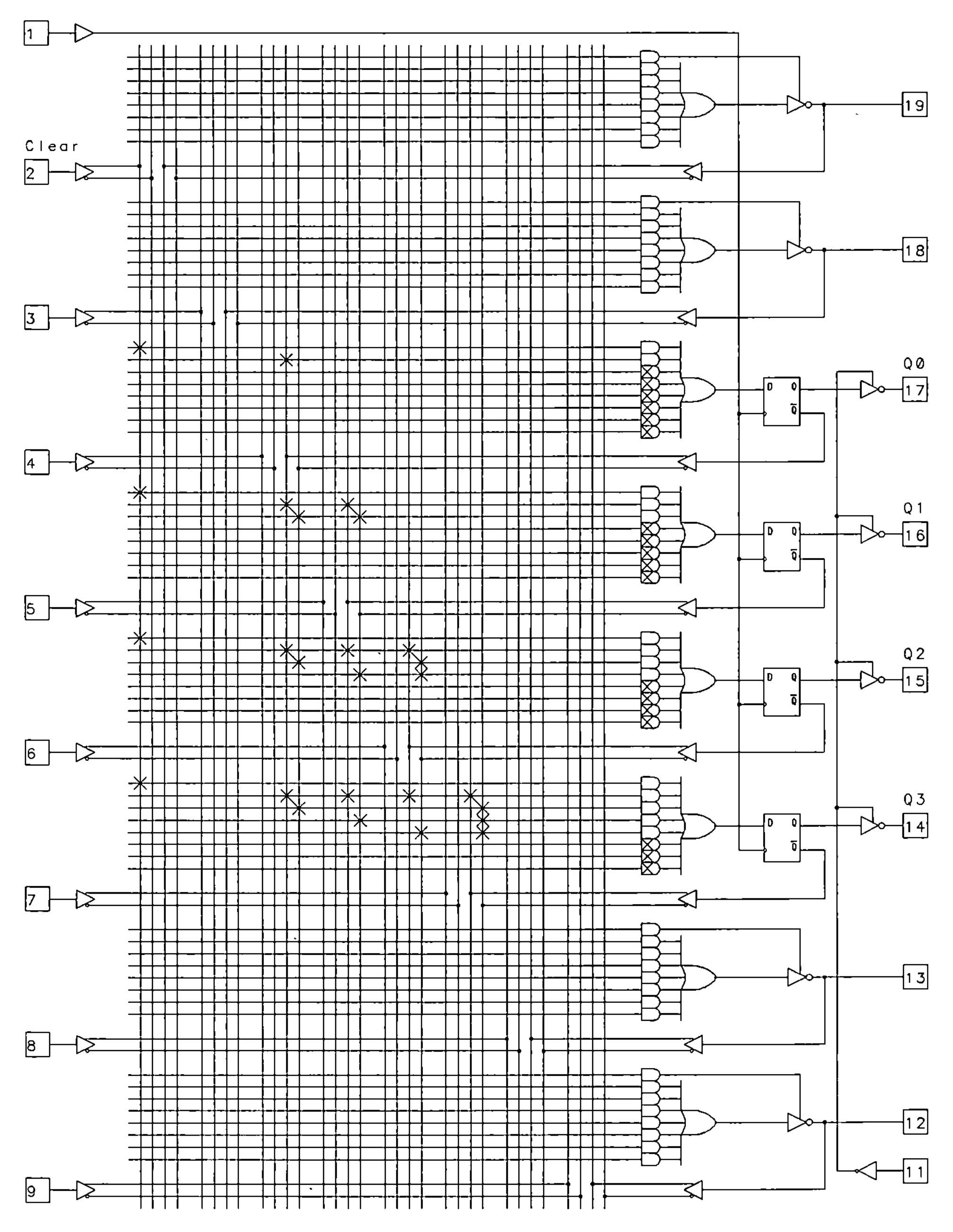

Figure 3.14 Counter design implemented in a 16R4 PAL

```
ABEL(tm) 3.20 Data I/O Corp.  JEDEC file for: P16R4 V8.0
4 bit counter with synchronous clear
Michael Holley and Dave Pellerin*
QP20* QF2048* QV6* F0*
NOTE Table of pin names and numbers*
NOTE PINS Q3:14 Q2:15 Q1:16 Q0:17 Clk:1 Clear:2 OE:11*
L0512 01111111111111111111111111111111*
L0544 11111111110111111111111111111111*
L0768 01111111111111111111111111111111*
L0800 11111111110111011111111111111111*
L0832 11111111111011101111111111111111*
L1024 01111111111111111111111111111111*
L1056 11111111110111011101111111111111*
L1088 11111111111011111110111111111111*
L1120 11111111111111101110111111111111*
L1280 01111111111111111111111111111111*
L1312 11111111110111011101110111111111*
L1344 11111111111011111111111011111111*
L1376 11111111111111101111111011111111*
L1408 11111111111111111110111011111111*
V0001 C1XXXXXXXXN0XXLLLLXXN*
V0002 C0XXXXXXXXN0XXLLLHXXN*
V0003 C0XXXXXXXXN0XXLLHLXXN*
V0004 C0XXXXXXXXN0XXLLHHXXN*
V0005 C0XXXXXXXXN0XXLHLLXXN*
V0006 C1XXXXXXXXN0XXLLLLXXN*
C337C*
```

Figure 3.15 JEDEC file

design can be operated (see Chapter 10). For most applications, though, the propagation times are short enough that this speed constraint is of little concern.

Other Devices of the 16R4 Variety

The 16R6 and 16R8 devices are identical to the 16R4, the only difference being in the number of registered outputs. The 16R6 has six registered outputs and two combinational outputs, while the 16R8 has eight registered outputs and no combinational outputs. Twenty-four-pin versions of these devices are also available (such as the 20R4, 20R6, and 20R8 devices) and are in all respects identical to the 20-pin devices with the addition of four extra input pins.

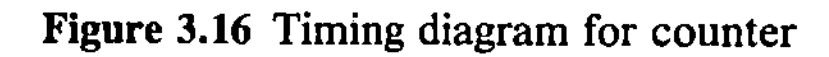
Figure 3.16 Timing diagram for counter

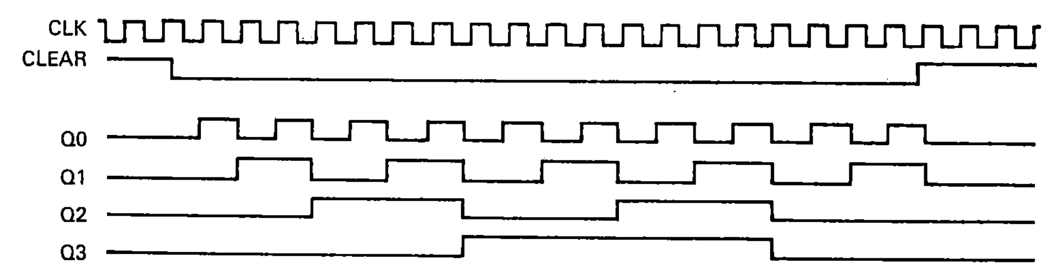

Configurable (Generic) PALs

In recent years, configurable (sometimes called generic) device architectures have become extremely popular. These devices simplify procurement, qualification and inventory requirements by replacing a large number of simpler PAL type devices with a "one size fits all" device. In addition, their flexible architectures allow designs to be implemented that are challenging or simply impossible for the simpler PAL devices to handle.

This architectural flexibility is provided by equipping the device with a variety of configuration fuses separate from those found in the programmable AND array. The 22V10 is one such device, and was designed by AMD to be a replacement for all of the 24-pin PALs of the architecture previously described. To meet this requirement, the 22V10 was designed with configurable outputs. These outputs are enhanced with special circuitry and are called *output macrocells*. Output macrocells are found on ten of the 22V10's pins, as shown in Figure 3.17.

The 22V10 has a total of twelve dedicated inputs, one of which (pin 1) also functions as the common clock input to the edge-triggered D-type flip-flop of each output macrocell. Any of the 22V10's ten output pins can be used as inputs, so the device is capable of supporting applications requiring up to 22 inputs (of course, if you use all ten I/O pins as dedicated inputs, there is no way to observe the results of your efforts, since there will be no output pins left).

We'll examine the structure of the individual output macrocells in a moment; but first, notice that the number of product terms available to the various OR gates in the device differs. The OR gate associated with pins 18 and 19, in fact, have sixteen product terms each available. This means that logic functions of significantly more complexity can be implemented in the 22V10. The irregular nature of the outputs does place more burden on the designer, though, since the design outputs may have to be assigned to device outputs based on their complexity.

The complexity of the 22V10 results in a rather unwieldy logic diagram. For larger devices, the logic diagram format becomes completely impractical, so these devices are often presented in the block diagram form that we used back in Chapter 1. Figure 3.18 illustrates the 22V10 in block diagram form.

As we said, the 22V10 has ten output macrocells, all of which are identical. Figure 3.19 shows the construction of one of these output macrocells.

Each macrocell contains an edge-triggered D-type flip-flop and a pair of fuse-configurable multiplexers (we have shown the 22V10s configuration multiplexers complete with their fuse interconnections and pull-up resistors; in subsequent figures we will omit the fuse and resistor). The two fuses that control the multiplexers can be configured in four different ways, as shown in Figure 3.20.

The 22V10 also has available two extra product terms that can be seen in Figure 3.17. These product terms can be used for synchronously presetting the 22V10 registers, or asynchronously resetting them. The feedback in the 22V10 can be configured to be from the register, the output pin, or from the OR gate

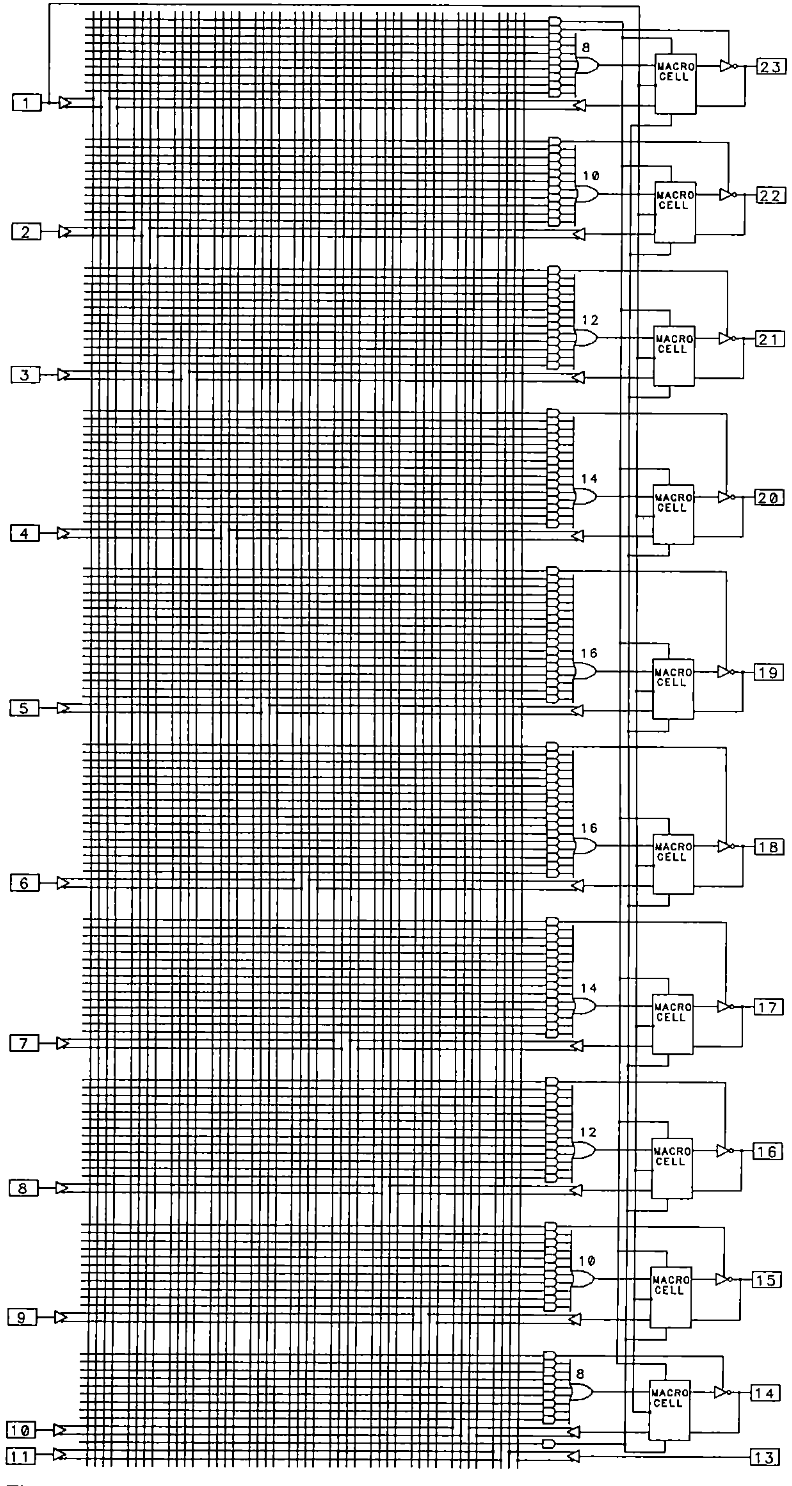

Figure 3.17 22V10 configurable PAL

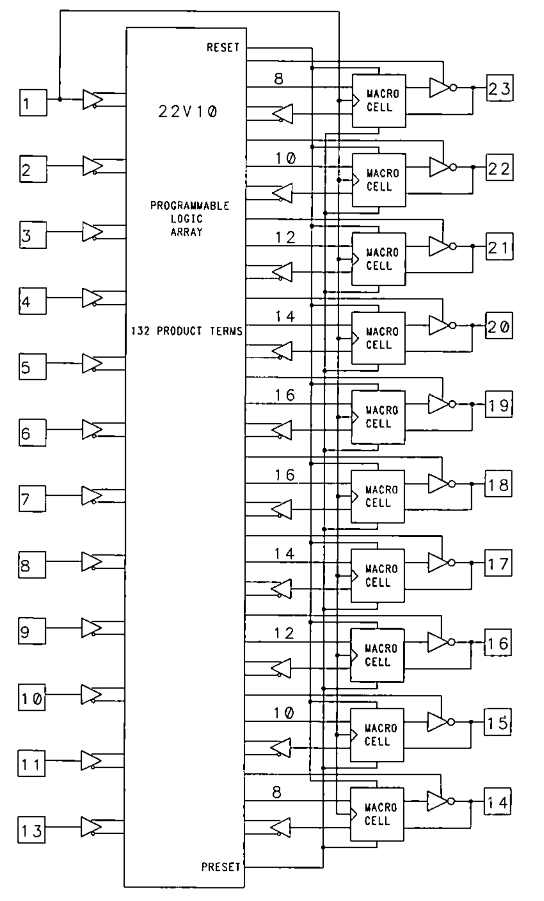

Figure 3.18 22V10 block diagram

Figure 3.19 22V10 output macrocell details

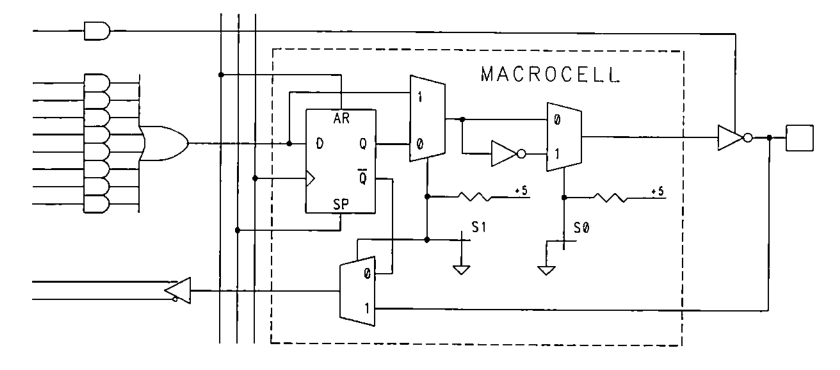

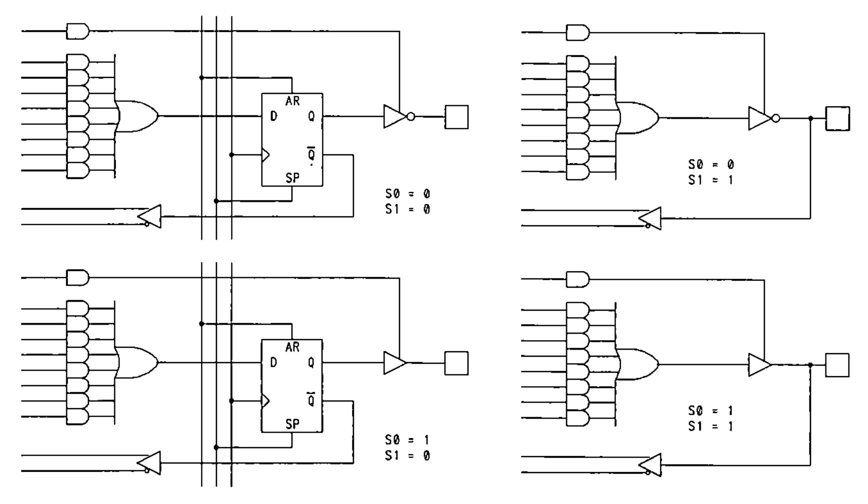

Figure 3.20 Four possible 22V10 macrocell configurations

output. While AMD's 22V10 has the feedback path and register bypass configurations controlled with a single fuse, TI's 22VP10, on the other hand, has independently configurable feedback and register bypass.

Another example of a configurable PAL is the Lattice GAL (generic array logic) device. These devices, more so even than the 22V10, are intended as pin-for-pin replacements for a wide variety of PAL devices. The GAL device, in fact, is designed to be compatible all the way to the fuse level—JEDEC format files for virtually any simpler PAL can be directly implemented in the GAL device.

Another major distinction between the GAL device and the original 22V10 is the fact that the GAL is electrically erasable (later CMOS versions of the 22V10 are available that are erasable). This makes the GAL particularly well suited for engineering prototype activities. We'll discuss the erasable CMOS technology later in this chapter.

The GAL comes in two basic versions. The GAL 16V8 device replaces most 20-pin PAL devices, while the 20V8 replaces most 24-pin PAL devices. The 16V8 device is shown in Figure 3.21.

Like the 22V10, the GAL devices utilize a configurable output macrocell. Although the function of the GAL macrocell is similar to the 22V10, the GAL macrocell differs from that of the 22V10 in a number of areas.

The GAL macrocell (which Lattice refers to as an *output logic macrocell*, or OLMC) is shown in Figure 3.22. The most important difference is that the OR gate is considered to be a part of the output macrocell. This is necessary because of the differing architectures of combinational and registered outputs in devices like the 16R4 described earlier. The control fuses for the GAL macrocells allow each macrocell to be configured in one of three basic configurations. These configurations correspond the various types of I/O configurations found in the PAL devices that the GAL is designed to replace.

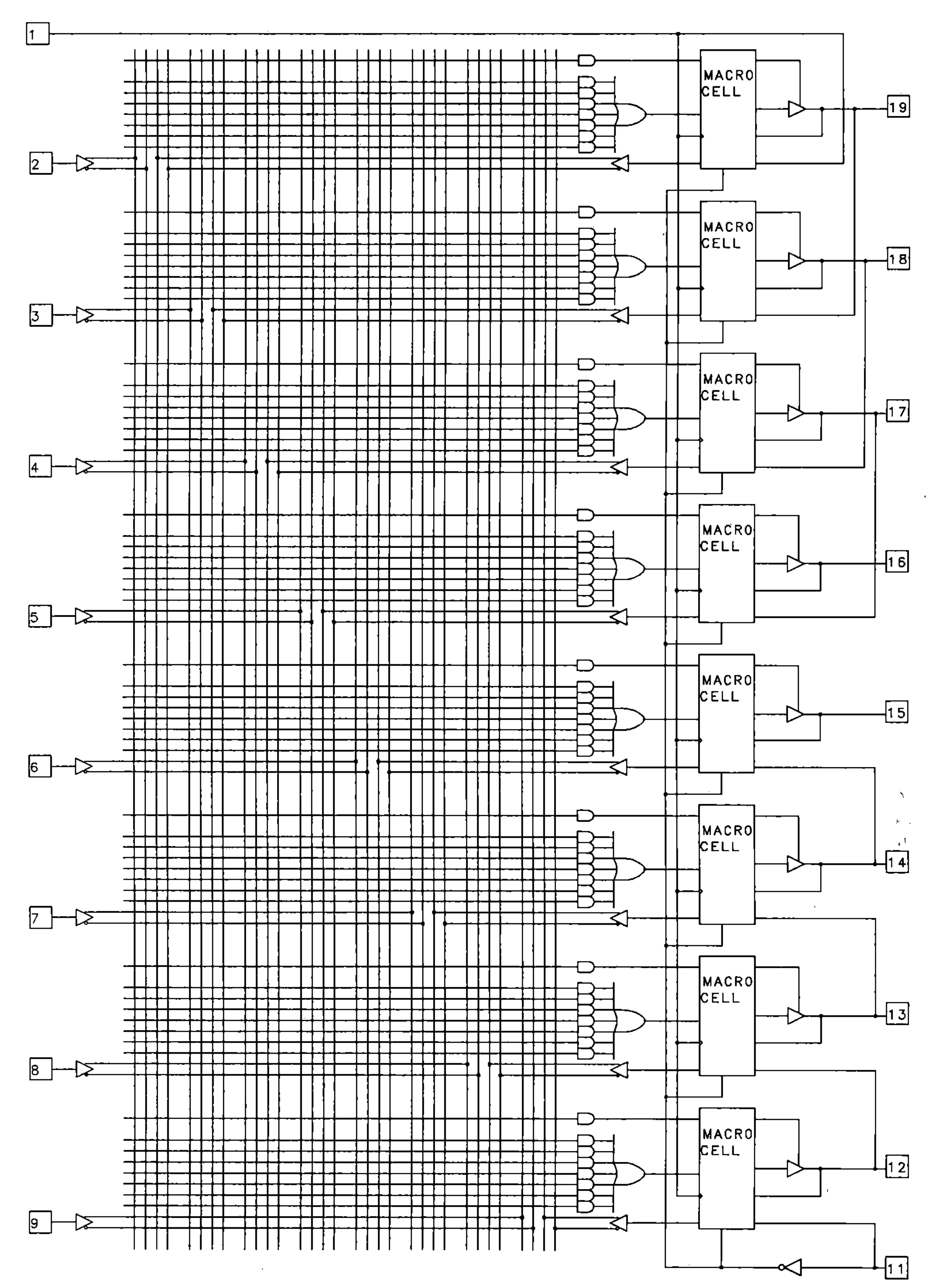

Figure 3.21 16V8 GAL device

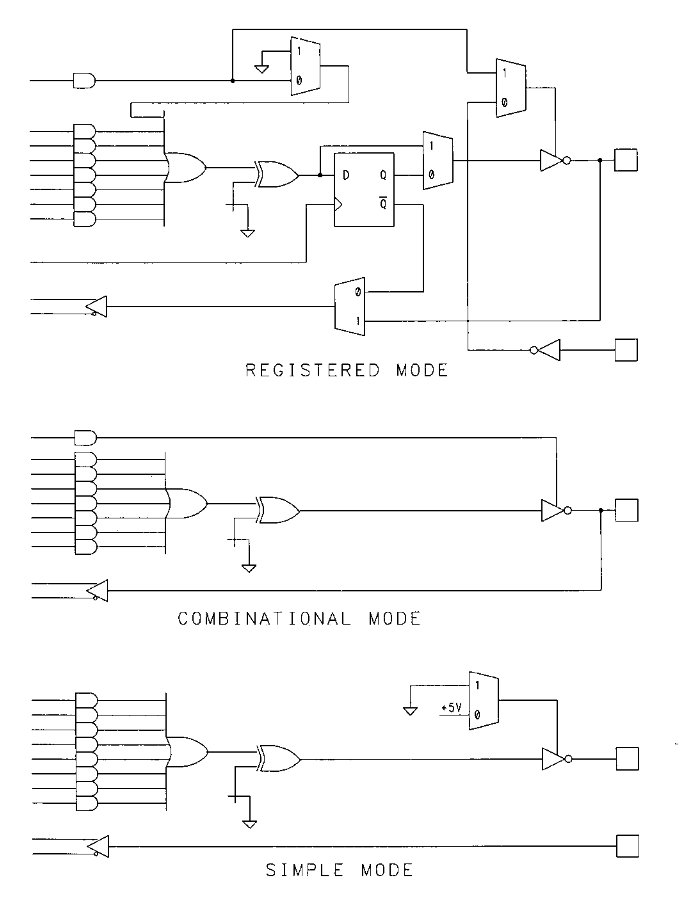

Figure 3.22 16V8 GAL output macrocell detail

In the registered PAL mode, each of the 16V8's eight outputs can be registered or combinational. Those outputs that are configured with registers have eight product terms and a fixed enable input (from pin 11), while those that are combinational have seven product terms with a term-controlled output enable. Clocking in this mode is from pin 1, which can not be used as an input to the logic array.

In the combinational mode, the clock and enable inputs (pins 1 and 11) are made available as array inputs. In this mode, output pins 12 and 19 are not available for use as inputs to the array. The combinational mode is intended for emulation of the 16L8-type combinational PALs.

In the GAL's third mode, there are eight product terms available to each of the eight outputs and no output enable feature is provided. This mode is intended for emulation of the simple PAL-type devices (14H4, for example) so as many as 16 inputs can be used with two outputs available (in this situation, the two outputs must be pins 15 and 16, since these pins can't be used as inputs in this mode.)

The original GAL devices did not have pin feedback available in this mode, but the newer 16V8A devices do support this feature.

Still another family of devices that are intended as PAL replacements are the PEEL devices from International CMOS Technology. The PEEL 18CV8, shown in Figure 3.23, features output macrocells that can be configured in any of twelve different ways.

The PEEL output macrocell is illustrated in Figure 3.24. The PEEL's macrocells provide a wider selection of feedback options than do the GAL's. Notice that there are four fuses used to select the macrocell configuration. Figure 3.25 shows the twelve different configurations possible for each macrocell using these fuses.

The PEEL 18CV8 architecture supports the use of up to 18 inputs to the array (including fedback outputs); the clock and enable pins can be used as inputs to the array even while they are being used for their primary functions. The same architectural features with 22 inputs are provided in the PEEL 20CG10, 22CV10, and 22CV10Z devices. Like the 22V10 device, the PEEL includes product-term controlled preset and reset functions. The 24-pin 22CV10Z, in fact, can be used as a functional replacement for the 22V10. Like the GAL, the PEEL device is erasable through the use of CMOS EEPROM technology. This technology will be discussed in more detail later in this chapter.

Exclusive-OR PALs

Exclusive-OR gates, or XORs, are found in PALs for a variety of purpose. We have already seen how XORs are used to implement fuse configurable output polarity. XORs have other applications as well.

Figure 3.26 is the logic diagram for a 20X8 PAL device. This device has an XOR feeding each of the device's flip-flops. Each of these XOR gates is fed in turn by two sum-of-products arrays of two product terms each. These XORs can be used to reduce the amount of logic required for many applications, particularly counters. The XORs not only allow dynamic polarity control of the outputs, but also allow surprisingly complex designs to be implemented using very few product terms.

XORs have been included in many PLDs over the years. In recent years, the value of these extra gates has become increasingly apparent to circuit designers. We will examine some of the uses for these XOR gates in Chapters 7 and 8.

Asynchronous PALs

Independently clocked flip-flops are useful for many applications (such as asynchronous state machine applications) and an increasing number of devices support this feature. The first device to have independently controlled clocks was MMI's 20RA10. The 20RA10's macrocell architecture is shown in Figure 3.27. This 24-pin device features product terms that control each of the device's ten D-type

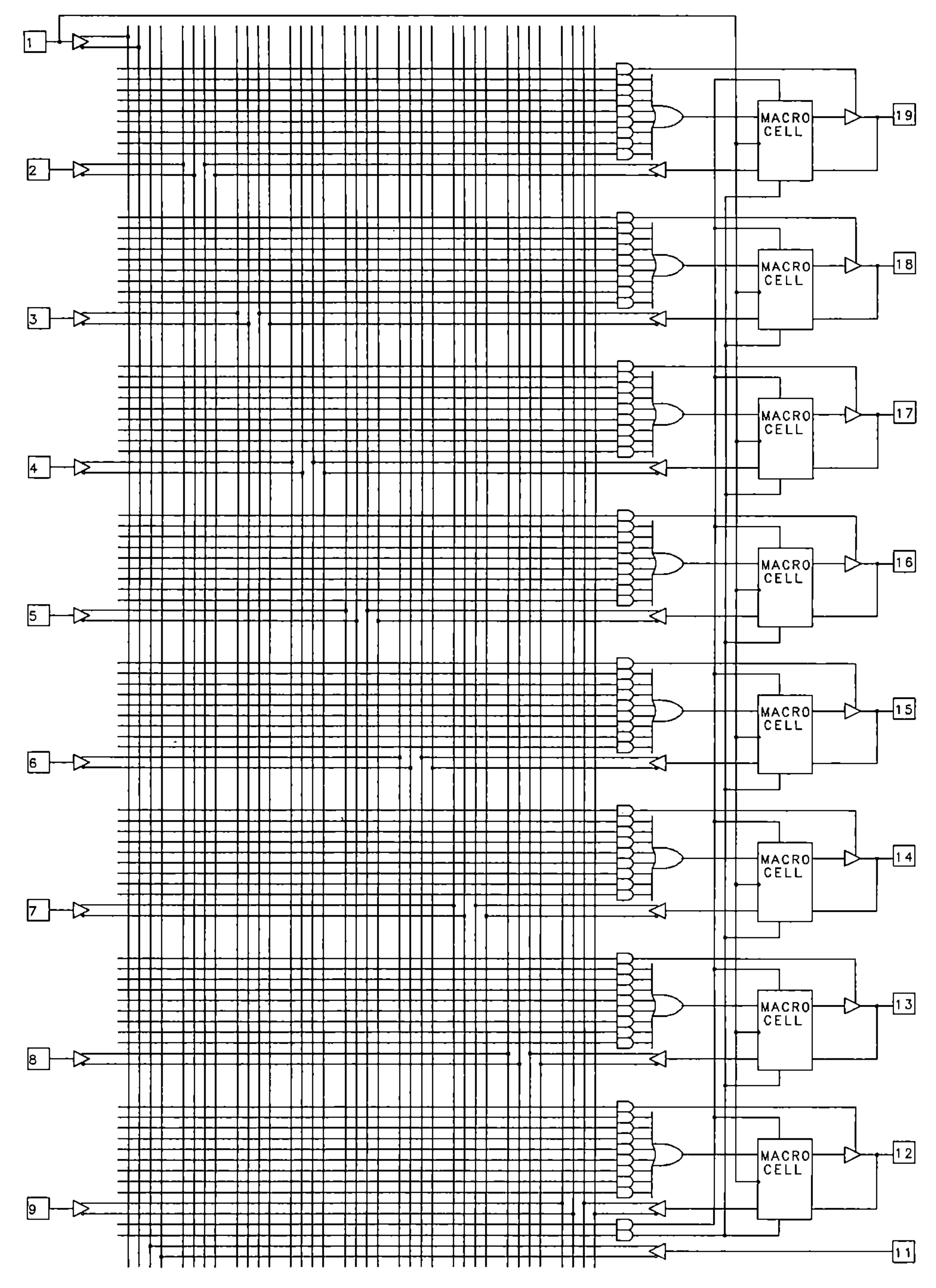

Figure 3.23 PEEL 18CV8 device

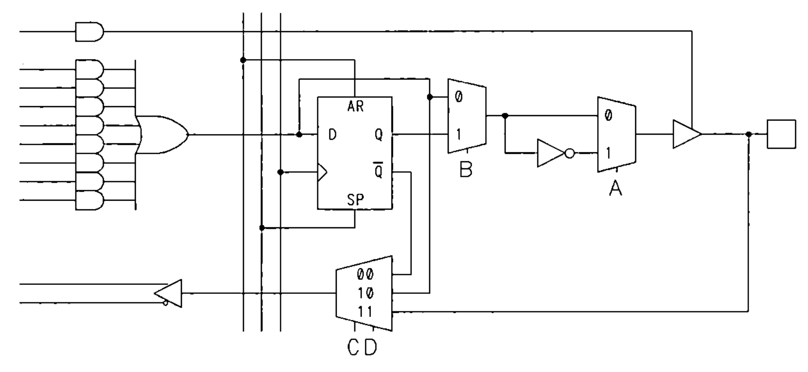

Figure 3.24 PEEL 18CV8 macrocell detail

Figure 3.25 PEEL 18CV8 macrocell configurations

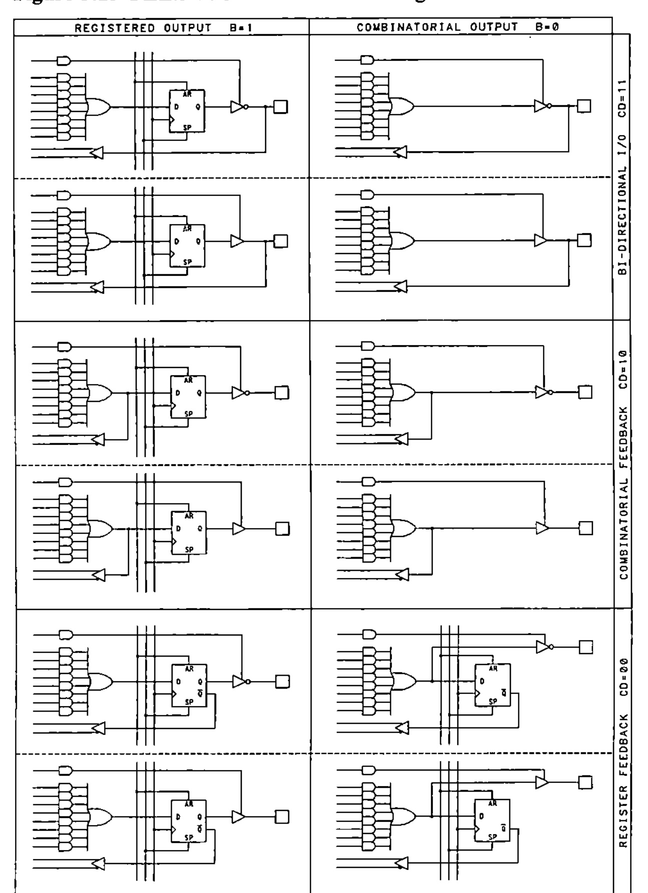

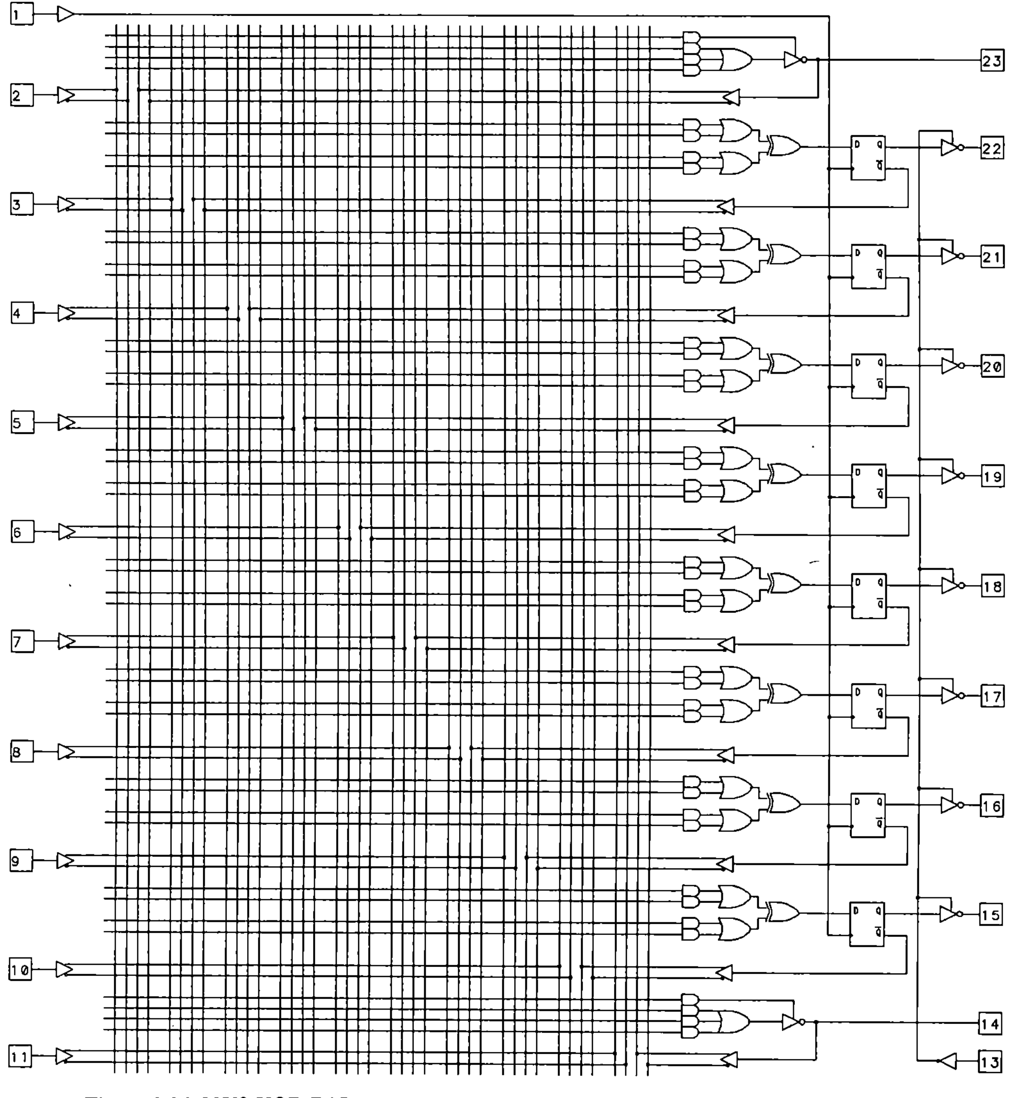

Figure 3.26 20X8 XOR PAL

flip-flops. The flip-flops also feature a transparent latch mode that is selected by asserting both the preset and reset signals high.

Complex PAL Devices

In the devices presented thus far, we've seen how configurable macrocells allow a single PAL type device to replace a number of different types of fixed output PALs. We have also seen how the adding of various output macrocell features

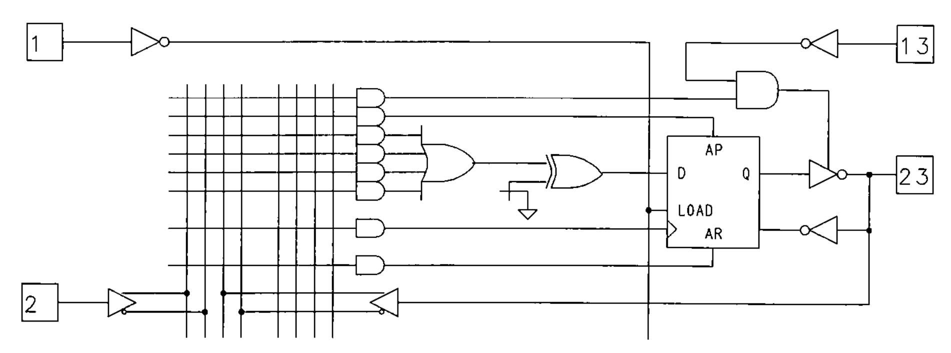

Figure 3.27 20RA10 output macrocell

such as XORs have extended the capabilities of the standard PAL architecture. Configurable output macrocells can also be used to increase the effective size of a device.

Examples of particularly powerful macrocells are found in the Altera EP600 and EP900 families of devices. The EP600 and EP900 families are popular UV-erasable devices of 24 and 40 pins (in DIP packages), respectively. Figure 3.28 illustrates the smaller EP600 device in block diagram form.

The macrocells found in these two families of devices are identical and have a number of useful features. As shown in Figure 3.29, the EP600 and EP900 macrocells feature the ability to be configured for operation as combinational inputs and/or outputs, or as configurable flip-flops that can be D-type or T-type. (JK-type or SR-type flip-flops can be emulated in the devices using Altera's design software.) These devices were designed with high-level design tools in mind, so they are regular in their construction and highly configurable. The flip-flop emulation features are designed so that no product terms are wasted in the device when alternative (non D-type) flip-flop types are used. Programmable inversion is provided between the PAL array and the flip-flop inputs, simplifying the use of programmable polarity.

When operated in the D-type or T-type flip-flop mode, the feedback path from each macrocell can be configured to route the feedback from the flip-flop, route it from the associated I/O pin, or disable feedback entirely. When JK-type or SR-type flip-flops are emulated, feedback from the I/O pin is not available.

Asynchronous operation is provided by a clock-select multiplexer. The multiplexer for each macrocell is fed by a single product term that can be routed to either the associated flip-flop's clock input, or to the output enable for the associated device pin. When this product term is routed to the output enable, the flip-flop is clocked from a global clock pin; when the product term is used for clocking, the outputs are always enabled. The EP900 family of devices have two clock pins, each of which can control up to twelve of the EP900's 24 macrocells. The flip-flops in the EP600 devices are similarly banked, with a total of sixteen macrocells clocked from two dedicated pins. In addition to the clock/enable product term, another product term is dedicated to each macrocell for clearing the macrocell's flip-flop.

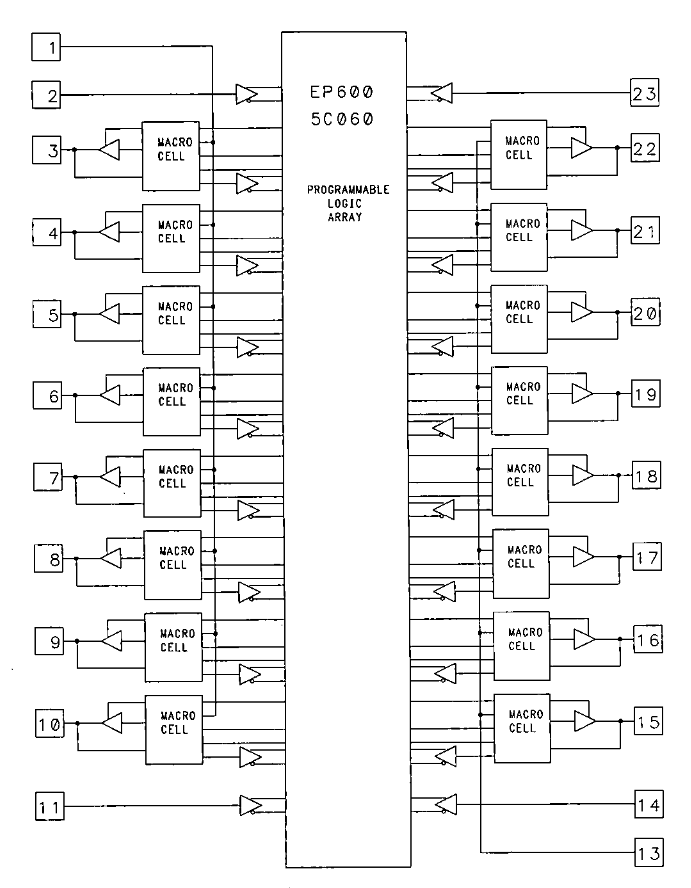

Figure 3.28 EP600 block diagram

Figure 3.29 EP600/EP900 output macrocell

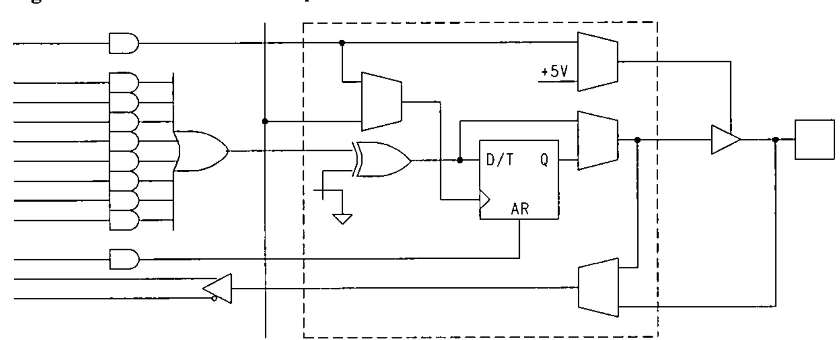

Another popular device that features complex macrocells with asynchronous capabilities is the AMD 29MA16. This device has sixteen complex macrocells (one of which is shown in Figure 3.30) that include configurable clock sources that can be either from a dedicated device pin or from a product term unique to each macrocell. Output enable is also selectable from multiple sources: either from a dedicated pin, from a product term, or fixed.

The 29MA16's registers are further configurable to operate as edge-triggered D-type flip-flops or as level sensitive latches. Dual register feedback provides the ability to bury the registers for state machine purposes while using the associated pin as an input. It's also possible to use the 29MA16's registers for input synchronization purposes by connecting the pin directly to the register's *D* input.

The trend toward more configurable output macrocells has continued to the point where the macrocells have grown so complex that it's almost impossible to comprehend them without the aid of software tools. One such complex macrocell is found in the Cypress 7C331. The device is shown in block diagram form in Figure 3.31. As the diagram shows, the macrocells are paired, and each pair of adjacent macrocells has a common feedback path, in addition to their own dedicated feedback paths.

The 7C331's macrocell architecture is shown in Figure 3.32. As the figure shows, the 7C331 has XOR gates associated with each macrocell. These XOR gates are arranged with one of their inputs fed by a single product term, while the other is fed by a normal sum-of-products array. The number of product terms allocated to each macrocell varies in the device—a situation common in complex PALs. The asymmetrical XOR architecture makes it easy to emulate T-type or JK-type flip-flops in the 7C331, as we'll describe in Chapters 7 and 8.

Figure 3.30 29MA16 output macrocell

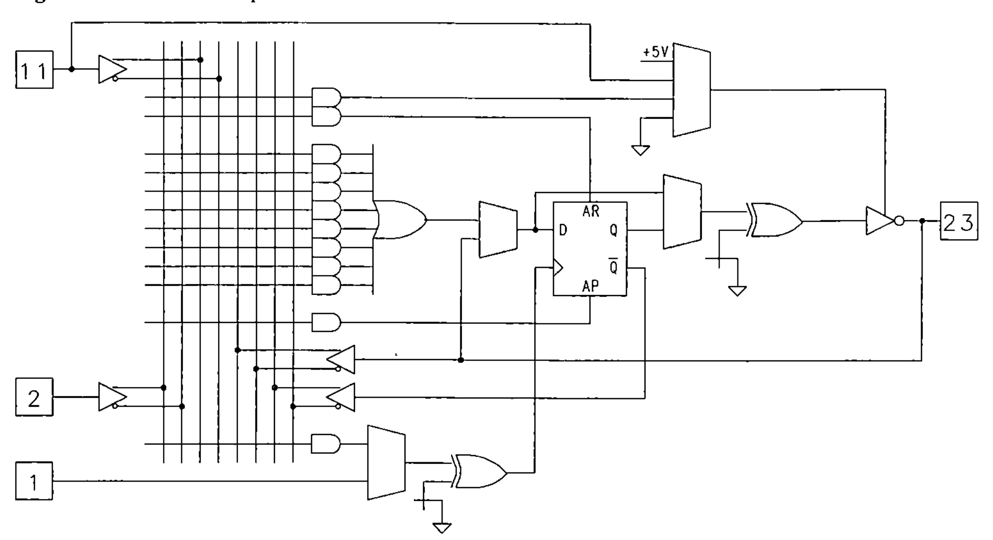

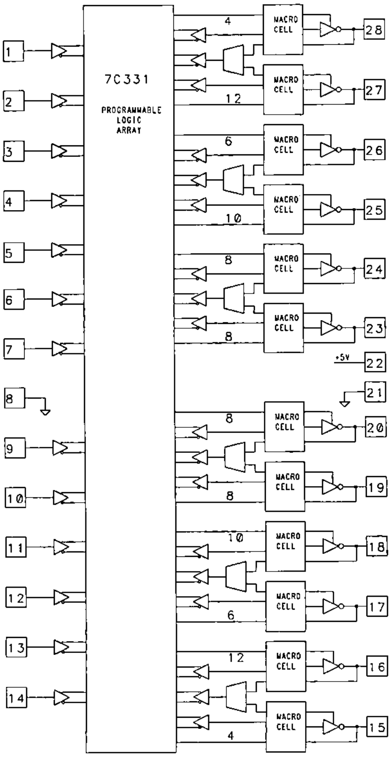

Figure 3.31 7C331 block diagram

Figure 3.32 7C331 macrocell

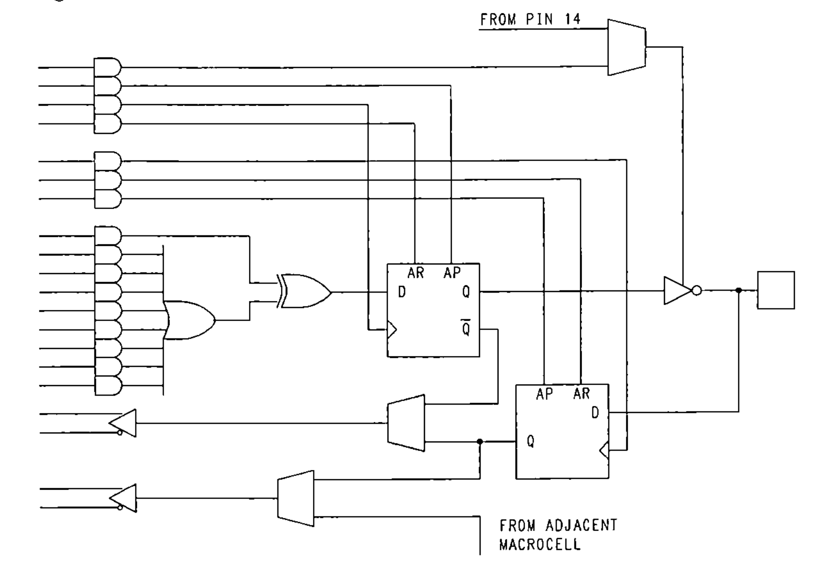

In addition to the primary D-type flip-flop, a second flip-flop is associated with each of the 7C331's twelve output macrocells. These secondary flip-flops can be used as input registers while the primary flip-flops are being used as buried state registers. Both flip-flops can be bypassed dynamically by asserting their preset and reset signals simultaneously. Notice that the presets and resets of both the 7C331 and 29MA16 are asynchronous, unlike in the 22V10 where the preset is synchronous and reset is asynchronous. Having both the preset and reset asynchronous allows polarity reversal (often necessary to fit a design equation into the device) to be accommodated easily through the swapping of preset and reset functions.

An architectural tradeoff is found in the 7C331's dual feedback scheme. Rather than providing dual feedback paths for all macrocells, the device instead has feedback paths that are shared between pairs of adjacent macrocells. This means that buried state registers must be assigned to pins based on the availability of feedback. Cypress parts are known for their high-speed operation, and the 7C331 is no exception. The device is packaged in a large 28-pin DIP, but has dual ground pins to help reduce ground bounce problems.

PLA Devices

As we said in Chapter 2, the PLA was the basis for the first programmable logic devices. Devices that include a complete PLA (with programmable AND and OR arrays) are more flexible in terms of the variety of circuits that can be implemented in them. Their drawback is that they tend to be slower and more expensive due to the second programmable array.

One of the simplest examples of the PLA architecture is found in the Signetics PLS100. This device (which is identical to the 82S100 chip described in Chapter 2) is diagrammed in Figure 3.33. The diagram orientation commonly used for PLA architecture looks somewhat different from the PAL diagrams we have previously presented. The biggest difference is that the product terms are represented by vertical (rather than horizontal) lines. This representation makes it easier to illustrate the programmable OR array, which is the lower of the two arrays.

As Figure 3.33 shows, any of the 48 product terms for this device can be accessed by any OR gate. This makes it possible to implement more complex logic functions than possible in a PAL device of similar pin configuration. The device shown has no provision for configurable I/O, aside from the ability to select an output inverter for each output pin. A 24-pin version of the PLS100 device is available and is called the PLS161.

The most popular PLA-type devices are designed for state machine applications. The first of these, which Signetics refers to as FPLS type devices, was the Signetics 82S105, which is now called the PLS105. This device is shown in Figure 3.34. The PLS105 has a number of interesting features that set it apart from the registered PALs in its ability to implement complex state machines.

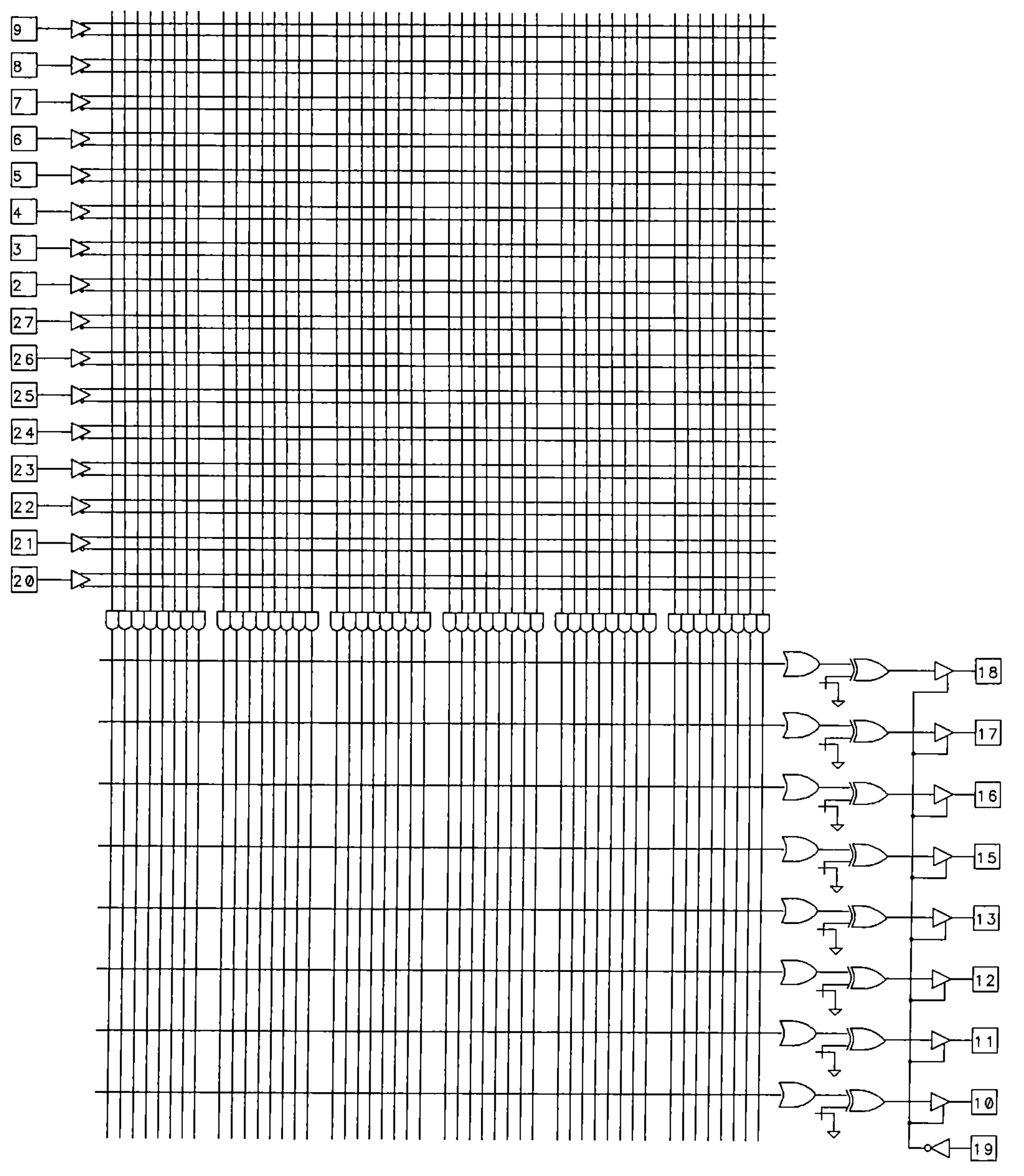

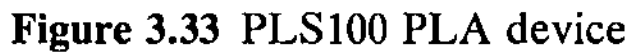

Figure 3.33 PLS100 PLA device

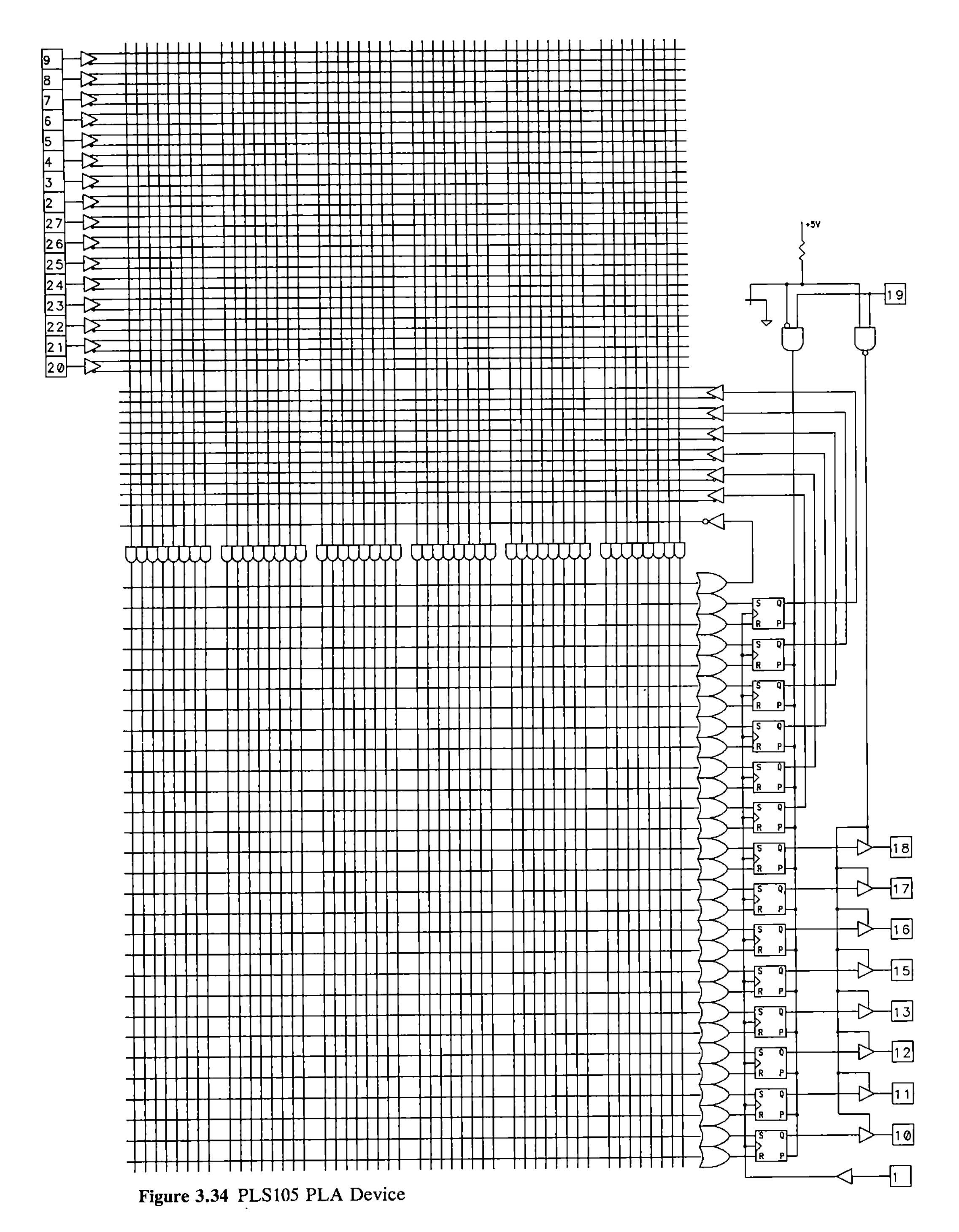

Figure 3.34 PLS105 PLA Device

The first unique feature found in the PLS105 is its use of SR-type flip-flops for memory elements. SR-type flip-flops are particularly well suited to state machine applications, often requiring far less logic to implement transitions than equivalent designs implemented using D-type flip-flops.

Six of the PLS105's SR-type flip-flops are buried within the device, and aren't accessible as outputs. They are, however, fed back to the AND array. This *buried register* feature is being found on an increasing number of devices, and will be discussed in more detail later. The remaining eight SR-type flip-flops are associated with device outputs, and are not fed back to the AND array. This configuration of internal state memory and output synchronization registers is ideally suited to complex state machines (state machine design will be described in later chapters).

Implementing complex state machines in PAL devices can be difficult because there are generally not enough flip-flops to provide both state memory and output synchronization. In a device such as the PLS105, however, such state machines are straightforward and easily achievable.

Another feature found in the PLS105 is the complement array. The complement array is a single OR array output that is inverted and fed back into the AND array to provide a very useful function, which is to detect undefined states and force the machine to return to a known state. The exact techniques for using this feature will be explored further in Chapter 8.

One attribute that limited the early success of the PLS105 is its size. The device has 28 pins, which is larger than the standard 20- or 24-pin PAL devices. In a DIP package, this device takes up quite a lot of board space. In the newer PLCC packages, however, 28 pins is a more reasonable number, so this decade-old design is gaining in popularity. The PLS105 device does have a smaller brother, the PLS167, that is supplied in a more conventional 24-pin package with corresponding reductions in the number of available inputs and outputs while retaining the same number of buried state registers.

TI's TIBPLS506 and TIBPSG507 are similar in architecture to the Signetics PLS105, featuring SR-type flip-flops for internal state registers and for output purposes. The TIBPLS506 feature an increased number of buried registers over the PLS105 and also includes a complement array. Unlike the PLS105, the output registers in the TI devices can be bypassed for combinational uses.

The TIBPSG507 is a specialized sequence generator device that includes an internal 6-bit counter. The device (diagrammed in Figure 3.35) is intended for applications such as waveform generators, dividers, timers, and counter-based state machines.

Another popular PLA-type device is the Signetics PLS159 device. This device, shown in Figure 3.36, extends the flexibility of the PLS105 architecture while at the same time reducing its size.

In the PLS159, there are eight registers, each of which can be used as either a state memory element or as a synchronized state machine output. In addition, four additional combinational I/O pins (of the sort found in the 16L8) are provided

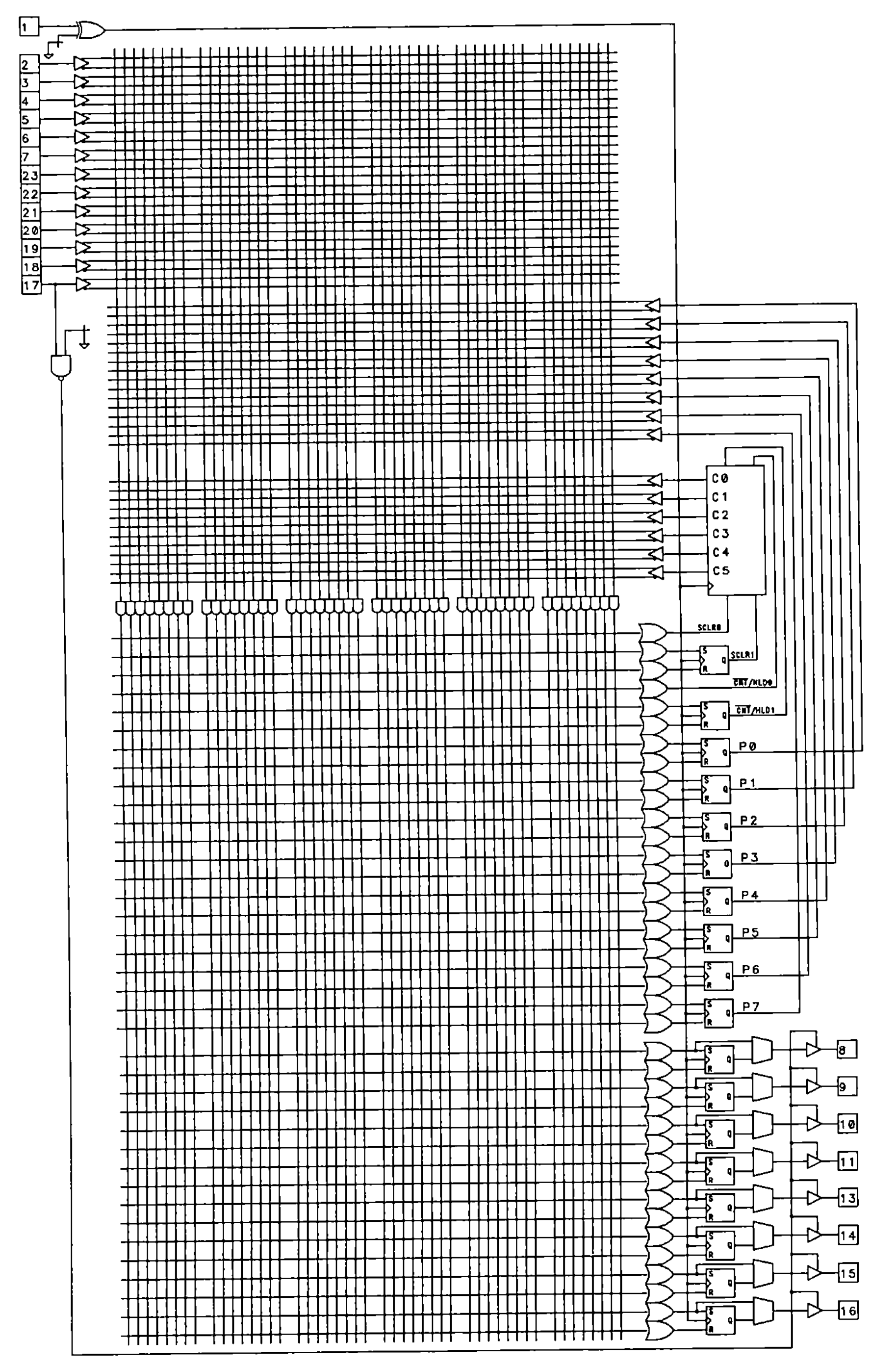

Figure 3.35 PSG507 Programmable Sequence Generator

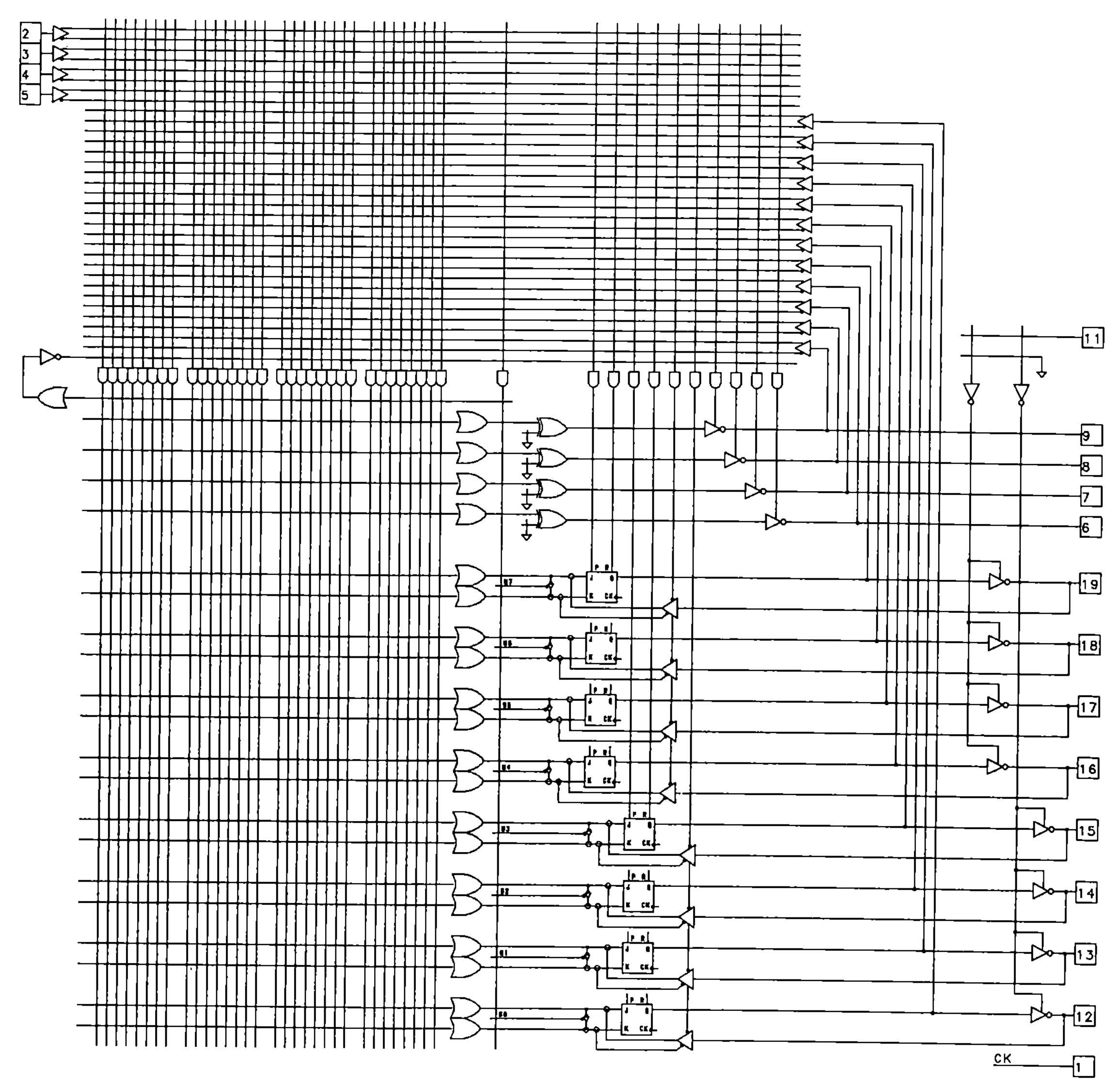

Figure 3.36 PLS159 PLA device

that can be used as inputs, outputs, or feedback to the AND array for multilevel logic purposes. These combinational I/O pins have fuse selectable inverters, further increasing their flexibility.

The registers used in the PLS159 are far more flexible than their counterparts in the PLS105. Rather than being composed of SR-type flip-flops, the registers are constructed from JK-type flip-flops. The use of JK-type flip-flops, coupled with the mode control product term (*Fc* on the diagram) and individually selectable mode control lines for each flip-flop (*M0* through *M7*), allows each register element to be operated either as a fixed JK-type, fixed D-type, or dynamically selectable D/JK-type flip-flop. The dynamically configurable flip-flop is detailed in Figure 3.37.

The availability of dynamically selectable register types gives increased power to the state machine designer. For example, the D mode can be used to load specific states into the machine and the JK mode can be used during normal machine operation. Signetics' newer PLC42VA12 device also features dynamically configurable flip-flops that can be bypassed if necessary.

Additional device features that aid in state machine design (and testability) include product-term controlled preset and reset and product-term enabled load functions. Like the PLS105, the PLS159 includes a complement array for escaping from undefined states.

One of the most complex PLA-type devices currently available is the 6001 device from Lattice. This device also features configurable flip-flops and eight buried macrocells that can be used for state machine purposes. A simplified version of the device is shown in Figure 3.38 (the actual device has eight buried macrocells and ten I/O macrocells). The device is significantly more complex than the earlier PLA-type devices. Not only does it have eighteen complex output macrocells, it has some other intriguing features. Before examining these additional features, let's look at the output macrocells.

Like the PLS105, the 6001 device has a set of buried macrocells. These eight macrocells are located on the left side of the diagram and are called the state logic macrocells, or SLMCs. The buried macrocells are identical in design and function to the ten output macrocells associated with pins 14 through 23.

Each of these eighteen macrocells is composed of a D-type flip-flop with a gated clock. When operated in a D mode, the *E* input to the macrocell can be used as a clock enable line. This is useful for data latching and holding operations.

Figure 3.37 PLS159 dynamically configurable flip-flop

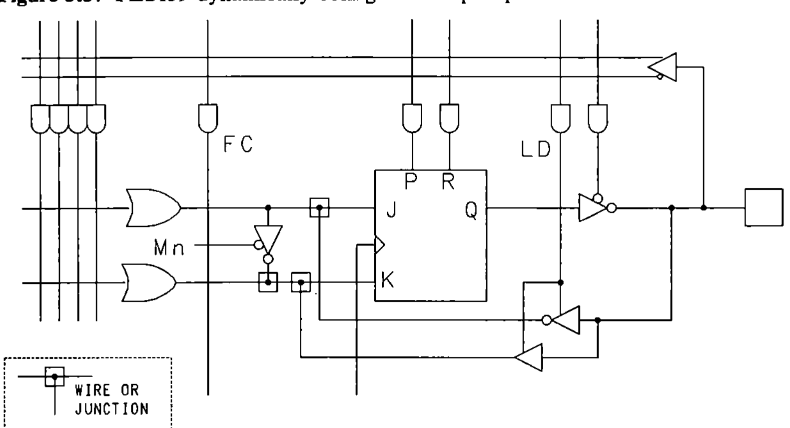

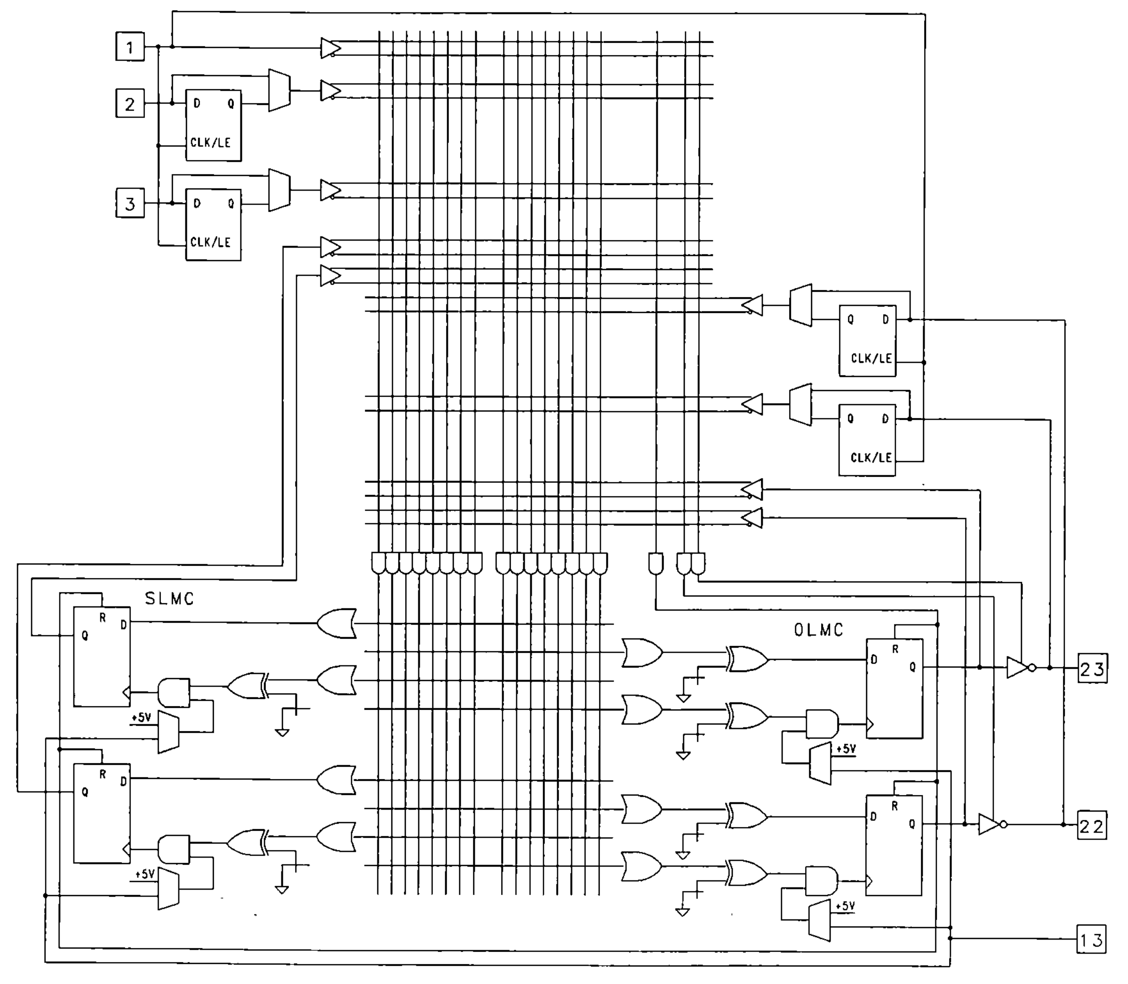

Figure 3.38 6001 PLA device (simplified)

Alternatively, the *E* input to the macrocell can be used for asynchronous clocking purposes or for emulation of alternate flip-flop types, as shown in Figure 3.39. Notice that, since the 6001 is a PLA, it's possible to emulate SR-type flip-flops with no waste of product terms.

As is the case in the 22V10 output macrocell, any of the 6001's flip-flops can be bypassed to provide purely combinational operation. Unlike most devices, it's possible to isolate any of the output macrocells from their associated output pins. This allows macrocells to be used as buried registers while at the same time using the corresponding pins as dedicated inputs.

The final feature that is noteworthy about this device is its set of banked input macrocells. These configurable macrocells are provided for all of the 6001's input and I/O pins, with the exception of pin 1. Each of the two input macrocell

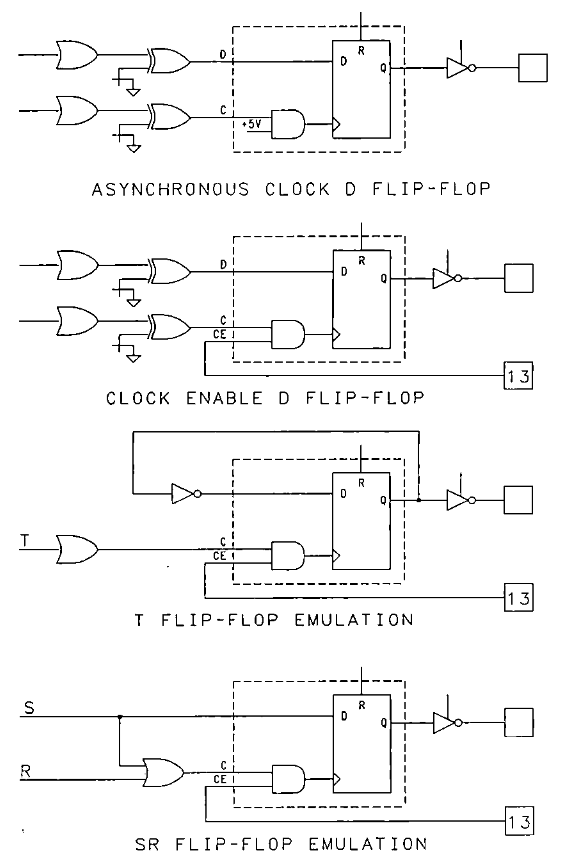

Figure 3.39 6001 PLA flip-flop emulation options

sections are configurable as a group to operate as either asynchronous, latched, or registered inputs. Configurable input macrocells such as these are appearing on more devices, and are useful for a variety of applications requiring input synchronization.

Product-term Steering

Product-term steering is a limited implementation of the PLA concept in a PAL type device. In its simplest form, a product-term steering device such as a 20S10 (shown in Figure 3.40) allows adjacent outputs to share terms. This means that those outputs provided with term steering can use as many as fourteen product terms each. This feature might be more appropriately called product-term stealing, since it isn't possible for both of the adjacent outputs to use the same product term.

Another form of product-term steering is found in the Atmel 750 and 2500

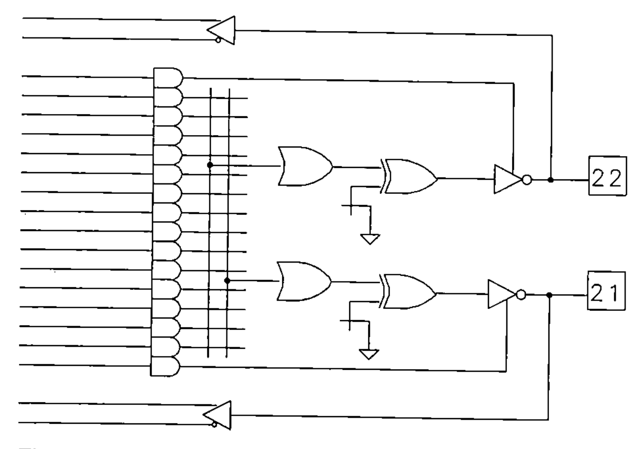

Figure 3.40 20S10 PAL device

devices. These devices feature buried macrocells associated with output macrocells. Figure 3.41 shows one output macrocell for the 24-pin Atmel 750 with its associated buried node. Typically, the buried macrocells are used for state registers but, if desired, the product terms feeding a buried node can be appended to the product terms of its primary macrocell. This effectively doubles the number of product terms available to the primary macrocell. The buried macrocell can still be used, assuming the shared product terms are appropriate for both functions.

Figure 3.41 750 macrocell pair

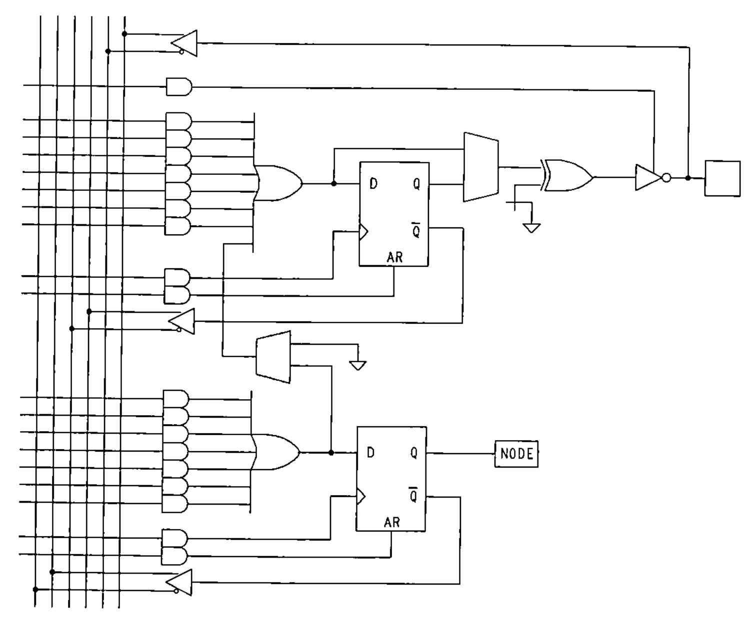

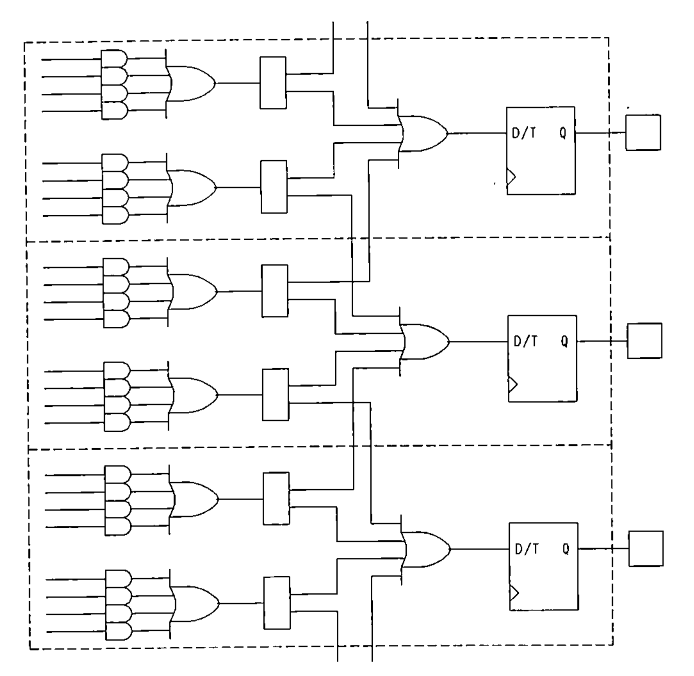

Figure 3.42 5AC312 product-term distribution

The 40-pin Atmel 2500 device is similar, but features two buried macrocells for every primary macrocell.

The Intel 5AC312 has another form of product-term steering. As shown in Figure 3.42, the device has its product terms segmented into groups of four. These groups are each allocated to a single output of the device, but if more product terms are required for a particular output, they can be reallocated. The reallocation is limited to nearby outputs, allowing one macrocell to have as many as sixteen product terms.

Still another product-term steering concept is found in the Altera MAX family of parts (which will be covered in more detail in the next chapter). These devices include a set of product terms that aren't allocated to any particular output. These product terms, referred to as expander terms, can be allocated as needed to outputs that require additional product terms. Use of expander terms has speed penalties, however, so their use must be weighed against the performance criteria for the design. Figure 3.43 illustrates the expander term concept.

Unlike other term steering schemes, the expander term architecture requires that the design equations be converted into an alternate (non sum-of-products) form.

Segmented Arrays

As PLDs increase in complexity, the size of the programmable array quickly becomes unmanageable. For this reason, some manufacturers have developed

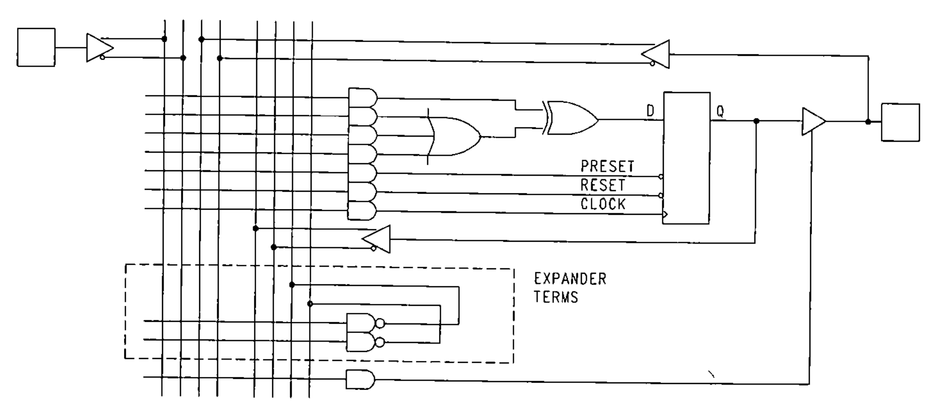

Figure 3.43 MAX expander terms

architectures that are segmented into smaller arrays with limited interconnections. One such device is the Altera EP1800 (shown partially in Figure 3.44).

The EP1800's programmable array is split into four identical quadrants that have limited interconnects. Portions of two quadrants are shown in the figure. This segmentation means that designs being implemented in the EP1800 must be partitioned and allocated to quadrants based on the amount of interconnection required. There are two types of feedback available in the EP1800: one feedback path is available to the local quadrant only, while the other path feeds signals globally around the device. This is illustrated in the diagram of Figure 3.45 which shows one output macrocell of the EP1800. The EP1800 macrocell architecture is identical to that of the EP600 and EP900 devices with the addition of the global feedback line on some (but not all) of the macrocells.

The EP1800 has some other interesting features, as well. The registers are configurable and may be used as T-type or D-type flip-flops. The clocking and output enable scheme is also interesting and allows either the clock or enable signal (but not both) to be controlled from a product term.

Folded PLAs

The sum-of-products array is, as we have seen, the mainstay of traditional PLDs. For large designs, however, the standard sum-of-products array can prove to be too restrictive. In particular, multilevel logic applications are difficult to implement in standard PLDs. To implement such designs in a PLD, succeeding logic levels must be fed back into the logic array, with often unacceptable speed penalties or wasted device resources.

The Folded-NOR Array

One architecture that is better suited to such multilevel applications is the folded-NOR architecture pictured in Figure 3.46. The *folded-NOR array* permits suc-

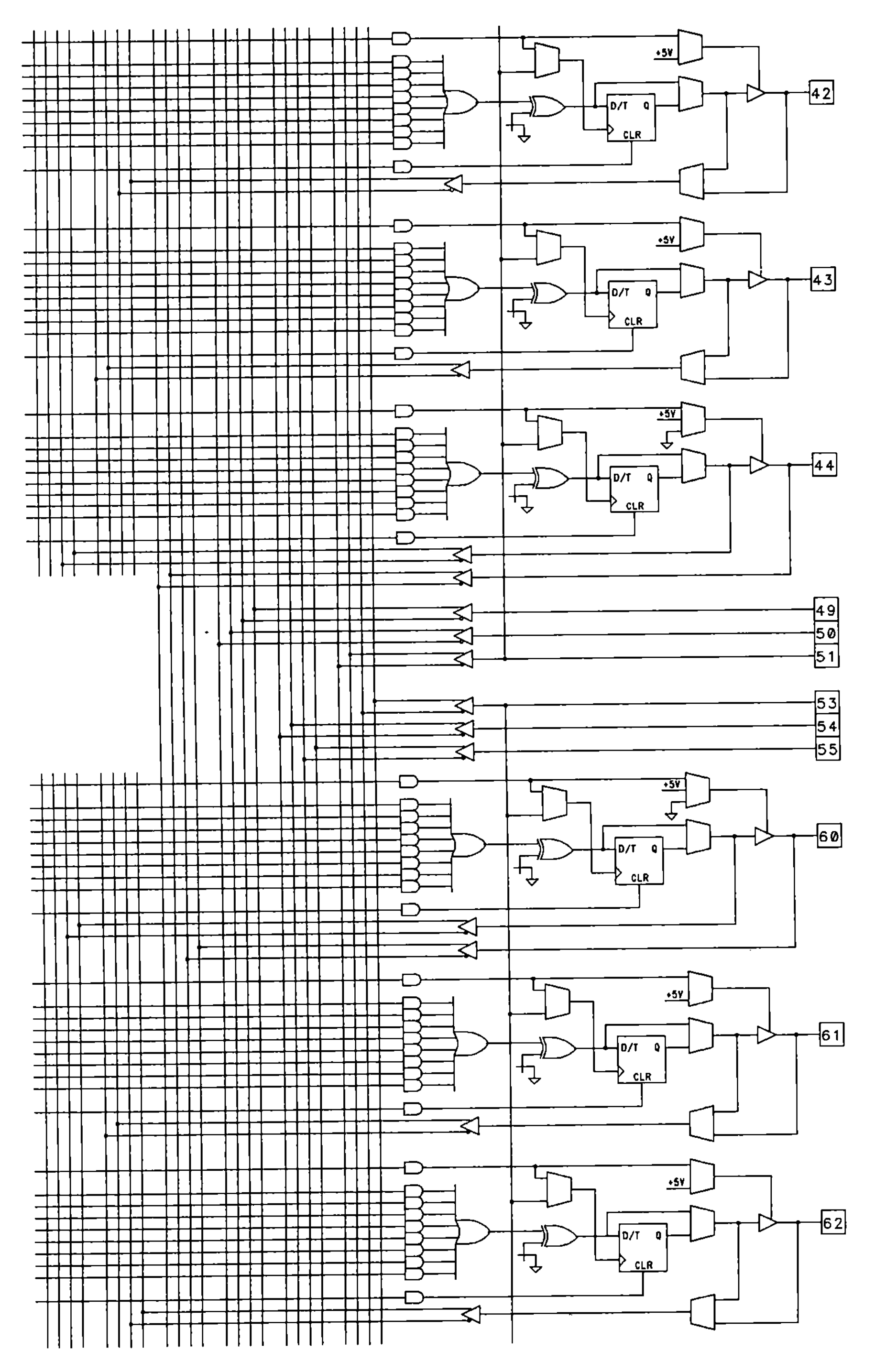

Figure 3.44 EP1800 segmented architecture (partial logic diagram)

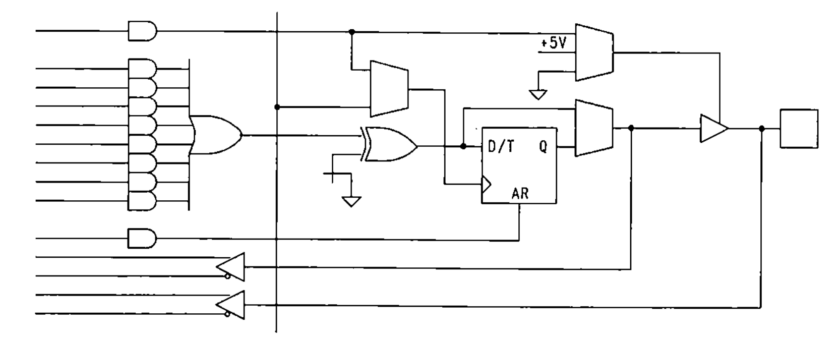

Figure 3.45 EP1800 output macrocell detail

cessive levels of NOR gates to be fed back and cascaded to implement multilevel logic designs. The array shown is a simplified version of Exel's ERASIC device.

The ERASIC is a good example of a folded-NOR PLA device. The XL78C800 features ten highly configurable output macrocells and eight dedicated input latches in addition to the folded-NOR array. The device consists of a single programmable NOR plane, which can be used to generate logic of as many levels as logic for a single output as you would ever practically require. The only limitation is the number of NOR terms provided in the array.

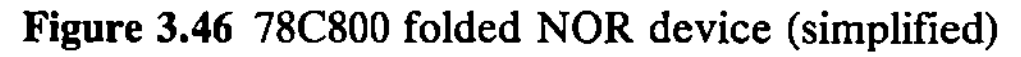

Figure 3.46 78C800 folded NOR device (simplified)

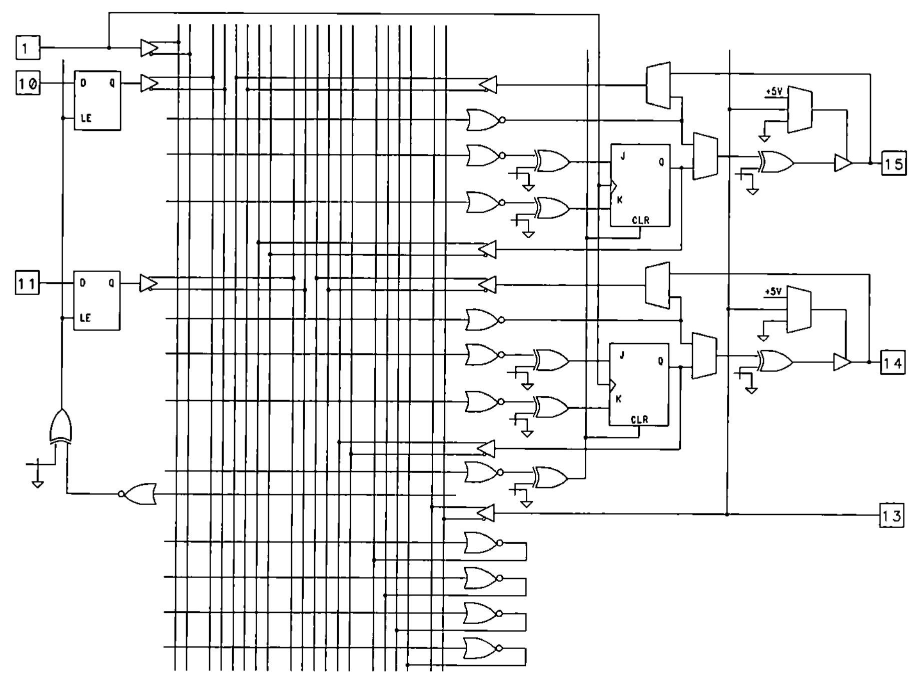

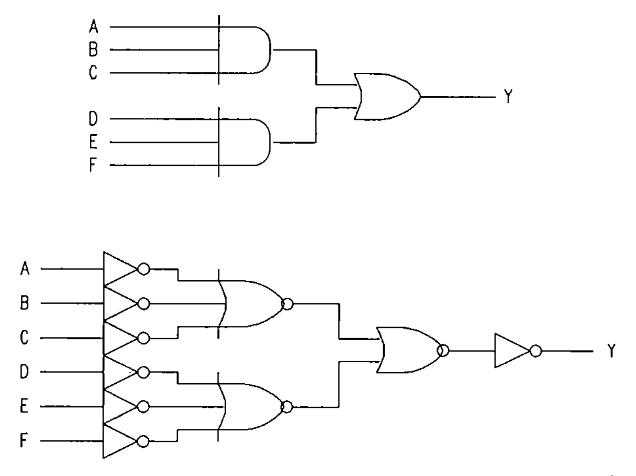

Figure 3.47 Implementing logic functions in NOR–NOR logic

How does this work? Consider the logic shown in Figure 3.47. As the figure illustrates, it's possible to implement either a product term (an AND operation) or a sum term (an OR operation) in a NOR term simply by choosing the appropriate polarity of inputs to the NOR gate.

From this, it can be shown that all possible logic functions can be implemented in NOR–NOR logic, just as they can be implemented in sum-of-products. Furthermore, the use of just one array allows the implementation of multilevel logic functions that are quite cumbersome and wasteful to implement in a sum-of-products array.

The folded-back NOR terms are referred to as *asynchronous feedback terms*. In the ERASIC device, there are two types of these terms available. The lower part of the array (shown in the logic diagram) is composed of these terms in their simplest form. These 32 terms are not associated with any device output and so, are roughly analogous to the AND terms in a standard PLA (remember, however, that these NOR terms can be used for OR functions in a multilevel application).

The upper, more complex part of the array is broken into groups of NOR terms that feed the ERASIC's ten complex output macrocells. Each macrocell is fed by three NOR terms. In addition to the NOR terms associated with output macrocell functions, there are two additional NOR terms that are used for asynchronously clearing the device's flip-flops.

The different configurations of this output macrocell are shown in Figure 3.48. Notice that it's possible to bury the JK-type flip-flop and free the associated output pin for use as an input. It's also possible to bury the combinational NOR gate and feed it back to the array as an asynchronous feedback term while still using the pin as an independent output.

An interesting feature of this macrocell is the method used to control the output polarity. Rather than having an XOR gate on the register output, this macrocell contains programmable polarity elements on the J and K inputs to the

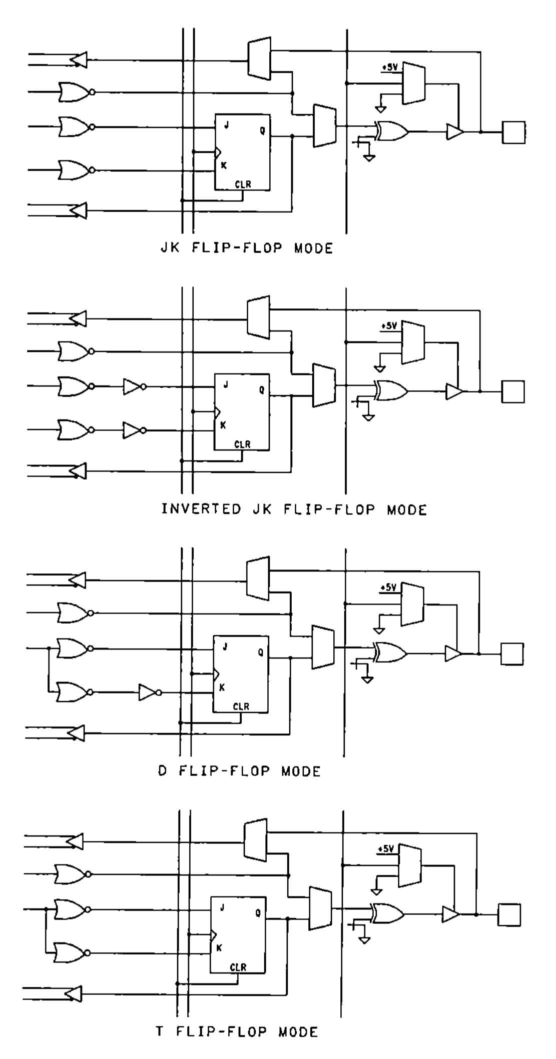

Figure 3.48 78C800 macrocell configurations

registers. In order to reverse the polarity of the macrocell output, you must swap the *J* and *K* inputs. In this situation, the register clear function becomes a register preset. Using the programmable polarity flip-flop inputs, the macrocell can be set up to emulate D-type or T-type flip-flops by supplying either the *D* and $\overline{D}$ inputs to the *J* and *K* inputs or by supplying the *T* inputs to both flip-flop inputs with no inversion.

Finally, the ERASIC device features latchable inputs on eight of the dedicated input pins. The latching of these inputs is controlled by two dedicated NOR terms.

The Folded-NAND Array

Another family of folded-PLA devices is produced by Signetics. These devices, which Signetics has dubbed PML (*Programmable Multilevel Logic*) devices, feature a *folded-NAND array* rather than a folded-NOR array. A simplified diagram of the PLHS502 is shown in Figure 3.49.

The 52-pin PLHS502 (and it's smaller brother the PLHS501) have a large number of buried logic elements. As the diagram shows, the device includes eight SR-type flip-flops intended for state machine applications, and eight D-type flip-flops useful for shifter or output synchronization purposes. In addition to the folded-NAND array structure, the PMLs have a powerful clocking array (not shown) that allows many different clocking strategies to be used.

Specialized PAL Devices

The versatility of the PAL architecture has led to a number of highly specialized programmable devices that are PAL-based. These devices are, by their very nature, difficult to categorize.

In-circuit Reprogrammable Devices

In-circuit reprogrammable PLDs are those can be reconfigured dynamically while they are operating in a system. One such device is the Lattice ISPGAL (*In-System*

Figure 3.49 PLHS502 folded-NAND device (simplified)

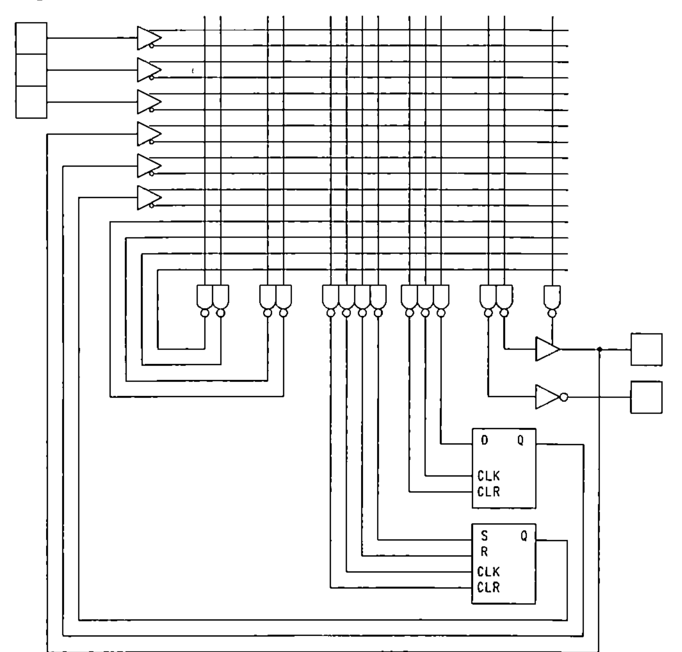

Programmable GAL). The 16Z8 ISPGAL is an in-system programmable version of the 16V8 GAL device. The 16Z8 has four extra pins that are used exclusively for programming operations. Having these extra pins means that the normal inputs and outputs of the device aren't interfered with during programming operations.

Asynchronous PLDs

Asynchronous applications have long been a source of trouble for designers using PLDs, and device manufacturers have responded by introducing PLDs with more flexible clocking schemes, or completely asynchronous output registers. The 20RA10 and 29MA16 are examples of devices designed for asynchronously clocked applications.

The problem of metastability has led to some new PLD architectures intended for asynchronous interface applications. One such device is the PAL 22IP6 produced by AMD. This device utilizes unusual output macrocells that include non-clocked output registers, three of which are SR-type flip-flops, while the other three are dual T-type flip-flops.

These output macrocells, which AMD refers to as *interface protocol asynchronous cells* (IPACs) aren't subject to the risks of clock setup and hold violations, although the input signals applied to the IPAC registers must still meet certain, less stringent timing requirements.

Programmable Bus Interfaces

Programmable bus interface devices fill a specific need for interface logic in bus-oriented systems. These devices are usually designed with a particular bus architecture in mind, such as the VMEbus or Micro Channel. Bus interface PLDs can be used to eliminate a large number of simpler devices normally required for such applications.

Altera's BUSTER is one device that is designed for bus interface applications. The device has input synchronization registers, latched outputs, and a large number of configurable macrocells which are found both on I/O pins and buried within the device.

Another PLD intended for bus interface applications is the Intel 85C960. This programmable bus control is targeted specifically at systems utilizing Intel's 80960 microprocessors. The device provides address decoding, wait state, and ready generation at the relatively high speeds typical of systems using the 80960.

Intel also has a more general bus interface PAL called the BIC device. The 5CBIC contains three octal latched transceivers that can be clocked in any of eight modes. Each I/O pin on the 5CBIC has a configurable bus-oriented macrocell. In addition, the 5CBIC has four buried macrocells.

PLX Technology has yet another of these devices. The PLX448 is similar to a 22V10 in construction, being based on the PAL structure. The device features ten inputs and eight bidirectional outputs. The PLX448's I/O drivers are compatible with the IEEE-448 interface standards possessing four outputs that can drive 48 mA loads, and four separately clocked outputs capable of driving 24 mA loads. Four of the outputs can be configured as open-collector type buffers. The PLX464 is similar and has outputs with 64 mA drive capability.

Programmable Sequencers

Programmable sequencers are hybrid devices that contain both a PAL array (used for state machine branch control) and a PROM-based microsequencer. MMI's PROSE device was one of the earliest of these devices and contains a bipolar PAL array (similar to the 14H2 PAL) of sixteen product terms along with a 128 word by 21-bit PROM array. Eight of the PROM outputs are connected to registered outputs, five are routed back into the PROM array, two are XORed with the two PAL outputs, and six are routed back into the PAL array as inputs. The PROSE device hasn't proven to be a popular part, partly because of the difficulty in using it (there is no high-level design tool support) and because of the limited advantages of using the PROSE versus implementing the same function in multiple smaller devices.

The Altera SAM device has had more success than the PROSE. This device is block diagrammed in Figure 3.50. The SAM is much larger than the PROSE, featuring a 768 product-term erasable PAL array and a 448-word by 36-bit EPROM. The SAM also features a 15-byte stack and an 8-bit loop counter. High-level design tool support is provided by Altera in the form of a microcode assembler and functional simulator.

AMD's 29PL141, like the Altera SAM device, is a microprogrammed device. Microinstructions are programmed into the chip's program memory. AMD's device has 29 microinstructions that include branching, subroutine calls, loops, and

Figure 3.50 SAM programmable sequencer

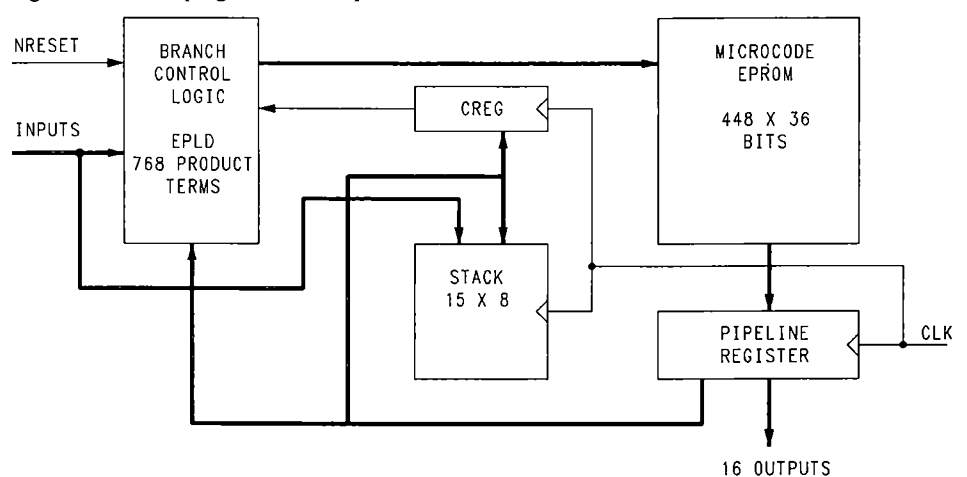

jumps. The device supports up to 64 states for state machine designs. Output data is stored in the device's PROM section. A block diagram of the 29PL141 is shown in Figure 3.51.

The AMD 29PL141 is supported with a microcode assembler that allows instructions to be written in a high-level language. These instructions are assembled into PROM data and microinstructions. Like a conventional PLD, the device uses JEDEC format files for fuse data and is programmed on a standard device programmer.

High-speed Architectures

The high clock speeds of many modern systems place heavy demands on the speed of the interface logic for those systems, much of which can be most efficiently implemented in PLDs. There are two approaches to speeding up PLDs. The traditional approach has been to use high-speed technologies (processes) such as ECL (emitter control logic). This is expensive and creates interface problems when the PLDs must be integrated with the rest of the circuit.

A second approach is to modify the architecture of the traditional PLD, optimizing it for high-speed operations. This architectural approach to high-speed devices is seen in the Intel 85C508. This device is intended specifically for high-speed address decoder applications in which OR gates may not be required. The AND gates of this device directly feed its latched outputs, eliminating the propagation delay normally associated with the OR gate.

Another device which uses an architectural approach to high-speed circuits is the Cypress CY7C361. This device is quite complex and is optimized for high-

Figure 3.51 29PL141 programmable sequencer

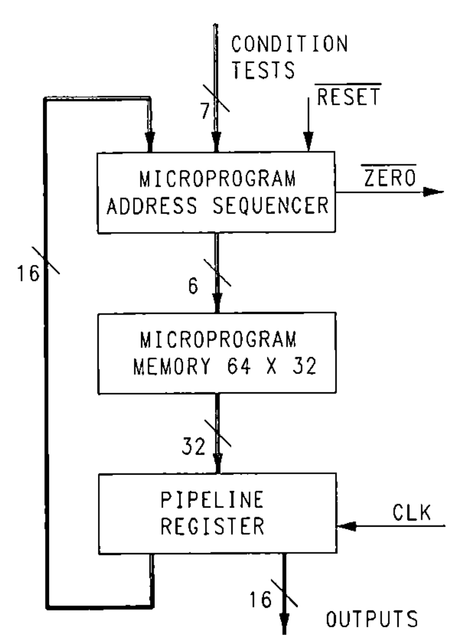

speed state machine applications. The CY7C361 has a bank of 32 state registers that are located between the AND and OR arrays in a structure that Cypress calls a *split-plane* logic array.

The central location of the CY7C361's registers means that signals can propagate from the device inputs to the state registers with less gate delay than in traditional PLDs. More significantly, the delay times of the clock and register inputs are more closely matched, and the OR array outputs are synchronized through the use of a special de-skewing buffer. This combination of features allows the device to be operated at very high clock speeds.

Security Fuse

An important architectural feature that is found on virtually all PLDs is not shown on logic diagrams. This feature is the *security fuse*. Normally, the fuse pattern programmed into a PLD can, like a PROM, be read and displayed or copied by programming hardware. Devices with a security fuse, however, provide the ability to disable this read function. This allows the design to be somewhat secure from attempts to copy or reverse engineer it.

In reality, it's relatively easy to shave the top off of a bipolar PLD and examine the programmed fuses with a microscope. For bipolar PLDs, then, would-be copiers are merely inconvenienced. Erasable CMOS PLDs are considerably more secure, since it's very difficult, if not impossible, to determine their function from examination.

Power/Speed Selection Fuses

Many devices feature additional fuses that can be used to select operating modes appropriate for the requirements of the circuit as a whole. For example, the Altera devices feature a set of configuration fuses that allow control of their autonomous power-down features. Disabling the low standby power results in a higher usage of power and a higher speed of operation.

The ICT 22CV10Z also features a power selection feature. Selecting the zero-power option for the 22CV10Z results in a zero standby current mode and correspondingly slower device operation.

3.2 DEVICE TECHNOLOGIES

The device features we've highlighted to this point have been primarily architectural features. As we stated earlier, there are some 300 different PLD architectures in existence. For any given architecture, however, there are possibly dozens (in some cases, even hundreds) of technologically different devices you can buy from one of many separate manufacturers. The devices you can buy will differ in many technology criteria. In this section we will describe some of these technology differences.

Process Technologies

The two dominant processing technologies found in PLD manufacturing are bipolar and CMOS. Bipolar devices have historically been faster in operation, with propagation delays now under seven nanoseconds. The bipolar devices are also cheaper than equivalent devices fabricated with other technologies. Bipolar devices are, however, larger users of power due to their need for standby power. Every transistor of a bipolar device requires power whether or not it's in use. Typical power consumption figures are in the 100–200 mA range for a 20- or 24-pin PAL device.

CMOS devices generally use much less power, due to the fact that they only consume power for switching operations. This fact means that the amount of power actually used by a CMOS PLD is dependent on the function the device is performing and the speed at which it operates. The larger power consumption of the bipolar technology places limitations on the density of bipolar PLDs. In general, CMOS PLDs can be created with much higher densities than possible in bipolar PLDs.

The most striking difference between bipolar PLDs and CMOS PLDs is the erasable nature of most CMOS devices. Erasable PLDs have many advantages over *one-time programmable* (OTP) devices. An erasable PLD can be tested more thoroughly than an OTP device, since a test pattern can be loaded into the device and verified. This level of testing can be done by the device manufacturer or by the customer. Other technologies in use include ECL and gallium arsenide, which are characterized by high speeds (see below) and higher costs.

Programming Technologies

There are a number of different programming technologies in use in PLDs. Reprogrammability aside, these different technologies won't normally affect the user of the devices, but can affect how easily the devices can be programmed.

The first PLDs were developed by adapting existing PROM technologies. These devices were (and still are) based on bipolar technologies which allowed reasonably high-speed operation and relatively simple and reliable programming. *Fusible-link* PLDs, like PROMs, contain an array of programmable interconnect points that each consist of a narrow bridge of metal that may be left intact or removed to enable or disable the interconnect. This bridge of metal is referred to as a fuse and may be made of a variety of materials ranging from nichrome to more exotic materials such as titanium-tungsten or platinum-silicide.

Different fusible-link PLDs have different requirements for programming. Some devices utilize elevated voltages on the V_{cc} pin to place the device in a programming mode, while others utilize a combination of voltage levels on various pins to indicate that the device is being programmed. Each fuse in the device has a specific address that is used to select the specific fuse to be blown in each programming sequence. The actual programming of the fuse is accomplished by applying a high enough current to the fuse that it actually melts.

Because of the higher voltages and currents involved in programming a fusible-link PLD, the programming process can be extremely stressful to a device. The major enemy of bipolar fusible-link devices is heat. In order to reduce the amount of heat generated during the programming process, the programming hardware must be carefully designed to limit the amount of time required for any programming pulse.

Many device programming algorithms take into account the physical location of fuses and spread the programming process around the device over time to avoid successively blowing fuses that are physically close to each other. This helps to eliminate excessive localized heating of the device.

Even with these precautions, there are a number of possible programming failures that can occur when fusible link technology is utilized. First, if the programming equipment is of poor quality or hasn't been properly calibrated, the critical voltages needed on various pins may not be correct. This may result in marginally programmed fuses that operate erratically. A typical situation is that of *underblow* in which a fuse has only partially melted and under differing operating conditions may or may not act as a programmed interconnect. In some cases, an underblown fuse can actually "grow back" over time.

If the programming equipment meets the device manufacturer's specifications and the device is manufactured properly, such failures are quite rare. Fusible link technology has proven to be a reliable method of programming not only in PLDs, but in many years of PROM use as well.

Avalanche induced migration, or AIM, is a programming technology pioneered by Intersil and used in their first PLA devices. The AIM programming element consists of a single open base NPN transistor. A transister is located at each programmable junction of two PLA signal paths. The transister's emmiter is connected to the one signal path of the PLA, while the collector is connected to the other signal path. In the unprogrammed state, the AIM element does not pass current between the two signal paths. To connect the two paths, a high current is forced between emitter and collector. This causes a breakdown of the transister and subsequent short between the emitter and base. As soon as the short circuit occurs, the current drops and the programming process halts. The short between the emitter and base leaves the transistor to operate as a diode between the emitter and collector.

Ultraviolet erasable PLDs, commonly called EPLDs, are based on technology similar to that of EPROMs. EPLDs typically utilize CMOS technology internally, while providing TTL interface circuitry for external pins.

The major programming consideration for EPLDs isn't programming reliability since the devices can be tested for programmability during incoming inspection or at the factory; rather, the primary consideration for these devices is cost. Since UV erasable parts are manufactured in ceramic packages (it's extremely difficult to design a plastic package with a quartz window) they are typically much more expensive than a comparable device in a plastic package. For this reason, most EPLDs are available in a windowless plastic package that isn't reprogrammable. The erasable version of the device can be used for engineering purposes, while the plastic packaged version is used for production. This, of course, means that the devices can't be tested for programmability prior to production.

Electrically erasable PLDs (EEPLDs) also utilize CMOS technology, but rather than having UV erasable cells, the EEPLDs contain programming cells that are erased electrically. As far as the user of these devices is concerned, there is no erase process to be concerned with at all; a previously programmed device can simply be reprogrammed at any time. In theory, there is no reason why EEPLDs can't be reprogrammed dynamically in-circuit, but in reality this is very rare. The real benefit to EEPLDs is their lower cost packages and their short erase and reprogram cycles. Because EEPLDs require so little time to erase, they can be tested very quickly by programming and verifying a variety of test patterns before programming the final pattern.

High-speed Technologies

Speed has often been cited as a limiting factor in moving designs from standard logic to PLDs. In recent years, however, new devices have been introduced that are just as fast as high-speed TTL devices and, in some cases, are faster. More complex bipolar PALs such as the 22V10 are now available with delay times as low as 15 ns. (clock to output). Simpler PAL type devices (the 16L8, 16R8, 16R6 and 16R4) are now available from various sources with speeds of 5 ns. or less.

Cypress has been producing high-speed CMOS PLDs for a number of years, and Lattice Semiconductor offers its customers a series of high-speed GAL devices. The GAL16V8A-12 and GAL20V8A-12 both have propagation delay times of 12 ns.

Real speed freaks, of course, can utilize ECL versions of the simple PAL devices. These devices have propagation delays of a little as 4 ns. but are expensive and consume a large amount of power (typically around 1.25 W for a simple PAL). AMD has recently announced an ECL device that is similar in functionality to the popular 22V10. National Semiconductor, having recently purchased Fairchild Semiconductor and its high-speed ECL process, is likely to produce ECL PALs in the 2- to 3-ns. speed range.

To overcome compatibility problems between TTL and ECL technologies, Gazelle Microcircuits produces a mask programmable 22V10 device that is implemented in gallium arsenide (GaAs) and operates at a speeds of 7 ns. As of this writing, however, the Gazelle device can't be programmed by end-users. Instead, designs must be prototyped using a standard 22V10 and the design then submitted to Gazelle for production. Although the turnaround time for this service is under a week using laser programming techniques, Gazelle is reported to be working on a field-programmable version of the part to satisfy the demands of users.

3.3 SUMMARY

PLDs have evolved from the simple PLAs and PALs of the late seventies into complex and highly configurable architectures. Although the devices covered in this chapter are a great advance over standard logic, fundamental limitations in the size of PLA-based architectures has led to completely new approaches to programmable logic. In the next chapter we'll continue our tour and explore some of these more advanced devices.

3.4 REFERENCES

Advanced Micro Devices Incorporated, *Programmable Array Logic Handbook,* Advanced Micro Devices, Santa Clara, CA, 1983.

Bursky, Dave. "Clock-Free Macrocells Simplify Asynchronous System Design." *Electronic Design* (July 1988):53–56.

Collett, Ronald. "Programmable Logic Declares War On Gate Arrays." *Digital Design* (July 1986):32–39.

Cypress Semiconductor Corporation, *CMOS Data Book,* Cypress Semiconductor Corp., San Jose, CA, 1986.

Exel Microelectronics Incorporated, *E^2 Data Book,* Exel Microelectronics Inc., San Jose, CA, 1988.

Meyer, Ernest L. "Programmable Logic Overview." *VLSI Systems Design* (November 1987):62–70.

Monolithic Memories Incorporated, *PAL/PLE Programmable Logic Array Handbook,* 5th edition, Monolithic Memories, Inc., Santa Clara, CA, 1986.

Rutledge, David. "E^2 CMOS Dominates PLDs In Near Term." *Electronic Engineering Times* (March 20, 1989):43–44.

Saffari, Bobby and Sasaki, Paul T. "Customizing Image Processor Chips With PLDs." *VLSI Systems Design* (November 1988):50–59.

Signetics Corporation, *Integrated Fuse Logic Data Manual,* Signetics Corporation, Sunnyvale, CA, 1984.

Texas Instruments Incorporated, *Programmable Logic Data Book,* Texas Instruments Inc., Dallas, TX, 1988.

Wilson, Ron. "Vendors Ponder Interconnect Schemes In Search For New PLD Architectures." *Computer Design* (May 1989):27–28.

Bursky, Dave. "Programmable Sequencer Hits 125-MHz Clock Speed." *Electronic Design* (September 28, 1989):43–46.

4

Beyond PLDs

The PLDs that we examined in the previous chapter represent the most widely used forms of user programmable ICs, even surpassing common programmable memory devices such as PROMs. These PLDs have all been based on a similar architecture: the programmable logic array.

Over the years, device manufacturers have added more functionality to the basic PLD architectures. Programmable arrays have increased in size and complexity, and highly configurable I/O macrocells have been added. There are limits to the flexibility and expandability of the programmable logic array, however. Recently, new devices have appeared that break away from previous PLD architectures and rival mask-programmed gate arrays in complexity.

4.1 THE FIELD PROGRAMMABLE GATE ARRAYS

In response to demands for more functionality in a single programmable device, manufacturers have begun developing alternative architectures. These devices have been dubbed *field programmable gate arrays,* or FPGAs. They are distinguished from PLDs by their use of non-PLA architectures and by their requirement for completely different implementation strategies. Specifically, FPGAs are composed of uncommitted device resources that must be selected, configured, and interconnected. A PLD does not require the interconnection step. Some of the more complex PLDs do require more complex resource selection and configuration than earlier devices, but the internal resources (registers, logic arrays, and I/O macrocells) of these devices are still essentially fixed in relation to one another.

What constitutes an FPGA? The FPGA acronym is a little misleading; the user-programmable devices that are included in this new category actually bear little resemblance to large, mask-programmed gate array devices. As FPGAs grow in size, however, they begin to appear in applications that formerly would have been implemented with a gate array.

The FPGAs currently available all share a few fundamental characteristics. First, they are all composed of some number of relatively independent logic modules that can be interconnected to form a larger circuit. These logic modules may be large configurable blocks of logic or smaller fixed-function logic modules consisting of just a few gates. The major distinction between complex PLDs and FPGAs is the constrained routing resources of the FPGAs. We'll explain this further as we examine the different FPGA families. Complex PLDs that feature segmented architectures and buried macrocells are also sometimes dubbed FPGAs, and we'll examine some of these devices in more detail later in the chapter.

FPGAs differ primarily in the size and configuration of their logic modules. FPGA designers have come up with different solutions to the problem of balancing logic module sizes with interconnection requirements. Each of these compromises has its advantages and disadvantages.

FPGAs with larger logic modules are difficult to efficiently utilize since entire logic modules (representing tens of logic gates) are often wasted performing simple logic functions. The use of smaller logic modules leads to interconnection problems, since a much larger number of interconnect paths must be provided in the device. Depending on their construction, these interconnections can create a significant delay factor, as well as consuming a large percentage of the FPGA's area.

The optimal logic module size and the interconnect requirements are, of course, highly dependent on the type of application that's being implemented in the device. For example, an application composed of regular or semi-independent structures may be better suited to an FPGA with larger logic modules, while an application composed of a variety of random interrelated logic functions may be better implemented in a device with smaller logic modules and more interconnect resources.

The logic modules of an FPGA are interconnected with configurable channels. This is done through a process known as *routing*. Routing is the determination (either manual or through the use of computer-aided tools) of an efficient interconnection strategy. For a single application and FPGA device, there are many possible ways to interconnect the resources required by the application.

Sizing Up FPGAs

The differences in architectures between PLDs and FPGAs often leads to confusion about the densities of these different devices. Equivalent gate count is one measurement that has been used by device manufacturers to describe the relative sizes of various types of devices including PLDs, gate arrays, and FPGAs. This measurement is an approximation of the number of two-input NAND gates that would be required to equal the functionality of the device in question. Unfortunately, such measurements of gate densities are highly subjective, being dependent on the types of circuits that are to be implemented, and on the structures of the devices themselves.

For FPGAs the concept of counting gates is less meaningful than for standard PLDs or gate arrays. While even the simplest FPGA may claim over 1,200 usable gates, in reality it's uncommon to find designs that can utilize all of an FPGA's circuitry. The difficulties of placing and routing the logic modules result in a much lower typical gate count for completed designs. In addition, FPGAs available today have widely varying architectures that make such comparisons difficult. Some FPGA manufacturers are now de-emphasizing gate count statistics, and instead publish data on the percentage of device resources required to implement various common circuits.

With these caveats, if we accept the gate count figures published by manufacturers for comparison purposes, we find that FPGAs range in size from 1,200 to 20,000 equivalent gates, while PLDs range in size from a few hundred to 2,000 gates. Since FPGAs are rapidly increasing in density, we can expect that typical FPGA applications will soon be an order of magnitude more complex than typical single-PLD applications.

Matching FPGA Architectures To Applications

Rather than simply accepting a published gate count as an indicator of a device's complexity and suitability for an application, it may be more appropriate to match the requirements of the application to the resources available in each type of device. The first step is to analyze the requirements of your application and tally up the resources that will be required. Determine how many primary inputs and outputs will be required and how many flip-flops will be required. These measurements of complexity will be approximate since design tradeoffs can be made to accommodate the chosen device architecture.

Also determine whether the application requires an unusually large amount of flip-flops compared to the amount of combinational circuitry and whether the combinational circuitry has high fan-out or gate-width requirements. Another factor that should be considered is whether the application is modular; does it consist of many small pieces with just a few interface signals, or is it a single mass of interconnected logic?

Armed with basic information such as this, you can begin to match the application to various PLD and FPGA types. Using the number of flip-flops and I/Os, for example, you can quickly determine which devices are either too small, or are overly complex. The information about fan-out and combinational logic complexity can help in determining the appropriateness of different types of FPGAs.

4.2 FPGA FAMILIES

There are a number of different types of FPGAs and completely new architectures can be expected to appear in the near future. Let's examine a few of the device families currently available. These families of FPGAs all have unique architectures, and also differ in their basic technologies. The higher densities of FPGAs have led to the development of new programming technologies for these devices. While nearly all PLDs have adopted the programming methods of PROMs and EPROMs (one-time-programmable fuses or erasable CMOS cells), all three of the FPGA families described below use alternative programming technologies.

The Logic Cell Array

The first family of FPGAs was announced in 1984 by Xilinx Corporation. The devices, called *logic cell arrays* (LCAs), are now available in a variety of sizes from Xilinx. AT&T has also signed an agreement with Xilinx to produce the LCA, using a higher-speed CMOS manufacturing process.

Like all FPGAs, the LCAs are not based on a single programmable logic array, being instead composed of many smaller logic modules. In an LCA, the logic modules are small programmable logic elements called *configurable logic blocks* (CLBs).

Figure 4.1 shows a block diagram of the LCA architecture. The simplest device in the LCA family is the Xilinx XC2064. The XC2064 is composed of 64 general purpose CLBs, each of which is capable of implementing either two three-input logic functions, or a single four- or five-input logic function. These CLBs are each similar in many respects to a small PROM with the addition of a flip-flop and feedback logic. Each CLB contains a memory element in the form of a

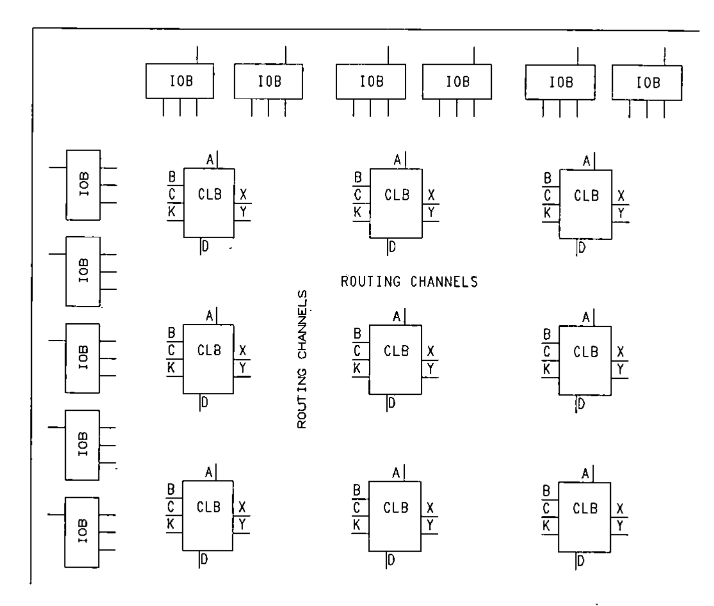

Figure 4.1 Xilinx XC2064 FPGA

configurable flip-flop that can be used as a transparent latch or edge-triggered D-type flip-flop.

Special-purpose logic cells, called *I/O blocks* (IOBs), are arranged around the device, and can each be configured for use as an input, three-stated output or bidirectional I/O pin. The IOBs can be further configured with flip-flops and can be accessed by the internal CLBs for non-I/O functions.

These configurable cells are arrayed as shown on the LCA chip and can be interconnected through the use of programmable routing channels. Since each CLB of the XC2064 LCA (shown in Figure 4.2) is capable of implementing functions of limited complexity, the interconnection resource requirements are correspondingly large.

Placement of logic functions in CLBs and determination of interconnect routing are a major focus of design efforts for users of LCAs. Automatic placement and routing software is provided by Xilinx that can perform most of this part of the design process. Most designs, however, require a certain amount of hand routing to efficiently utilize the LCA (this is expected to improve with newer LCAs that feature more complex logic modules and improved interconnection schemes).

The programming method used in the LCA is also a departure from traditional PLDs. An LCA device isn't programmed in the same way that PLDs are programmed (with fuses or erasable PROM technology). Instead, the LCA's programming scheme is based on static RAM technology. The device is loaded with a configuration pattern while in-circuit. This means that new configurations can

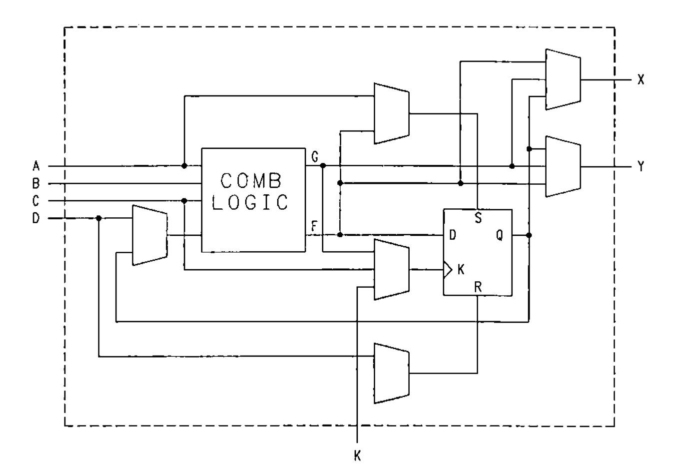

Figure 4.2 XC2064 configurable logic block (CLB)

easily be loaded into the chip even while it's operating in the system. In most cases, the LCA programming pattern is stored in a configuration PROM that is installed on the board with the LCA device. The configuration PROM is read automatically by the LCA upon power-up.

While the XC2064 device shown is comparable to a complex PLD in density and gate count, larger LCAs from Xilinx are closer to gate arrays in size and complexity. In addition to the 2000 series of LCAs (encompassing parts up to 1,800 usable gates in size), Xilinx offers the 3000 family of LCAs, which includes parts ranging from 2,000 to 9,000 usable gates in size. In late 1989, Xilinx announced its 4000 family with a more generalized architecture and simpler interconnect strategy.

The LCA devices boast high speeds for flip-flops (up to 100 MHz toggle rates are claimed) but the devices are generally not suitable for high-speed applications due to the delay inherent in the interconnection matrix. Interconnections in the LCA devices include non-metal elements, resulting in delayed signals. Total circuit delay is correspondingly difficult to predict, being dependent on the routing of interconnect signals.

The Actel ACT

In 1988, Actel announced its new family of CMOS-based devices, which range in size from 3,000 to 6,000 equivalent gates. The Actel FPGAs, dubbed ACT devices, are composed of rows of uncommitted logic blocks separated by routing channels as shown in Figure 4.3.

The logic modules of the Actel FPGAs consist of three two-into-one multiplexers and an OR gate arranged as shown in Figure 4.4. By virtue of their simplicity, the Actel logic modules are more general that the CLBs found in the

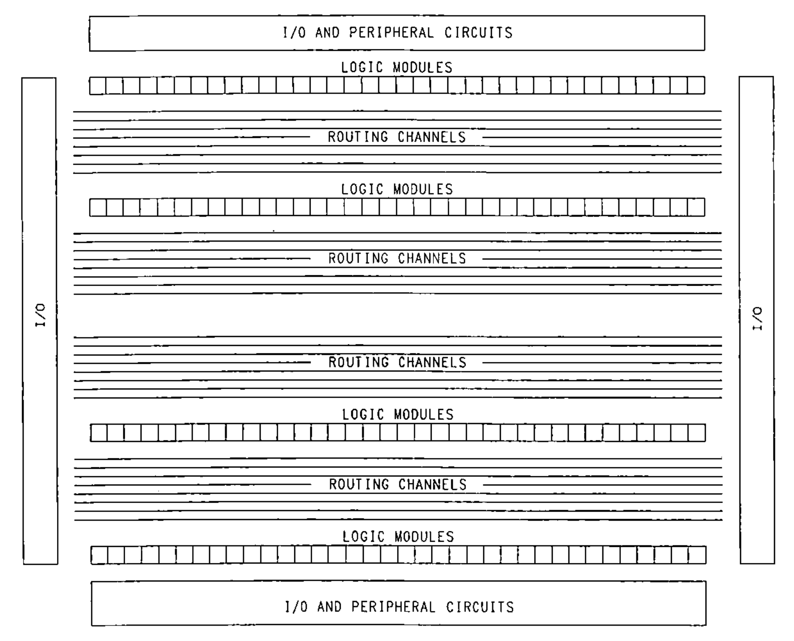

Figure 4.3 Actel ACT FPGA

LCAs, having eight inputs each and no built-in registers. Since registers are constructed from the basic logic modules (one or two logic modules can theoretically implement a flip-flop of any type) they can be applied where needed and no flip-flops are wasted in circuits with large amounts of combinational logic.

Unlike the LCA's logic modules, which are internally configurable, Actel's logic modules are configured by connecting the appropriate logic module inputs to the interconnect channels and tying the remaining logic module inputs high or low. This architecture is more similar to that of mask-programmed gate arrays and gives the Actel devices the power to implement a much wider variety of circuits.

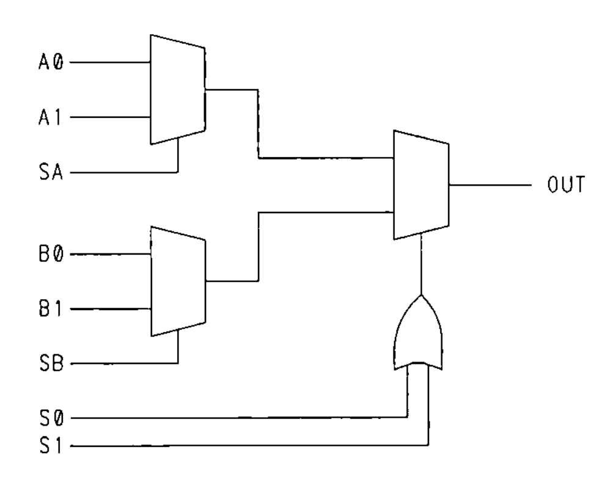

Figure 4.4 The Actel logic module

As we pointed out earlier, the use of less complex logic modules has a price. Since the Actel logic modules are of a relatively small size and are configured purely through their interconnection with other modules, the Actel devices must devote a large part of their area to interconnect resources. To overcome the problem of programmable interconnect area, the Actel FPGAs utilize a unique and extremely compact programming element.

Actel calls their new programmable element the *programmable low impedance circuit element,* or PLICE. Unlike the fuses found in PROMs and PLDs, the PLICE is unconnected in its virgin state and is then connected when programmed. This is similar in some respects to the vertical AIM programming technology pioneered by Intersil, but uses a dielectric insulator rather than a transistor as the programming element. Actel has dubbed their new PLICE programming element an *antifuse.*

The antifuse technology is significant because of the extremely small size of the programming elements. An Actel antifuse is 50 times smaller than the static RAM elements found in the LCAs and ten times smaller than a CMOS EPROM cell. The Actel antifuses aren't erasable and, therefore, cannot be tested before being programmed. For this reason the device contains special test circuitry that allows every logic and I/O module, every interconnect channel, and all other device circuitry to be tested in addition to verifying the integrity of the antifuse connections.

The nonreprogrammable nature of these devices, coupled with their relatively high price, has proven to be a barrier to many new users who are comparing FPGAs. Designs being implemented in Actel devices must be complete and functionally verified before programming. While different programming patterns can be quickly loaded into an LCA and tried out, this form of firmware breadboarding isn't possible in an Actel device. This places a greater burden on the designer and on the design tools.

In Actel's favor is the simpler routing process created by the generous and more generalized routing resources. Most designs implemented in an Actel device can be automatically routed with much higher utilization than possible in an LCA, and in less time. This means that designs that have been functionally verified (with a simulator, for example) can be automatically mapped into the device with a high degree of success. Another advantage is the intrinsic capability for higher device densities brought about by the PLICE antifuse technology.

For many applications, speed is the primary factor affecting the decision and, in this area, the Actel devices have the advantage. The interconnect paths are all metal and experience less inherent delay than the signal paths of the LCAs.

The Plessey ERA

A recent entrant into the FPGA field is the *electrically reconfigurable array* (ERA) chip offered by Plessey Semiconductors. This device is similar in some respects to the logic cell array. The first of the ERA devices claims 10,000 equivalent gates and Plessey has indicated that it will be offering devices of up to 100,000 gates by the end of 1991.

Like the LCA, the ERA is based on static RAM technology. The ERAs have a configuration feature not found in the LCA: they can be partially reprogrammed while they are operating. This means that one segment of logic within the ERA could, theoretically, control the dynamic reconfiguration of a different segment in the device. Unlike the LCA, the Plessey device is composed of a large number of simple two-input NAND gates instead of more complex logic blocks. If the application requires more complex functions, such as flip-flops, these functions must be constructed out of the basic NAND structure.

The ERA has, in addition to its central array of NAND gates, a series of 84 configurable cells for I/O purposes. Each of the NAND gate logic modules can be used for either logic purposes or for interconnection routing. To provide more efficient routing of signals in the NAND array, the device also has a ten-line configurable bus that surrounds the main array. This bus can be used to route signals from one section of the array to another or to and from the configurable I/O cells or fixed I/O pads.

The simplicity of the ERA device provides the potential for more efficient device utilization than possible in the LCA, but this efficiency advantage may be lost in the greater need for circuit interconnects in the device. Interconnections in the ERA are routed through the logic modules themselves, so greater interconnection requirements lead to correspondingly fewer available NAND gates. A large design with primarily combinational logic and relatively few feedback paths will be better suited to the ERA than to the register-rich LCA, but a design that is modular and register-intensive will be better suited to an LCA.

4.3 FPGA OR PLD?

In the previous chapter, we examined many devices that featured configurable macrocell features and buried resources. Many of the devices presented were of sufficient complexity to require more comprehensive design tools and techniques than simple PAL-type devices, but all of the devices presented in Chapter 3 are promoted as PLDs by their manufacturers.

Some of the more complex configurable PLDs have been marketed as FPGAs, reflecting the fuzzy nature of the FPGA category. The simpler routing process and predictable timing characteristics of these "composite PLD" devices have led many observers to conclude that they are not FPGAs at all. While it's true that the design process is simplified with these devices (routing is trivial, with no affect on timing), their architectures are sufficiently modular to warrant special consideration. As to whether or not these complex PLDs belong in the FPGA category: we'll leave that debate to those who find such arguments interesting.

Altera's MAX

Altera's *multiple array matrix* (MAX) device is a complex erasable PLD that features the expander term array described in the previous chapter. The architecture of the 5,000-gate equivalent MAX EPM5128 is diagrammed in Figure 4.5. This 68-pin device is composed of relatively large PAL-based logic modules that are interconnected with fixed-delay interconnections.

The EPM5128 MAX device is composed of eight logic modules that Altera has dubbed *logic array blocks*, or LABs. The logic array blocks are interconnected by a *programmable interconnect array*, or PIA. The PIA is what distinguishes the larger MAX devices (such as the EPM5128) from the smaller MAX devices such as the EPM5032. This interconnect matrix provides the capability to route signals from one section of the device to another. Any macrocell (there are sixteen found in each logic array block, for a total of 128 macrocells) can be connected through the PIA to any other macrocell.

The macrocells themselves are illustrated in Figure 4.6. Each macrocell (there are two shown in the figure) has a pair of expander terms that can be used to increase the effective size of the programmable PAL-type array. The expander terms are all connected to the global interconnection bus, so each expander term can be shared between multiple macrocells. An XOR gate found in each macrocell also increases the effective size of the MAX's array. The XOR gates can be used to control polarities, emulate flip-flops, or implement complex logic functions more efficiently.

Of the two macrocells shown, one is capable of driving an output directly, while the other is buried. When the lower I/O macrocell is used as an input, the expander terms of the upper buried macrocell are not accessible. The macrocells all feature product-term controlled clocks and asynchronous resets and presets. The I/O macrocells also feature product-term controlled output enables.

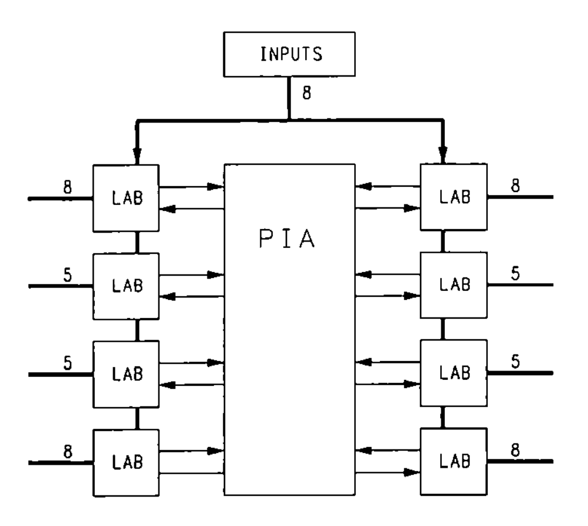

Figure 4.5 Altera MAX EPM5128 block diagram

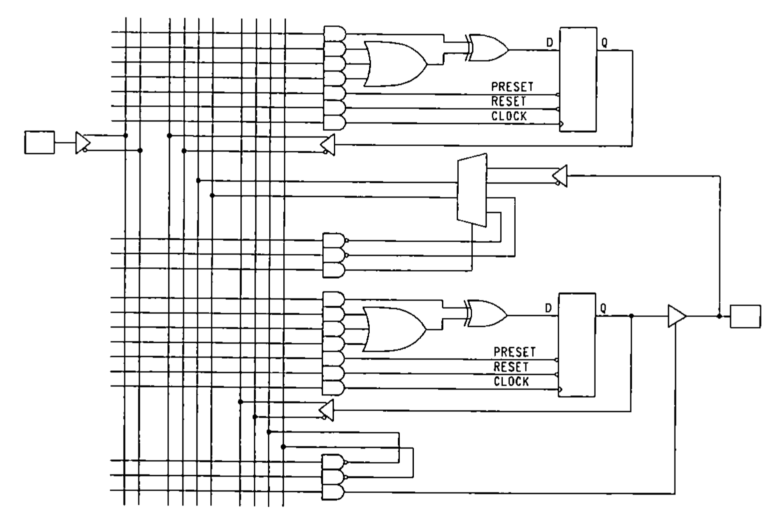

Figure 4.6 Altera MAX macrocell detail

The macrocells all feature configurable register elements that can be programmed for use as either D-type flip-flops or as latches. XOR gates associated with each macrocell also allow the emulation of T-type, JK-type, or SR-type flip-flops. If desired, the registers can be clocked from a common global input for increased speed of operation over the product-term controlled clocking scheme.

AMD's Mach

The Mach family of devices, announced in early 1990 by AMD, are similar to Altera's MAX devices. Like the MAX, the Mach devices are composed of a bank of interconnected logic blocks. Each block can be viewed as an independent PAL device that can be connected to other blocks through the Mach's central switch matrix. Figure 4.7 illustrates the Mach architecture.

The Mach's macrocells (a number of which are associated with each logic block) are conventional, and feature registers that can be bypassed for combinational uses, or configured for use as either D-type or T-type flip-flops. Unlike the MAX, the AMD parts have up to 16 product terms directly usable for each macrocell. Product-term steering (discussed in the previous chapter) is used in the Mach devices for variable product-term distribution, so there is no speed penalty when a larger number of product terms are allocated to a macrocell. This architecture, combined with the faster process technology, makes the AMD parts attractive for high speed applications.

There are two types of Mach parts available. The three devices in the Mach 1 series feature 32, 48, and 64 macrocells, respectively, with pin counts of 44, 48, and 68 pins. The larger Mach 2 devices feature buried macrocells allowing up to

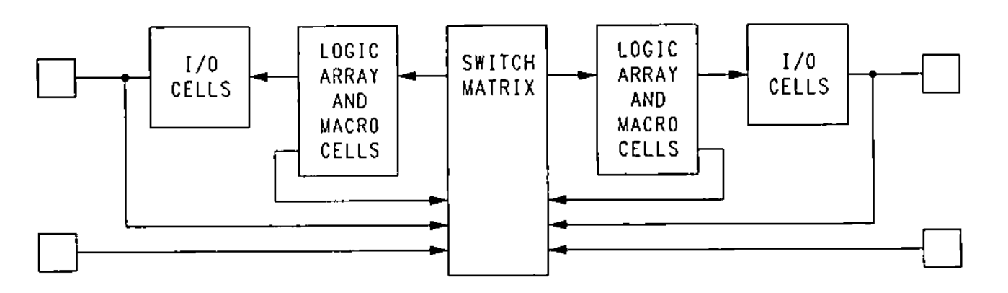

Figure 4.7 AMD Mach block diagram

128 macrocells with pin counts similar to the Mach 1 devices. All of the devices in the Mach 1 and Mach 2 series feature similar architectures, so designs can be easily migrated to larger devices when needed.

Plus Logic

Another complex PLD that is similar to Altera's MAX was announced at the end of 1989 by Plus Logic, Incorporated. This family of devices, which range in size from 1,000 to 6,000 gates, is designed for higher-speed applications than are possible with routing-constrained FPGAs. Like the MAX, the Plus Logic devices have a central matrix of interconnections (crossbar switches) that are configured using UV-erasable cells like those found in standard EPROMs. Figure 4.8 illustrates the architecture of the FPGA2020 device.

In the Plus Logic FPGA2020 device, the interconnect matrix is surrounded by a battery of function blocks, each of which has nine outputs and 21 inputs, as well as a carry input and clock line. These function blocks may be thought of as interconnected PLDs, since they each contain a sum-of-products array. The function blocks are all connected directly to the interconnect matrix and to the configurable I/O blocks.

In addition to the function blocks (eight of which are found in the FPGA2020 device), the devices include high-speed comparators useful for address decoding

Figure 4.8 Plus Logic FPGA2020 device

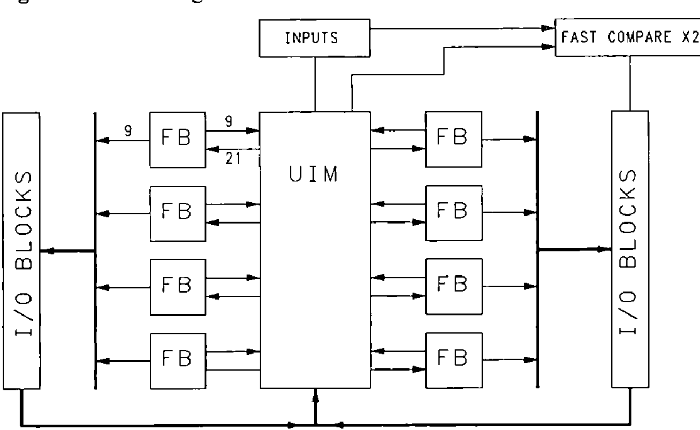

and other purposes. The interconnect matrix itself may be used for logic purposes, further increasing the device's effective size.

In addition to general purpose devices, Plus Logic has announced their intention to produce a family of application-specific FPGAs that will contain specialized logic for system applications.

ICT's High-density PEEL

ICT's PA7024 (illustrated in Figure 4.9), is a complex, electrically erasable CMOS PLD with 20 I/O pins and two independent clock inputs. Each I/O has a selectable input register/latch, a general purpose register, and a variety of configuration options.

The PA7024 device features highly configurable macrocells associated with, but not dedicated to, each of 20 I/O pins on the device. These macrocells (which ICT refers to as logic control cells, or LCCs) are capable of being isolated from their associated I/O pins for use as buried register or multilevel combinational purposes and each contains a dynamically configurable register element that can be used as a D-type, T-type, or JK-type flip-flop. A diagram of a logic control cell and it's associated I/O cell are shown in Figure 4.10.

The PA7024 is based on a large sum-of-products logic array (four sum-of-products functions are available to each macrocell) and, like the Plus Logic and

Figure 4.9 ICT PA7024 device

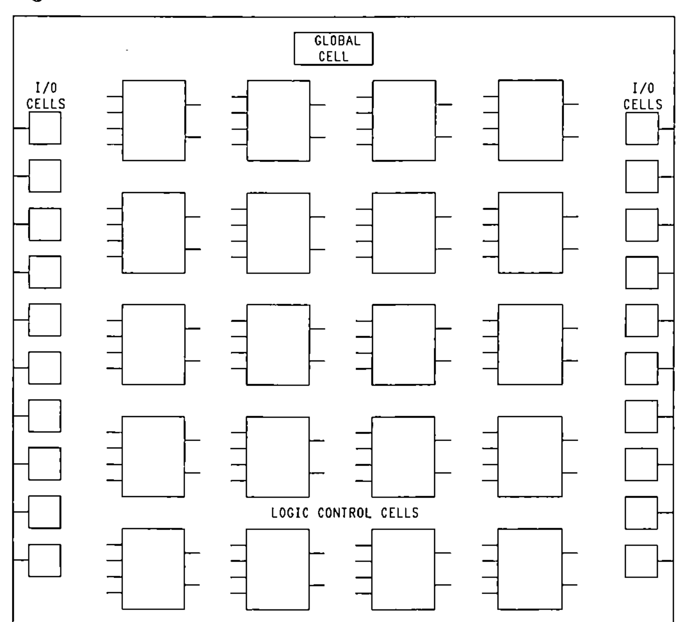

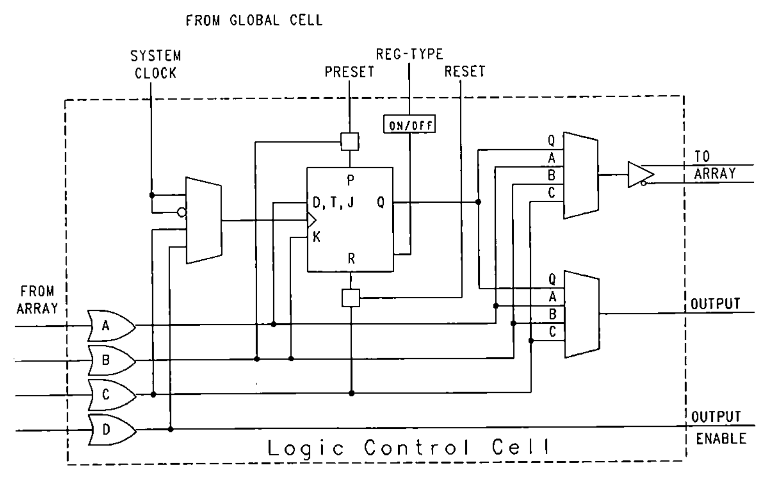

Figure 4.10 PA7024 logic control cell

MAX devices, does not have complex routing requirements that affect the timing behavior of the programmed device.

4.4 MIGRATING DESIGNS FROM PLDs TO FPGAs

The high densities of FPGA devices, coupled with their user programmability, makes them attractive choices for value engineering efforts for PLD-based designs. There are some fundamental differences between PLDs and FPGAs that should be considered before undertaking such a task.

First, it's important to understand the limitations of each architecture and how these limitations affect the designs that are typically created for each architecture. The most limiting factor when designing for PLDs is the relatively small number of outputs and flip-flops available. We might think of PLDs as generally being short on registers, and rich in input forming logic (although if the design requires multilevel logic, the simpler PLDs are quite limited even in this respect).

An FPGA, on the other hand, may have lots of registers, while input forming logic for these registers is limited. This is particularly true of the LCA devices. While one output of a PLD may have the ability to implement an equation with 20 or more inputs, that same equation implemented in an LCA may require as many as seven of the device's CLBs. In the smaller LCAs, this can represent more than ten percent of the CLBs available in the device.

What this means is that a design that has complex input forming logic, a large number of inputs and relatively few registers may fit quite comfortably in a small number of PLDs, but may not fit efficiently into a device such as an LCA.

Efficiently moving a design from PLDs to an FPGA may require a major redesign effort to exploit the FPGA's specific set of features.

Timing is another area where PLDs and FPGAs differ dramatically. If you require high speed operation, it may be impossible to utilize an FPGA for a design that is implemented in PLDs.

4.5 SUMMARY

The FPGA families described in this chapter all have unique design characteristics, both architecturally and in their programming technologies. From the early success of these devices, we can expect to see other manufacturers announce completely new FPGAs. As PLDs become more complex, the distinction between PLDs and FPGAs may become increasingly blurred. The complex PLDs now available from Altera, Plus Logic, and ICT are prime examples of devices that fall somewhere between PLDs and FPGAs in complexity.

4.6 REFERENCES

Collett, Ronald. "Programmable Logic Declares War On Gate Arrays." *Digital Design* (July 1986):32–39.

Collett, Ronald. "A Programmable Gate Array: Can It Be True?" *Electronic System Design* (April 1988):28.

Conner, Doug. "PLD Architectures Require Scrutiny." *Electronic Design News* (September 28, 1989):91–100.

El Gamal, Abbas. et al. "An Architecture for Electrically Configurable Gate Arrays." *IEEE Journal of Solid-state Circuits* (April 1989):394–98.

Haines, Andrew. "Gate Utilization Goes up with Antifuse FPGAs." *High Performance Systems* (October 1989):47–57.

Harbert, Tammi. "Speed, Density Improvements Continue for FPGAs." *Electronic Design News* (January 11, 1990):8–14.

Merrill, H. Wayne. "Evaluating PLD Implementations Through Design Synthesis." *Computer Design* (July 1, 1989):56.

Mohsen, Amr. "Desktop-configurable Channeled Gate Arrays." *VLSI Systems Design* (August, 1988):24–33.

Novellino, John. "Development Tool Trouble-Shoots PGAs In The Target System." *Electronic Design* (January 1989):64–70.

Small, Charles H. "User-programmable Gate Arrays." *Engineering Design News* (April 27, 1989):146–58.

Wirbel, Loring. "Plus Logic Rethinks PLD Approach." *EE Times* (October 30, 1989):46.

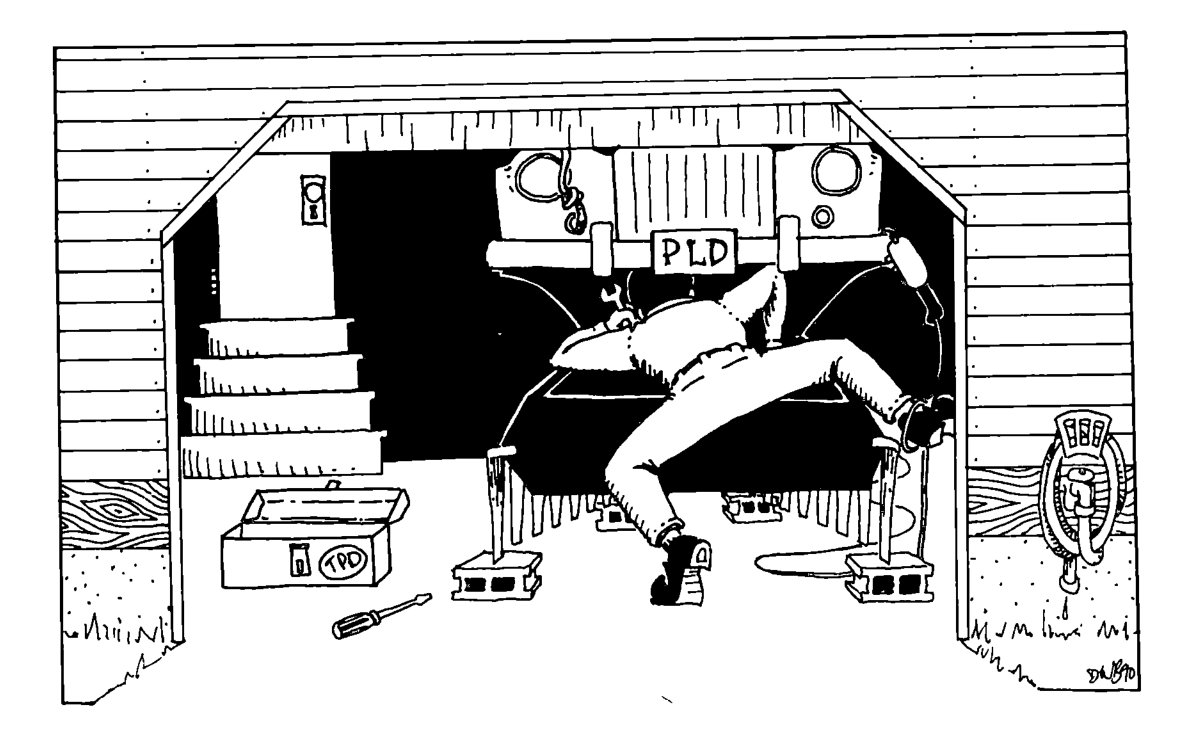

5

Development Tools

In the past decade, a wide variety of design tools have appeared that aid in the development of PLD-based designs. In this chapter, we'll describe the general features of PLD design tools and survey the tools available by comparing their various features and approximate prices.

Before describing each of these tools, let's examine where these computer-aided design tools fit in the overall design process. There are two phases to the design process in which *computer-aided engineering* (CAE) tools can help. The first phase is the creation and description of a design concept. The second is the conversion of that conceptual description into an actual circuit that meets the design's physical and electrical constraints. CAE tools that aid in the first phase, the *design entry* phase, include *hardware description languages* (HDLs), design entry tools such as schematic and text editors, and behavioral simulation and design analysis tools.

The second phase, the conversion of a design description into a physical circuit, can be automated with currently available CAE tools. This process is called *logic synthesis* and can involve many different levels of automation. Automated tools such as simulators can also help to verify that a synthesized circuit actually performs the required function.

5.1 DESIGN DESCRIPTION METHODS

There are two fundamental approaches that can be taken when describing a logic circuit. The first approach, *structural design,* is the method in which various logic gates and functional blocks are connected together with wires to create a circuit. This approach is typified by the user of discrete TTL parts. TTL users frequently design at the level of an actual circuit and may begin the process of verifying and optimizing that circuit by flipping a switch and looking for smoke.

Computer-aided structural design tools include graphic editors for schematic entry, design rule checking programs, fault simulators, and, for VLSI designers, hardware description languages such as VHDL. These tools are often used as front-end design systems for circuit simulators, circuit board layout systems, and automatic wire wrapping systems. For the majority of design engineers, however, such tools are used for documentation purposes only, since the engineer in all likelihood has a working circuit prototype sitting on his workbench long before the circuit is ever entered into a computer.

The second approach is to use *behavioral methods* of design description. In this approach, the design is conceptualized at a higher level and the resulting high-level description, called the *behavioral description,* is then processed into some intermediate form that can be manipulated for the purposes of optimization and physical implementation. The behavioral description of a circuit can take many forms, including Boolean equations, flow diagrams, waveforms, truth tables, and combinations of these representations. Schematics can also be used as vehicles for behavioral descriptions. To synthesize actual logic from the behavioral description, these various forms of representation are translated to a common intermediate form that can be processed by logic synthesis tools.

The behavioral approach is becoming widespread for the design of VLSI devices such as gate arrays and standard cell ICs and is the method of choice for experienced PLD users. Computer-aided tools for behavioral design include high-level design languages, silicon compilers for standard cell IC design, PLD assemblers and compilers, and graphic design entry tools such as flow diagram, schematic, and waveform entry systems.

To summarize, a structural design description is an exact representation of a circuit (the most accurate structural representation is the working circuit itself) while a behavioral description is a higher level representation of the circuit's function. CAE tools exist that aid both design methods.

Choosing a Design Method

The method used during the design of a circuit is affected by the type of hardware into which the design will be implemented and by the nature of the circuit itself. The more generalized and less specialized an architecture is, the more appropriate are behavioral design techniques. For example, we can think of many of the larger standard TTL devices as examples of specialized architectures. If your circuit concept includes an adder segment and there is a standard, specialized adder device available that matches that requirement, you can proceed directly to the implementation of that segment using structural design techniques.

This isn't to say that behavioral design techniques are appropriate only for ASIC-based designs. On the contrary, many of the simpler fixed-function devices (such as the NAND and NOR gates of the 7400 series of devices) are generalized by virtue of their simplicity and can lend themselves well to behavioral design techniques.

PLDs are highly generalized devices. There is rarely a situation in which a circuit concept maps directly and naturally into a PLD architecture, so behavioral design techniques are almost universally used (computer-based structural design tools for PLDs do exist in the form of fuse editors, but these tools are extremely cumbersome to use for anything but minor design corrections).

The type of circuit being designed is an important factor in determining how to proceed with the design process. Generally, logic circuits fall into three broad groups: data paths, glue logic, and control paths. Structural design methods are most appropriate for *data path circuitry,* including such things as register arrays, pipelined registers, multiplexers, and other designs in which the actual structure of the circuit directly follows its function. If you know immediately what a circuit will look like, there is no benefit to describing it behaviorally unless you need such a description for simulation purposes. Behavioral design methods are most useful for creating glue logic and control paths, since the actual circuitry for these types of circuits isn't usually obvious from their function.

Glue logic is used to provide the miscellaneous interface requirements of system circuitry. Typical glue logic circuits include address decoders, wait-state generators, and a wide variety of other interface circuits. Glue logic circuits can usually be described best using Boolean equations or truth tables, or may be described graphically using functional blocks representing standard logic functions.

Control path applications include decision-making circuits and circuits that detect or generate sequences of events. These circuits are often described as state machines and are the most difficult to design. Since control path circuits are usually sequential and frequently have to contend with asynchronous signals, they often have critical timing requirements.

5.2 DESIGN ENTRY FORMATS

There are a variety of methods that can be used to describe a circuit, with or without a computer. Let's explore how these design description formats are utilized in PLD design tools.

Equations

Equations have come a long way since the early days of PALASM. Modern PLD design languages incorporate high-level equation features such as sets, arithmetic and relational operators, hierarchy, and modularity. These features allow far more abstract forms of expression than the simple sum-of-products Boolean equations of the past.

Equations are most useful when the design to be described has some underlying pattern or regularity. Multiplexers, shift registers, and counters are all examples of circuits that have these attributes. There are as many ways of implementing high-level equation features as there are PLD design tools. Most PLD design tools support set notation, which is useful for grouping related signals that are operated on as one design element. Some PLD design languages support higher level arithmetic operators and relational operators in addition to the standard Boolean operators.

Truth Tables

Truth tables are most useful for those designs that have no underlying pattern or order. A typical example of such a circuit is a decoder for a seven-segment display. The truth table is a natural way to describe partially specified functions in which there are don't-care conditions. When truth tables are used for state machine applications, they are referred to as *state tables*. State tables are a convenient way in which to describe the behavior of sequential circuits that contain a large number of state transitions.

Truth table features are found in many of the design tools mentioned in this chapter. You should be aware, however, that if a design tool doesn't include support for don't-care optimization, the benefits of truth table representations over equations are reduced.

State Diagrams

State diagrams, either graphic or text based, are used to describe the behavior of state machines. The choice of whether to use state diagrams or state tables for descriptions of these circuits is largely a matter of personal taste. State diagrams tend to be more lengthy than equivalent state table descriptions, but are usually more readable if there are a large number of states. *Textual state diagram* features are found in many PLD design tools and *graphic state machine* design tools are beginning to appear as well.

Schematics

Schematics are considered by many to be the purest form of structural description. Structural design description isn't the only use for schematics, however. Schematics can also be a highly effective method of behaviorally describing a circuit. There are many schematic translation tools available that will read a computerized schematic or netlist representation of a circuit and generate behavioral circuit information. One popular use for such tools is to convert existing designs composed of discrete TTL blocks into PLD implementations. In this capacity, the formerly structural schematic is being used to describe the behavior of the circuit in terms of a discrete TTL representation.

Another behavioral application for schematics is to create circuit descriptions composed of block diagrams. The individual functional blocks appearing on the schematic may be represented by lower level logic in the form of logic gates or textual description methods such as those already discussed.

Waveforms

Waveforms are another method for describing circuit behavior for PLDs. A waveform can be used to specify relationships between input signal events (edges or values) and corresponding output signal events. Waveforms allow these relationships to be expressed in terms of timing relationships, rather than as simple switching functions. *Waveform editors* can be used to draw these relationships for either synchronous or asynchronous circuits.

5.3 AUTOMATED LOGIC SYNTHESIS FOR PLDS

Logic synthesis is a phrase that has been appearing more frequently in descriptions of commercially available PLD design tools. If we wish to see through such marketing jargon, it's important for us to understand just what this term means. As we said earlier, logic synthesis is the process of converting a circuit description into a form appropriate for hardware implementation by optimizing that form to create a physical circuit that meets specific constraints. These constraints may be weighted and may include such things as the total circuit size, operating speed, power usage, or testability. Logic minimization and multilevel logic optimization are examples of specific logic synthesis techniques.

Device mapping is the process of implementing the optimized circuit in hardware. The mapping process may involve further refinement and optimization of the circuit to meet the specific constraints of a selected target architecture. These constraints may include such things as gate sizes, fan-in and fan-out restrictions, and device resource constraints. The actual mapping process may include partitioning, resource allocation, routing, and circuit translation.

Computer-assisted logic synthesis isn't always necessary, but as designs and devices become more complex and less reliance is placed on fixed-function devices and simple PLDs, an understanding of available logic synthesis options becomes increasingly beneficial.

The Importance of Effective Logic Synthesis

When behavioral design methods are used, effective logic synthesis techniques are the key to a successful and efficient final circuit. Since the goal of logic synthesis is to massage a circuit representation into a form that meets the specified constraints of the design, it's important to decide what constraints are most important for each design. The constraints that affect logic synthesis include operating constraints, cost constraints, and architectural constraints. Other indirect constraints include such things as interface constraints or testability requirements.

For most designs being implemented with PLDs, the most direct constraint is architectural. Automated logic synthesis tools that help to optimize combinational circuits for various cost and operational constraints have existed for some time. Automated tools that help with logic synthesis for constrained architectures, however, are still in their infancy.

Logic Synthesis Features

There are many levels of logic synthesis features found in today's PLD design tools. We'll quickly described these features to help you compare the tools available. The first logic synthesis features to appear in PLD design tools were algorithms that perform logic minimization. *Logic minimization* (described in detail in Chapter 7) is used to reduce the amount of circuitry required to implement a given logic function. The first logic minimization algorithms used in PLD design tools were limited in their features and weren't capable of fully optimizing circuits that contained don't-care conditions. Modern logic reduction algorithms such as Espresso (developed at the University of California at Berkeley) have recently appeared in PLD design tools. The Espresso algorithm supports a large number of options, including don't-care optimizations, and is quickly becoming the algorithm of choice for experienced logic designers.

Multilevel logic optimization algorithms are just beginning to appear in PLD design tools. Complex multilevel devices (such as the FPGAs described in the previous chapter) require optimizations beyond the scope of sum-of-products-oriented logic minimization algorithms. *Equation factoring* is one multilevel synthesis feature that is useful for these devices. Factoring allows large combinational functions to be fractured into smaller units that are more appropriate for constrained architectures. As we will describe in Chapter 7, multilevel optimizations can be used for reducing the overall amount of circuitry required for a circuit or for tailoring a design to a specific target architecture.

As PLDs become more complex, specialized *device fitting* algorithms are becoming more common. These algorithms are used to optimize a logic function for the particular constraints of a target architecture. Device fitting features may include automatic pin and resource assignment, flip-flop type conversion, preset and reset emulation, and automatic polarity control. Devices with segmented architectures may require *logic partitioning* features.

In addition to their use for device fitting, automatic partitioning algorithms can also be used to break large designs into smaller units that are appropriate for multiple-PLD implementations. The automatic partitioning feature found in the PLDesigner product, for example, accepts partitioning constraints (such as the maximum number of partitions and the maximum delay) and device selection criteria (such as preferred vendors and package types) and can automatically generate an optimized circuit partitioned into multiple PLDs.

Limitations of Logic Synthesis

Logic synthesis features available in today's PLD design tools are certainly useful for that part of the design process in which logic synthesis can benefit. It's important to realize, however, that the currently available PLD synthesis tools, no matter how highly automated or tightly integrated they may be, cannot be used without an awareness of their limitations.

In particular, you must always be aware of how basic design decisions can affect the final circuit produced by the tools. Automated logic synthesis tools can only help to optimize a logic design as you have specified it. They cannot explore design-level alternatives and make intelligent decisions about about those alternatives. In later chapters, we'll explore some of these design-level optimizations and alternatives.

5.4 SIMULATION OF PLD-BASED DESIGNS

To describe the entire field of simulation tools that can be applied to PLD designs is beyond the scope of this book. It's important to understand, however, that there are two fundamental approaches to design simulation that are useful to PLD users. These two approaches are *functional* simulation and *device* simulation. Both of these simulation methods have important advantages and disadvantages.

Functional Simulation

Design-level functional simulation is a method of determining whether your design will operate as intended, based on its high-level behavioral description. Functional simulation can be performed at any phase of the design entry process, if the simulator can process the design data at that level. For example, a design entered in the form of Boolean equations could be functionally simulated directly from those equations, if the simulator was capable of reading and calculating circuit behavior from an equation form.

Functional simulation is particularly useful as a tool for verifying small sections of a design as they are entered. This allows design entry to proceed in a step-wise and modular way, simplifying the overall design process. Functional simulation also allow you to compare the results of various circuit optimizations if the simulator supports delay calculation or hazard detection.

While functional simulators are useful for verifying higher-level circuit behavior (ensuring that your 252-state counter doesn't reset after only 251 states, for example) they usually can't help with circuit problems that are related directly to PLD architectural differences. That is the realm of the device simulators.

Device Simulation

Device simulation is normally done as the last step in the conversion of a design description into a specific PLD implementation. A device simulator will construct a software model of the target PLD and will then "program" this PLD model using the programming data provided (normally in the form of a JEDEC file). Test stimulus can then be applied to the simulator to determine if the device will function correctly.

Device simulators may or may not include device timing information, but all include information about specific architectural details in the device (macrocell configurations, dynamic flip-flop control, output enables, and so on). A device simulator is the only way to determine, with a high level of confidence, that a real device programmed with the specified design will function as expected. Examples of device simulators are found in many PLD design tools. Device simulation is also available from many companies producing board-level simulation tools. For example, the SUSIE simulator from Automated Logic Design (ALDEC) includes models for popular PLDs and can simulate programmed PLDs within a larger circuit.

The drawback to device simulation is that you must enter your design, optimize it, and map it to one or more specific devices before you can perform the simulation. For this reason, it's useful to have both functional simulation and device simulation tools available.

Test Vectors

The test stimulus used for simulation of PLD-based designs, and testing of actual devices, is most often provided in the form of *test vectors*. Test vectors are sequences of input stimulus values and expected output values. The input stimulus is applied to the inputs of the PLD (either a real PLD or a software model of a PLD) and, after enough time has passed for the signals to propogate through the

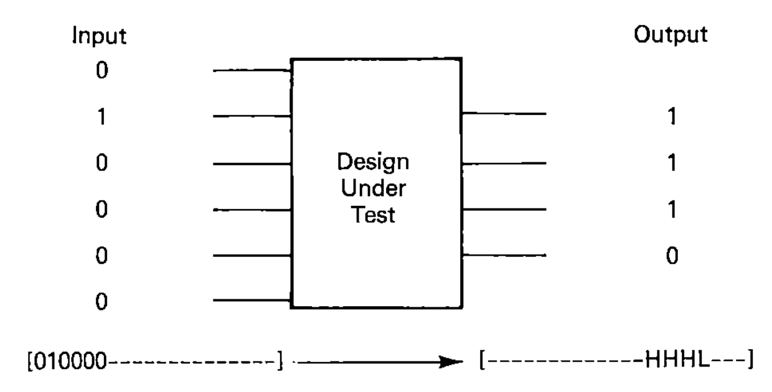

Figure 5.1 Test vector applied to a PLD

device, the resulting values appearing on the device's outputs are compared against the expected outputs indicated in the output portion of the test vector (Figure 5.1).

Many PLD design languages include language features that help simplify the creation of test vectors for design verification. *Design verification vectors* are an important part of any complex PLD-based design. Without them, it can be very difficult to determine whether a complex behavioral design description will actually result in the desired circuit.

Other tools are available that can automatically generate test vectors that are used in production to verify that each PLD programmed from a master design is identical to that master. These test vectors, called *device verification vectors,* should not be confused with design verification vectors. If you are attempting to verify that your design concept was correctly translated into an equivilant circuit, then you clearly can't create test vectors automatically from the circuit you are trying to verify. We will examine the creation and use of device verification vectors in a later chapter.

5.5 STANDARD FORMATS FOR DESIGN TRANSFER

While there is no standard design entry language that has emerged for PLDs, there are a variety of data transfer formats that have become real or defacto standards.

The JEDEC Format

The first standard format to be accepted in the PLD industry was *JEDEC standard 3*. This standard format is used to transfer device programming data to device programmers. Fields in the file specify the condition of actual device fuses within the target device. A sample file (often simply referred to as a *JEDEC file*) is shown in Figure 5.2.

The first few lines of the file are a comment field which is terminated by the

```
ABEL 4.00 Data I/O Corp. JEDEC file for: P16R8 V9.0
Created on: Sat Aug 25 07:47:34 1990
Decade counter *
QP20* QF2048* QV15* F0*
X0*
NOTE Table of pin names and numbers*
NOTE PINS Clock:1 Reset:2 D3:14 D2:15 D1:16 D0:17*
L0512 11111111111111111101110111111111*
L0544 01111111111111111111111111111111*
L0576 11111111111111011111110111111111*
L0608 11111111110111111111111111111111*
L0768 01111111111111111111111111111111*
L0800 11111111111111111111110111111111*
L0832 11111111111011101111111111111111*
L0864 11111111110111011111111111111111*
L1024 01111111111111111111111111111111*
L1056 11111111110111011101111111111111*
L1088 11111111111011111110111111111111*
L1120 11111111111111011101111111111111*
L1152 11111111111111111111110111111111*
L1280 11111111111011111111111011111111*
L1312 11111111110111101111111111111111*
L1344 11111111111111011110111111111111*
L1376 11111111111111111101110111111111*
L1408 01111111111111111111111111111111*
V0001 C1XXXXXXXNXXXLLLLXXN*
V0002 C0XXXXXXXNXXXLLLHXXN*
V0003 C0XXXXXXXNXXXLLHLXXN*
V0004 C0XXXXXXXNXXXLLHHXXN*
V0005 C0XXXXXXXNXXXLHLLXXN*
V0006 C0XXXXXXXNXXXLHLHXXN*
V0007 C0XXXXXXXNXXXLHHLXXN*
V0008 C0XXXXXXXNXXXLHHHXXN*
V0009 C0XXXXXXXNXXXHLLLXXN*
V0010 C0XXXXXXXNXXXHLLHXXN*
V0011 C0XXXXXXXNXXXLLLLXXN*
V0012 C0XXXXXXXNXXXLLLHXXN*
V0013 C0XXXXXXXNXXXLLHLXXN*
V0014 C0XXXXXXXNXXXLLHHXXN*
V0015 C1XXXXXXXNXXXLLLLXXN*
C4328*
```

Figure 5.2 JEDEC standard 3 file

first asterisk (*) character. The following line is composed of four optional fields which specify the number of entries in the file, the total number of fuses in the target device, the number of test vectors included in the file, and the default fuse value for unspecified fuses.

The actual fuse data is contained in the lines of ones and zeroes that are prefixed by the "L" character and a fuse offset. A fuse value of one indicates that the specified fuse is to disconnected ("blown"), while a value of zero indicates that the specified fuse is to remain intact. Since this file specifies a default value of zero (indicated by the *F*0 entry on the fourth line) all fuses not specified in the file will remain intact. The JEDEC format includes not only device programming data, but test stimulus information as well. Test vectors are an important part of device programming. Most PLD programmers are capable of applying the specified vectors to the programmed device to ensure that it functions properly.

In addition to fuse programming data, this sample file also contains six test vectors. Each test vector is prefixed by the character "V" and a vector number, and each vector has exactly as many characters as there are pins on the target device. In this case, there are a total of 20 characters in each test vector. Each character corresponds to one pin on the device, with the left-most character representing pin number 1. The following characters are valid in JEDEC standard format test vectors:

0	Drive input low
1	Drive input high
2–9	Drive input to supervoltage (2–9)
B	Buried register preload
C	Drive input low, high, low (clock pulse)
F	Float input or output
H	Test output for high state
K	Drive input high, low, high (inverted clock)
L	Test output for low state
N	Unused pin (including power and ground)
P	Preload registers
X	Output not tested; input driven to default value
Z	Test output for high impedence

The final data field shown in this sample file is the *fuse checksum*. The fuse checksum is used to detect transmission errors occuring when the file is transferred between a host computer and peripheral device programmer. The checksum (which appears in the file as the character "C" followed by a hexadecimal value) is calculated from the fuse data for the entire device (including fuses not specified directly in the file) and is the 16-bit sum (modulo 65,535) of consecutive 8-bit words of programming data.

The JEDEC format file can be used to transfer device programming data from a design tool to a device (via a programmer), or from a programmed master device to another device. The JEDEC file can also be used to generate simulation models for programmed PLDs.

EDIF

Electronic design information interchange format (EDIF) is supported in some PLD design tools as an input or output option. EDIF was developed in 1983 and 1984 as a joint effort by a consortium of CAE workstation and semiconductor vendors. Version 2.00 of EDIF has been adopted as a standard by both the American National Standards Institute (ANSI) and by the Electronics Industries Association (EIA).

The intent of the EDIF standard is to provide a common data format in which to express netlists, schematic data, and other electrical design data. Detailed information about the EDIF standard can be obtained by contacting:

Electronic Industries Association
Standards Sales Office
2001 Eye Street, NW
Washington, D.C. 20006

EDIF Users Group
2222 South Dobson Road, Bldg. 5
Tempe, AZ 85285-5542

VHDL

The *very high-speed integrated circuit* (VHSIC) project, funded by the Department of Defense, includes a development program for a standard hardware description language. The result, *VHSIC hardware description language* (VHDL), is a language initially developed for the purpose of modeling large integrated circuits. While the first applications of VHDL were for circuit simulation, the VHDL design environment is intended to support a variety of system design tools and methods, including logic synthesis.

The VHDL design system supports hierarchical design methods, and includes a number of design entry and analysis tools. The language itself supports a variety of different circuit abstractions, ranging from purely structural descriptions (layout of transistors on a chip, for example) to high-level behavioral descriptions of circuit functions.

It's unclear at this point whether VHDL (or some subset of VHDL) is appropriate for a standard PLD design language, but we can expect that future PLD development tools will utilize VHDL as a vehicle for interface to other tools in the CAE environment.

The Berkeley PLA Format

The Berkeley PLA format was developed at the University of California at Berkeley as a means of transferring design information between various logic synthesis tools. A logic function specified in the PLA file is expressed in a tabular form, with each line of the table representing one product term in a sum-of-products logic function (these terms are described in detail in Chapter 7). A sample PLA file is shown in Figure 5.3. The Berkeley PLA format is beginning to appear as

```
#$ TOOL ABEL 4.00
#$ TITLE Decade counter
#$ MODULE cnt10
#$ DEVICE P16R8
#$ JEDECFILE cnt10
#$ VECTORFILE cnt10.tmv
#$ PINS 6 Clock+:1 Reset+:2 D3-:14 D2-:15 D1-:16 D0-:17
.i 6
.o 8
.type fr
.ilb Clock Reset D3.FB D2.FB D1.FB D0.FB
.ob D3.REG- D2.REG- D1.REG- D0.REG- D3.C D2.C D1.C D0.C
.phase 11111111
.p 24
--0--0 1~~~~~~~
----01 1~~~~~~~
---01- 1~~~~~~~
--11-- 1~~1~~~~
-1---- 1111~~~~
---111 ~1~~~~~~
---0-0 ~1~~~~~~
---00- ~1~~~~~~
--1--- ~11~~~~~
----00 ~~1~~~~~
----11 ~~1~~~~~
--1-1- ~~~1~~~~
-----1 ~~~1~~~~
1----- ~~~~1111
-01000 0~~~~~~~
-00111 0~~~~~~~
-00011 ~0~~~~~~
-001-0 ~0~~~~~~
-0010- ~0~~~~~~
-00-10 ~~0~~~~~
-00-01 ~~0~~~~~
-0-000 ~~~0~~~~
-00--0 ~~~0~~~~
0----- ~~~~0000
.e
```

Figure 5.3 Berkeley PLA file

a data transfer format between different synthesis modules found in commercial logic synthesis and PLD design tools.

5.6 DESIGN TOOLS

There are a large number of CAE tools available for PLD development. These tools can be lumped into three categories roughly corresponding to the source of the tools. The *vendor specific* tools are those developed by device manufacturers, while the *universal* and *internally developed* tools are those developed by third-party companies and end-users, respectively.

Vendor Specific Tools

Ever since MMI's success with PALASM, PLD manufacturers have recognized the need to have software support available for their devices before the devices are available to end-users. The success or failure of a new PLD architecture is often affected by the usability of the design tools for that device and by the availability of example applications. In order to ensure that software support is available in a timely manner, many manufacturers have chosen to develop and distribute design tools specific to their devices. These design tools range from relatively simple giveaway design compilers to very complex design entry and synthesis tools costing thousands of dollars.

At the low end of the spectrum, programs like PALASM and PLPL from Advanced Micro Devices, and PLAN from National Semiconductor are available free of charge and provide limited design entry options and no logic synthesis features. Some device manufacturers, such as Exel Microelectronics, have chosen to create add-on modules to established third-party PLD design tools in order to get satisfactory support for their devices.

At the high end of the spectrum, companies like Altera and Xilinx produce design entry tools that rival universal PLD design tools in complexity. For a startup company with a radically different device architecture, the revenues from the sales of their proprietary design tools can help to offset the cost of developing new devices.

Vendor specific tools can, in general, provide excellent support for a limited number of PLDs. The limitation in the number of devices supported can be a major disadvantage to the user, however.

Third-party Universal Design Tools

Third-party PLD design tools are those that have been developed by companies not directly involved in the production of devices. This includes companies like Data I/O, (whose primary business is device programmers) and CAE companies like Mentor, Daisy, and HP (who develop PLD design tools to complete their CAE product lines). In addition, a number of smaller companies, including Assisted Technology and Minc, have been formed specifically to develop and market design tools for PLDs.

The advantages of third-party software are many when compared to vendor-specific tools. The most obvious advantage is that designs are easily converted from one device architecture to another. If different vendor-specific tools are used for different devices, it may be necessary to re-enter your design when a new device is chosen.

Another advantage to using one third-party PLD tool is that there is little agreement between different tool developers on how various design tool and design language features should be implemented. The format for entering Boolean equations, for example, may be quite different from one design language to another. Using a single design tool for all of your PLD designs helps to shorten the amount of time required to learn the specific features of the tools.

A third, less obvious advantage to the third-party tools is the benefit of having a central source of information about new devices. The third-party tools are constantly being updated to reflect new device types and every software release gives you a new set of devices to experiment with.

Internally Developed Tools

Many companies and individuals have, for various reasons, developed their own tools for PLD development. Developing your own PLD design system isn't a difficult task, if you aren't concerned with supporting a large number of devices. All that is required is that you acquire some of the public domain logic synthesis tools already in existence and combine these tools with a PLD assembler such as PALASM (described in a following section), using the PLD assembler as a fuse pattern generator. Ideally, you will also develop some sort of design entry tool, such as a design language processor, that will allow users to describe their designs to the design system in some form other than simple Boolean equations.

This will require a certain amount of software savvy, but is not beyond the abilities of the average computer-literate engineer. The resulting set of design tools may not be well integrated, but will allow you to use high-level design techniques such as those described later in this book.

The real difficulty in developing such a PLD design system is the device support problem. Supporting the hundreds of distinct device architectures in a single design tool is a Herculean task. New devices are constantly being introduced, many of which have features never seen in previous devices. If you have any hope of using these new devices, you must first do whatever is necessary to add them to your design tool. If you don't have access to advanced information from the device manufacturers, your support for the new devices can lag their introduction by months.

We will discuss internally developed design tools very little in this book, since these tools are generally not accessible to the average user. It's important to note, however, that there are a number of quite powerful tools that have been developed for internal use at large companies. Some of these tools have directly influenced the development of commercially available PLD tools. One early example of this was Douglas Brown's SMS system, developed for internal use at Tektronix, which had a direct influence on the design of Data I/O's ABEL (see Chapter 2).

Assemblers, Compilers, and Architecture Independent PLD Tools

In terms of feature support, PLD design tools of today can be grouped into three major categories; PLD assemblers, PLD compilers, and architecture independent design tools. These categories are sometimes referred to as the first, second, and third generations of PLD design tools.

PLD assemblers are those design tools which convert sum-of-products design equations into fuse programming data. Assembler-level PLD tools are distinguished from second and third generation design tools by their lack of high-level design entry or logic synthesis features. PLD assemblers are characterized by sum-of-products Boolean equation entry and no logic minimization features.

PLD compilers are those design tools that provide support for a variety of high-level design entry options and include logic synthesis features. Typically, a PLD compiler will include support for other than sum-of-products forms of equations, as well as state machine description formats and other design entry options. Compilers are distinguished from higher-level design tools by the requirement that a specific PLD (or general class of PLD) be specified up-front, rather than after the design has been entered and verified.

PLD compilers offer a great deal of flexibility when it comes to design entry options and device support, but they also have their weaknesses. Most of the design tools that fall into the compiler category require that the user know in advance which PLD (out of the hundreds of unique architectures available) is going to be used for the design. While many of the modern PLD compilers do not require that the user specify the specific device to be used, these tools still require that a number of architecture-related design decisions be made up front.

The need for separation of the design description process from actual device architectures becomes more important as devices and designs become more complex, and the sheer number of different architectures grows to the point where an engineer simply can't keep up with the specific features of all devices.

There are a few design tools now available that, to some extent, address the need for *architecture-independent design* and it's expected that more will soon appear. These tools generally include more advanced logic synthesis features than the compiler tools, and have more output options, reflecting their appropriateness for target architectures other than PLDs.

5.7 SURVEY OF AVAILABLE TOOLS

In this section, we'll briefly describe some of the many PLD design tools that are available as of this writing. There are new PLD design tools coming into existence every year and new features are appearing in established tools, so this section cannot hope to cover all of the tools and features available. The vendors of these design tools are constantly improving their products, so it's important to get up-to-date information from the vendors before making a decision on which to use. Appendix B contains addresses and phone numbers for all companies mentioned in this chapter.

ABEL—Data I/O Corporation

The ABEL design system provides a variety of input language features that allow designs to be described in ways familiar to design engineers. These input forms include high-level equations, truth tables, and state diagrams. The ABEL-HDL language includes many features found in high-level programming languages and yet is easy for hardware-oriented engineers to learn and use. A sample ABEL-HDL design file is shown in Figure 5.4. This file describes a simple address decoder using ABEL's high-level equation syntax.

When processed by ABEL, these design equations are processed into sum-of-products Boolean form and then minimized and converted into a PLD fuse map. This fuse map is then written to a JEDEC standard file that can be downloaded to a device programmer. The design file also includes test vectors which have been written using the same set notation as the design equations. These test vectors are used by the ABEL simulator to verify that the final circuit will perform as expected and are also written to the JEDEC file for use by device programmers.

Figure 5.4 ABEL-HDL design file

```
module DECODE
title 'Address Decoder Example'

        decode device 'P16L8';

        !SRAM,!PORT,!UART,!PROM          pin 14,15,16,17;
        A15,A14,A13,A12,A11,A10,A9       pin 3,4,5,6,7,8,9;

        H,L,X,Z,C = 1,0,.X.,.Z.,.C.;

        Address = [A15..A9,X, X,X,X,X, X,X,X,X];

Equations
        SRAM = (Address <  ^h8000);
        PORT = (Address >= ^h8000) & (Address <= ^h81FF);
        UART = (Address >= ^h8200) & (Address <= ^h83FF);
        PROM = (Address >= ^h8400);

Test_Vectors
       (Address -> [!SRAM,!PORT,!UART,!PROM])
        ^h0000  -> [   L ,   H ,   H ,   H  ];
        ^h1000  -> [   L ,   H ,   H ,   H  ];
        ^h4000  -> [   L ,   H ,   H ,   H  ];
        ^h7000  -> [   L ,   H ,   H ,   H  ];
        ^h7FFF  -> [   L ,   H ,   H ,   H  ];
        ^h8000  -> [   H ,   L ,   H ,   H  ];
        ^h81FF  -> [   H ,   L ,   H ,   H  ];
        ^h8200  -> [   H ,   H ,   L ,   H  ];
        ^h83FF  -> [   H ,   H ,   L ,   H  ];
        ^h8400  -> [   H ,   H ,   H ,   L  ];
        ^hC000  -> [   H ,   H ,   H ,   L  ];
        ^hFFFF  -> [   H ,   H ,   H ,   L  ];
end
```

ABEL includes a variety of logic minimization options based on the well known Espresso logic minimization algorithm. Optimization of don't-cares is performed if requested by the user. ABEL also includes a PLD disassembler that reads JEDEC format files and produces an ABEL-HDL file. This is useful for documentation, testability analysis, or design translation. The ABEL-HDL language is architecture-independent; there is no need to specify an actual PLD in an ABEL-HDL design. ABEL's optional device selection and fitting features can be used to implement a design in a variety of different PLDs.

An optional module, PLDlinx, allows designs to be entered using Data I/O's FutureNet schematic capture package. Using this module, a design can be expressed in schematic form or in a mix of schematics, equations, truth tables, and state diagrams. PLDlinx is capable of extracting circuit logic from a schematic composed of standard logic devices so existing designs can be quickly converted to PLD implementations.

In terms of device support, ABEL has historically supported more PLD architectures that any other PLD design tool. ABEL is available on the IBM personal computer and on a variety of engineering workstations including the Sun and VAX. The IBM personal computer version sells for around $2,000.

AMAZE—Signetics Corporation

Automated map and zap equations (AMAZE) is Signetics' giveaway package for PLD design. AMAZE includes the BLAST assembler, which is designed to handle the PLA architecture of the Signetics devices. The language, called *Signetics standard program,* accepts Boolean equations, state tables, and schematic netlists (FutureNet or OrCAD formats). AMAZE includes a device simulator, and is capable of generating test vectors automatically. The package includes a rudimentary optimization module that can detect and combine shared product terms (see Chapter 7) for design outputs. Also included in AMAZE is a JEDEC file disassembler program that will convert JEDEC files into signetics standard program files.

APEEL—International CMOS Technology

The APEEL package, developed and distributed free of charge from ICT, is similar to PALASM (see below) in its feature set. The package supports only ICT's EEPLDs and outputs a JEDEC standard file that can be downloaded to any PLD programmer.

The APEEL language (which was initially developed for ICT by John Birkner) is actually a limited subset of the ABEL language (described in a previous section). The language also supports the use of PALASM-style equations. APEEL supports only sum-of-products Boolean equation entry and has no logic minimization features. The package does include a text editor that is integrated with the language processor for error checking and a PLD simulator.

Since all APEEL source files are ABEL compatible, the APEEL system is one of the few vendor-specific tools that allows designs to be migrated to a more powerful universal design tool without the need for source file translation or re-entry.

A+PLUS—Altera Corporation

Altera offers development tools that are designed specifically for users of their devices. The A+PLUS system is designed for the EP300 through EP1800 family of devices, while the MAX+PLUS package (described later) is intended for users of the MAX family of devices. The SAM+PLUS package is designed for users of the Altera SAM sequencer device. All of these packages support multiple design entry formats which include Boolean equations, state descriptions, and schematics (FutureNet, P-CAD, or Altera LogiCaps formats).

Unlike most device vendors' offerings, the Altera packages are quite sophisticated and are among the most advanced PLD design tools available. Since these design tools were created for a relatively small set of devices, Altera has been able to focus their efforts on design entry and synthesis for the specific architectures of their devices. In addition, the Altera devices were designed in parallel with the tools, so the architecture of the devices could be tailored for high-level design methods.

The Altera design systems support hierarchy features: designs may be specified in a modular fashion and individual modules can be simulated before they are incorporated into the larger design. The Altera systems accept design descriptions in a variety of forms, one of which is the *Altera design language* (ADL). A sample ADL design file is shown in Figure 5.5.

The logic minimization and device fitting modules of the Altera design systems translate designs into a form appropriate for the target device and automatically assign pins and other device resources as needed to implement the design. The packages also include an event-driven simulator that can verify the timing behavior of the resulting PLDs. The simulator interface has a waveform interface and other logic analyzer type features.

All of the features of the Altera products are packaged in an easy to use user interface. The A+PLUS product is available for use on IBM personal computers and sells for $4,995 (including a simple device programmer).

CUPL—Logical Devices Incorporated

CUPL was originally developed by Bob Osann, then president and founder of Assisted Technology, and was actually the first PLD compiler product, predating ABEL by about six months. The CUPL product was later bought by P-CAD, a supplier of PC-based schematic capture tools. In 1987, CUPL was sold yet again to Logical Devices, a PLD programmer manufacturer, when P-CAD re-focused its efforts on printed circuit layout tools.

```
Michael Holley and Dave Pellerin
Aug 25, 1990
1.00
A
EP600
8-bit binary counter

PART: EP600

INPUTS: ENABLE, RESET, CLOCK

OUTPUTS: Q0, Q1, Q2, Q3, Q4, Q5, Q6, Q7

NETWORK:
        ENABLE  = INP(ENABLE)
        CLR     = INP(RESET)
        CK      = INP(CLOCK)

        Q0,Q0   = TOTF(Q0t,CK,CLR,,)
        Q1,Q1   = TOTF(Q1t,CK,CLR,,)
        Q2,Q2   = TOTF(Q2t,CK,CLR,,)
        Q3,Q3   = TOTF(Q3t,CK,CLR,,)
        Q4,Q4   = TOTF(Q4t,CK,CLR,,)
        Q5,Q5   = TOTF(Q5t,CK,CLR,,)
        Q6,Q6   = TOTF(Q6t,CK,CLR,,)
        Q7,Q7   = TOTF(Q7t,CK,CLR,,)

EQUATIONS:

        Q7t = ENABLE * Q0 * Q1 * Q2 * Q3 * Q4 * Q5 * Q6;

        Q6t = ENABLE * Q0 * Q1 * Q2 * Q3 * Q4 * Q5;

        Q5t = ENABLE * Q0 * Q1 * Q2 * Q3 * Q4;

        Q4t = ENABLE * Q0 * Q1 * Q2 * Q3;

        Q3t = ENABLE * Q0 * Q1 * Q2;

        Q2t = ENABLE * Q0 * Q1;

        Q1t = ENABLE * Q0;

        Q0t = ENABLE;

END$
```

Figure 5.5 A+PLUS ADL design file

CUPL provides the user with a variety of design entry options, including Boolean equations, truth tables, and state diagrams. CUPL also supports set notation similar to ABEL's, although the design language doesn't support as many high-level equation operations as available in ABEL. A sample CUPL design file is shown in Figure 5.6. This design utilizes CUPL's set notation and set range operators to implement the same address decoder described previously in ABEL.

```
                PARTNO    decode;
                NAME      Memory Decoder;
                REV       01 ;
/*******************************************************/
/*  Target Device Types: 14L4, 16L8, 82S153, 16H8      */
/*******************************************************/

        pin [3..9]        = [a15..a9];
        pin 14            = !SRAM;
        pin 15            = !PORT;
        pin 16            = !UART;
        pin 17            = !PROM;

        field address     = [a15..a9];

/**  Logic Equations  **/

         SRAM    = address:[0000..7FFF];

         PORT    = address:[8000..81FF];

         UART    = address:[8200..83FF];

         PROM    = address:[8400..FFFF];
```

Figure 5.6 CUPL design file

CUPL includes a number of logic minimization options and can be equipped with an interactive device selector that accepts various selection criteria to help choose an appropriate device for a specified design. In the CUPL product, test vectors aren't included in the design file, but are written separately. These test vectors can be processed by the CUPL's CSIM device simulator and written to the JEDEC file for program-and-test purposes.

CUPL's device support is quite good and includes complex PLDs. An option to the CUPL product allows designs to be converted from schematic forms. CUPL is available on the IBM personal computer and on various engineering workstations. The IBM PC version is sold for around $1,500.

FutureDesigner—Data I/O Corporation

FutureDesigner is an architecture-independent design entry package developed at Data I/O that allows large designs to be specified in a language similar to that of ABEL. Using FutureDesigner, a design can be entered and functionally simulated before any logic synthesis steps are performed. All of FutureDesigner's many features are invoked from a highly interactive design environment that is similar in operation to a spreadsheet program.

The FutureDesigner system consists of a set of interactive design entry forms that allow designs to be entered using high-level equations, truth tables, and state descriptions. Since the system is forms based, design entry is quick and the system is easy to learn. Error messages are displayed immediately, without the tedious edit-compile-edit loop that is typical of compiler-style tools.

Designs entered in these forms can be simulated behaviorally (at the equation level) and logic synthesis features, such as logic minimization and equation factoring, can be invoked to tailor the logic for a specific type of architecture. For example, a design could be minimized and partitioned for implementation into PLDs or could be factored into a multilevel representation for implementation in other technologies such as FPGAs or gate arrays. Like other PLD design systems, the FutureDesigner system produces JEDEC files for the devices selected. FutureDesigner is also capable of automatically generating schematics and netlists that are useful for implementing designs in other types of device technologies.

FutureDesigner has a particularly powerful set of logic synthesis features. The Espresso algorithm is included in the product and the design language supports the specification of don't-care conditions (see Chapter 7), resulting in more optimal minimizations, particularly for state machines. Another feature not found in earlier tools is the factoring module. This module is used to generate multilevel logic such as that required for gate array and FPGA implementations. Factoring parameters such as desired gate widths can be specified as needed for the target technology. FutureDesigner is available on 80386-based MS-DOS computers and on Sun workstations. The 80386 MS-DOS version sells for about $4,500.

iPLDS II—Intel Corporation

The iPLDS II system from Intel is similar to the Altera design software, having the same origins. iPLDS II does support more devices, however, than it's Altera counterpart. iPLDS includes support for Boolean equations and state machine descriptions and allows schematic designs to be entered using FutureNet's DASH schematic editor, P-CAD's PC-CAPS, or Omation's Schema editor.

iPLDS has powerful device fitting features that can automatically assign pins based on the design requirements, optimizing the utilization of more complex devices. Intel also offers a logic simulator and timing analyzer (OEM versions of Viewlogic's ViewSim product). Available on the IBM personal computer, the iPLDS II tools begin at $895, with a full-featured product priced at $5,995.

LOG/iC—Adams-MacDonald Enterprises Incorporated

LOG/iC was developed by ISDATA, a West German company, soon after ABEL and CUPL were released. This product has evolved over the years and is now quite popular in Europe. LOG/iC includes support for Boolean equations, truth tables, and state flow table descriptions.

LOG/iC is sold as a variety of modules that can be selected to meet the needs (and budget) of the user. LOG/iC's Boolean equation language is less powerful than that of some other PLD languages, but their state machine flow table syntax is quite powerful. The logic minimization module is also above average and can utilize don't-cares for more complete minimization. The basic algorithms of LOG/iC were developed at the University of Karlsruhe in West Germany.

The product has many extra-cost options including a functional simulator, schematic entry, and a comprehensive database of PLDs that can be accessed through a menu interface for device selection purposes.

One reason for LOG/iC's success in Europe is the syntax of the design language. LOG/iC's language has fewer keywords than most PLD design languages. Instead, the language is more mathematical in its grammar. This helps users of the product who are not native English (or German) speakers. A sample LOG/iC design file is shown in Figure 5.7.

LOG/iC is available on the IBM personal computers and on a variety of CAE workstations, including the Apollo and DEC VAX. The price of LOG/iC depends on the options selected beginning with $1,480 for the entry-level package.

MAX+PLUS—Altera Corporation

In order to support the MAX family of devices, Altera offers a design system called MAX+PLUS. MAX+PLUS is similar to A+PLUS in its design entry options, but includes synthesis features designed specifically for the MAX's modular architecture. In conjunction with the developement of the MAX+PLUS software, Altera made some changes to the Altera design language that allow higher-level design descriptions to be used. These changes make the language (which Altera now calls AHDL) look more like ABEL, as shown in Figure 5.8. The MAX+PLUS software is available as a $3,900 add-on to the A+PLUS software or as a stand-alone package (with a simple device programmer) for $4,995.

MultiMap and MultiSim—EXEL Microelectronics Inc.

In order to support its ERASIC PLDs, Exel developed MultiMap and MultiSim. These modules replace the FuseMap and Simulate modules of the ABEL product and allow ABEL to be used for the unusual NOR/NOR architecture of the ERASIC devices. The MultiMap and MultiSim modules are designed to be invoked automatically when an ERASIC device is specified in an ABEL design. MultiMap is capable of performing simple multilevel logic optimizations to efficiently utilize the ERASIC's folded array. The MultiMap and MultiSim modules are priced at $295.

OrCAD/PLD—OrCAD Systems

OrCAD, a producer of PC-based schematic capture tools, has a product that supports Boolean equations, truth tables, state descriptions, and schematic entry for PLDs. The language features are, in general, less powerful than those of ABEL

```
;
;          a
;         ---        BCD-to-seven-segment decoder
;       f| g |b
;         ---        segment identification
;       e| d |c
;         ---
;
*IDENTIFICATION
 Seven Segment Display Decoder
*DECLARATIONS
 X-VAR = 4
 Y-VAR = 7
*X-NAMES
 BCD[3..0];
*Y-NAMES
 Aseg, Bseg, Cseg, Dseg, Eseg, Fseg, Gseg;
*RUN-CONTROL
 CPU-TIME = 1000;
 PROGFORMAT = JEDEC;
 LISTING = FUSE-PLOT, PINOUT, EQUATIONS;
 TEST-VECTORS = READ;
*STRING
 ON  = 1;
 OFF = 0;
*FUNCTION-TABLE
$ (BCD[3..0]) : ((Aseg, Bseg, Cseg, Dseg, Eseg, Fseg, Gseg));
       0H     :    'ON'   'ON'   'ON'   'ON'   'ON'   'ON'  'OFF' ;
       1H     :   'OFF'   'ON'   'ON'  'OFF'  'OFF'  'OFF'  'OFF' ;
       2H     :    'ON'   'ON'  'OFF'   'ON'   'ON'  'OFF'   'ON' ;
       3H     :    'ON'   'ON'   'ON'   'ON'  'OFF'  'OFF'   'ON' ;
       4H     :   'OFF'   'ON'   'ON'  'OFF'  'OFF'   'ON'   'ON' ;
       5H     :    'ON'  'OFF'   'ON'   'ON'  'OFF'   'ON'   'ON' ;
       6H     :    'ON'  'OFF'   'ON'   'ON'   'ON'   'ON'   'ON' ;
       7H     :    'ON'   'ON'   'ON'  'OFF'  'OFF'  'OFF'  'OFF' ;
       8H     :    'ON'   'ON'   'ON'   'ON'   'ON'   'ON'   'ON' ;
       9H     :    'ON'   'ON'   'ON'   'ON'  'OFF'   'ON'   'ON' ;
      REST    :     -      -      -      -      -      -      -  ;
*PAL
 TYPE = PAL16L8
*PINS
 BCD3 = 5, BCD2 = 2, BCD1 = 3, BCD0 = 4,
 Aseg = 18, Bseg = 12, Cseg = 13, Dseg = 14,
 Eseg = 15, Fseg = 16, Gseg = 17;
*END
```

Figure 5.7 LOG/iC design file

and CUPL. One unique language feature that OrCAD/PLD offers is indexed equations. Figure 5.9 shows a sample of the OrCAD/PLD indexed equation syntax. The design file shown describes a counter.

OrCAD/PLD includes a logic minimization module and an interface to the OrCAD/VST simulator. This interface, OrCAD/MOD, reads JEDEC format files and creates PLD timing models that can be simulated with other parts of a circuit. Since this interface reads standard JEDEC files, designs developed using other

```
Title "Arbitrary length counter with Carry Out";
%-----------------------------------------------
Clyde Cowpoke    Altered States Corp
-----------------------------------------------%
DESIGN IS Count116
BEGIN
    DEVICE IS "EPM5032"
    BEGIN
        Osc @ 1, Inc @ 2, Clear @ 12 : INPUT;
        A6 @ 17, A5 @ 18, A4 @ 19, A3 @ 20: OUTPUT;
        A2 @ 23, A1 @ 24, A0 @ 25, CarryA @ 26: OUTPUT;
    END;
END;

SUBDESIGN Count116
(
    Osc, Inc, Clear      : INPUT;
    A[6..0], CarryA      : OUTPUT;
)
VARIABLE
    CarryQ, CountQ[6..0] : DFF;

BEGIN

    CountQ[]      = (CountQ[] + 1) &  Inc & !CarryQ
                  #  CountQ[]      & !Inc;

    CountQ[].clk = Osc;
    CountQ[].clrn = Clear;

    A[]           = CountQ[];

    CarryQ        = (CountQ[] == 115) &  Inc & Clear
                  #  CarryQ            & !Inc & Clear;

    CarryQ.clk = Osc;
    CarryQ.clrn = Clear;

    CarryA        = CarryQ;

END;
```

Figure 5.8 AHDL design file

PLD design tools can be simulated using OrCAD/MOD and the OrCAD/VST simulator.

The list of devices supported by the OrCAD/PLD product is limited, generally lacking support for the more complex PLDs. OrCAD/PLD sells for $495 on the MS-DOS platform.

PALASM—Advanced Micro Devices

The first PLD assembler, PALASM, was in large part responsible for the rapid acceptance of MMI's PAL in the first years of its introduction and remains, to

```
Decimal Counter   26 Aug, 1990
Michael Holley and Dave Pellerin

|PAL16R6 in:(HOLD, RESET), out:Q[3~0], io:HOLDO, clock:CLK
|
| Active-Low:   "All"
| Registers:    CLK // Q[3~0]
|
| Q[3~0] = (Q[3~0]+1 & RESET' & HOLD' & Q[3~0]/=9)
|        # (Q[3~0]   & RESET' & HOLD )
|
| HOLDO = Q[3~0]/=9


|Vectors:
|{ Display CLK, (Q[3~0])d, HOLDO
|  Test RESET=1; CLK
|  Test RESET=0; CLK=15(0,1)
|  End }
```

Figure 5.9 OrCAD/PLD design file

this day, the single most used design tool for PLDs. PALASM is available free of charge from AMD and is also available as a built-in option to a number of PLD programmers.

PALASM converts design descriptions into JEDEC download files for a wide variety of devices. The original PALASM language has been enhanced over the years, but still features the same basic syntax. A representative design file is shown in Figure 5.10.

When processed by the PALASM software, the sum-of-products Boolean equations that describe the function are converted to a fuse pattern and written to a JEDEC standard data transfer file. The JEDEC format file is directly downloadable to most PLD programmers. Test vectors in the source file (the "FUNCTION TABLE" section of the listed file) are converted by PALASM into completely specified test vectors for all device pins. These test vectors are then written to the JEDEC format output file along with the calculated fuse information.

PALASM is an established and well known design tool for PLDs and was, in large part, responsible for their initial success. PALASM does suffer, however, from many limitations. First, designs entered using PALASM must be in the appropriate form for direct implementation in the device. This form (sum-of-products Boolean equations) is rarely the form in which a design is first conceived. Second, the device and pin assignments must be fully specified in the design file. This limits the flexibility for moving from one device type to another. Third, PALASM only supports a limited number of devices: generally those available from AMD.

PALASM 2—Advanced Micro Devices

A more recent version of PALASM, called PALASM 2, supports state descriptions in addition to Boolean equations. The PALASM 2 language (called PDS for

```
PAL16R4 PAL              PAL DESIGN SPECIFICATION
CNT4SC
4 bit counter with synchronous clear
Michael Holley and Dave Pellerin
Clk  Clear  NC  NC  NC  NC  NC  NC  NC  GND
 OE  NC     NC /Q3 /Q2 /Q1 /Q0  NC  NC  VCC

  Q3 :=  Clear
       + /Q3 * /Q2 * /Q1 * /Q0
       +  Q3 *  Q0
       +  Q3 *  Q1
       +  Q3 *  Q2

  Q2 :=  Clear
       + /Q2 * /Q1 * /Q0
       +  Q2 *  Q0
       +  Q2 *  Q1

  Q1 :=  Clear
       + /Q1 * /Q0
       +  Q1 *  Q0

  Q0 :=  Clear
       + /Q0

FUNCTION TABLE
OE Clear Clk   /Q0 /Q1 /Q2 /Q3
------------------------------
 L  H     C     L   L   L   L
 L  L     C     H   L   L   L
 L  L     C     L   H   L   L
 L  L     C     H   H   L   L
 L  L     C     L   L   H   L
 L  H     C     L   L   L   L
------------------------------
```

Figure 5.10 PALASM design file

PAL design specification) is an extended superset of the original PALASM language. PALASM 2 includes a logic minimization module that can be used for a limited number of devices. PALASM 2 also includes additional language features that provide for control of the PALASM device simulator. This section of a PALASM 2 source file contains statements that allow the setting of device input values and checking of device outputs. The simulation language includes PASCAL style looping constructs that help when dealing with sequential circuits.

PLAN—National Semiconductor

PLAN is a package developed and distributed by National Semiconductor in support of their devices. The PLAN language has a syntax that is based on PALASM, but is somewhat more flexible. In particular, the language supports device independent design methods. Designs can be entered and manipulated without the need to specify the target device. PLAN includes an automatic device selection module.

PLAN includes support for sum-of-products Boolean equations, state table entry, and schematic netlist entry. PLAN also includes a PAL disassembler that is useful for converting existing designs into PLAN design files. This is useful for documentation and test suite generation purposes and is also useful for moving designs to different devices. PLAN is available free of charge from National and runs on IBM or compatible personal computer systems.

PLDesigner—Minc Incorporated

Minc, Incorporated was founded in 1987 by four ex-employees of Hewlett-Packard. Their goal was to produce a design entry system for PLDs that was simpler to use than earlier PLD design tools. The resulting product, named PLDesigner, includes a variety of design entry formats, including Boolean equations, truth tables, and state descriptions. In PLDesigner, there is no need to specify a target PLD before entering and simulating a design.

In addition to the usual design entry formats, PLDesigner also includes a waveform entry feature that allows synchronous designs to be described in terms of timing relationships. The product also includes an event-driven functional simulator with a waveform interface. The PLDesigner simulator is capable of detecting critical race conditions and other timing discrepancies.

An EDIF netlist translation module is also available with PLDesigner. This allows schematic designs to be entered into the system from a variety of schematic capture products.

PLDesigner's most distinguishing feature is its ability to automatically partition a large design into multiple PLDs, based on user selected and weighted criteria such as the maximum number of devices permissible, cost and timing criteria, and preferred device vendors and device types. Using these criteria, the PLDesigner system prunes down a list of possible devices and device combinations and then chops the design into a variety of different partitioning configurations that the user can choose from.

PLDesigner is available on the IBM personal computer and on a variety of CAE workstations. The price of PLDesigner ranges from $1,000 to $5,000 (for the IBM PC) depending on the partitioning options chosen.

PLD Design System—Hewlett-Packard Company

Hewlett-Packard's PLD Design System (PLDDS) is similar in many respects to Minc's PLDesigner and runs on HP 9000 Series 300 workstations. PLDDS was originally developed in Japan by HP's YHP division and is intended for use as part of the HP Design Center Environment (a complete system including tools for all phases of electrical engineering).

The PLDDS system is integrated with HP's CAE environment and supports graphic state diagram input, as well as the Boolean equation, truth table, and state description formats. HP also supplies a waveform entry tool, but it differs from

Minc's in that it's used for asynchronous circuits. A waveform debugger is supplied that helps to detect timing and logic problems before the logic is synthesized into equations.

Designs entered using the PLDDS system can be simulated at the board level using other elements of the HP Electronic Design System. The product also creates simulation models compatible with GenRad's HILO simulator. Like PLDesigner, the PLD Design System Automatic includes automatic partitioning and device selection features. The product also includes a test vector generation package, timing analysis, and glitch detection features.

PLD Logic Assembler—Avocet Systems

A lesser known product is the PLD Logic Assembler sold by Avocet Systems. This assembler (which was developed by Columbine Software of Broomfield, Colorado) is similar in many respects to PALASM but has a more flexible language and supports a much wider range of devices, including the Signetics FPLA devices. The language is designed to support a much greater range of complex architectures than typical PLD assemblers, and the documentation (particularly the device logic diagrams) is above average. The PLD Assembler includes a tutorial disk, supports over 120 different device architectures, and runs on the IBM PC. It sells for $295.

PLD Master—Daisy Systems

The PLD Master system is a highly integrated design environment that utilizes Daisy's window interface. It combines Daisy's schematic capture and simulation tools with ABEL, PALASM, and AMAZE to provide users with a comprehensive set of design options. The Daisy system includes design entry methods corresponding to those in ABEL and also supports a graphic state diagram entry format.

PLDSynthesis—Mentor Graphics Corporation

The PLDSynthesis package is based on technology purchased from Minc and the product is very similar to PLDesigner in its feature set. Mentor has put considerable work into making the original Minc software fit into its design environment. Designs can be entered using Mentor's schematic editor and completed designs can be simulated at the board level using the QuickSim, QuickFault, and QuickGrade simulation tools.

SNAP—Signetics Corporation

The *Synthesis, netlist, analysis, and programming* (SNAP) system is Signetics' third-generation system targeted specifically at their more advanced PML devices. The package supports a number of design entry formats including Boolean equations, state equations, or schematics. The package is a modified version of gate array software that was originally developed by Philips (Signetics' parent company). SNAP allows designs to be entered in a variety of formats with no specific PLD architecture in mind. Schematic entry is supported with either the FutureNet or OrCAD schematic editors.

Since SNAP is based on a gate array design system, it may provide a path for migration of designs from PLDs to gate arrays. SNAP can output designs in netlist format and includes a logic simulator that will perform timing analysis and fault simulation. The product sells for $695.

SP11—Pistohl Electronics Tool Company

Pistohl sells a complete design environment for CMOS erasable PLDs. The system includes a device programmer that attaches to an IBM personal computer and a set of highly integrated design tools. Designs can be entered into the system through the use of a text editor that automatically detects syntax errors during the entry process. The designs are then processed by the SP11 assembler and loaded into the target device.

The Pistohl assembler accepts designs in Boolean equation form only. At last report, Pistohl was planning to release a version of the SP11 assembler that includes logic minimization and PLD simulation features. The complete system, including the device programmer and software, costs approximately $800.

Tango-PLD—Accel Technologies, Incorporated

Another maker of schematic entry tools, Accel Technologies, has released a PLD design tool. Although the initial version doesn't support the more complex PLDs, the Tango-PLD product is actually quite sophisticated considering its low price.

Tango-PLD's design language, called TDL, is similar to the C programming language and features looping constructs for state machine design and hierarchy features in the form of user-definable logic functions. Engineers who have experience with C will find the TDL syntax easy to learn (although the fundamental differences between software and hardware design may lead to some confusion). A sample TDL source file is shown in Figure 5.11.

The set notation and other high-level language features provided in TDL are as powerful as those in any PLD design language and allow logic designs to be developed independent of a specific PLD. In addition to logic minimization, Tango-PLD contains algorithms that can perform automatic flip-flop emulations and other technology mapping functions.

Tango-PLD utilizes the Espresso logic minimization algorithm and is one of the few PLD design tools that exploits Espresso's capabilities for don't-care optimizations. As in ABEL, the utilization of don't-cares for this purpose is selectable by the user.

Also included in Tango-PLD is a functional simulator that is controlled with

```
/*
**  COUNT4.TDL
**  Binary Counter
*/
count4 (in Clk,Clear,OE;
        reg Q3..0)
{
    group Q[Q3..0];         /* counter group */

    Q[].ck = Clk;           /* hook up clocks */
    Q[].oe = !OE;           /* hook up enables */

    if(Clear)
        Q[] = 0;            /* sync clear */
    else
        Q[]++;              /* count up */

    putpart("P16R4", "count4",
            Clk,Clear, _, _, _, _, _, _, _,GND,
            OE, _, _,Q3,Q2,Q1,Q0, _, _,VCC);

    #define L 0
    #define H 1

    /* test counter sequence */
    test (Clk => Q3..0) {
        tracef("%w %w  %d%d%d%d", Clk, Clear, Q3..0);
        Clear = 1;
        OE = 0;
        ( \C => L,L,L,L );        /* synchronous clear */
        Clear = 0;
        ( \C => L,L,L,H );        /* count */
        ( \C => L,L,H,L );
        ( \C => L,L,H,H );
        ( \C => L,H,L,L );
        ( \C => L,H,L,H );
        ( \C => L,H,H,L );
        ( \C => L,H,H,H );
        ( \C => H,L,L,L );
        ( \C => H,L,L,H );
        ( \C => H,L,H,L );
        ( \C => H,L,H,H );
        ( \C => H,H,L,L );
        ( \C => H,H,L,H );
        ( \C => H,H,H,L );
        ( \C => H,H,H,H );
        ( \C => L,L,L,L );        /* roll over */
    }

}
```

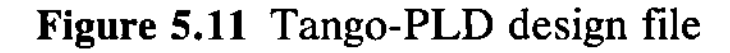

Figure 5.11 Tango-PLD design file

functions written in the TDL language. Tango-PLD doesn't have an advanced user interface; all features of the product (including the functional simulator) are invoked with a single command and operate in much the same way as a software compiler. Simulation results and documentation files are produced, as well as

JEDEC files (Tango-PLD can be invoked without a specified device, however, resulting in simulation results and documentation only).

For users of the BRIEF text editor, Accel supplies a set of macros that allow the compilation process to be invoked from within BRIEF. Tango-PLD is available on the IBM personal computer and sells for $495.

5.8 FPGA DESIGN TOOLS

Because of their unique architectures, FPGAs require a different approach to design than PLDs. Design tools for FPGAs include many functions that may be unfamiliar to PLD users. The most striking difference is the need to place and route the logic of an application into the various elements of the devices. In a PLD, it's a relatively simple matter to convert a logic circuit (normally expressed as an optimized set of Boolean equations) into fuse programming data. The PLA structure is generalized and little or no specialized technology mapping is required.

To implement a large circuit in an FPGA, however, it's often necessary to break the circuit up into small pieces that can be assigned to constrained resources within the device. These resources must then be interconnected through the process of routing. The placement and routing of circuit elements can have a critical impact on the timing characteristics of the device.

Each family of FPGAs has a unique set of placement and routing constraints and requirements, so FPGA vendors have had to shoulder the burden of supplying this piece of the design software. No third-party tool vendor has yet announced a willingness to tackle the problem of universal FPGA placement and routing software.

Xilinx supports its devices with a development system known as XACT that includes a graphic LCA editor and automatic optimization, placement, and routing software. XACT also has interfaces to schematic editors such as Data I/O's FutureNet system. The software runs on the IBM PC/AT and can also be equipped with an LCA simulator and in-circuit verifier. A complete XACT system with FutureNet or Omation schematic capture sells for approximately $5,000.

Actel supplies its customers with a development system called the *action logic system* (ALS) that allows designs to be entered in schematic form (using the ViewLogic schematic editor) and processed by automatic place and route software. ALS also includes an electrical design rule checker and an Actel-specific device programmer and sells for around $11,000 including the schematic capture interface.

Plessey provides its users with similar software that allows designs to be entered schematically (again using the ViewLogic schematic editor). Designs can be expressed with higher-level macros (standard logic equivalents, for example) and are automatically converted to a form appropriate for the ERA device architecture. An automatic routing program and circuit simulator are also supplied. The ERA Design kit with Viewlogic schematic capture sells for about $10,000.

Calculating Timing Characteristics of FPGAs

The timing behavior of an FPGA circuit is dependent on the placement and routing of the logic within it. For this reason, FPGA vendors are cautious about publishing timing specifications for their devices. Development systems such as Xilinx's XACT and Actel's ALS provide delay calculation features to aid in timing analysis. If the delay is unacceptable for certain critical circuit paths, those paths can be routed by hand and preserved during subsequent automatic routing.

In-circuit Verification

In-circuit verification has proven to be a useful approach to debugging FPGA devices. Both Xilinx and Actel offer their users in-circuit diagnostic tools. Actel provides a probe attachment to its Activator programming module that allows programmed ACT devices to be observed while they are operating in-system. This is done by utilizing special pins on the devices called *actionprobes*. These special pins may be interrogated on request by diagnostic software resulting in the display (on a logic analyzer) of internal device states. The diagnostics are, however, limited to frequencies below about 10MHz.

Xilinx offers the XACTOR in-circuit verifier is an emulator system that allows a microprocessor-based controller to be substituted for a programmed LCA in the target system. The system allows the internal nodes of the LCA to be examined and allows examination of asynchronous changes in device signals.

Data I/O's Mesa in-circuit verifier (see Figure 5.12) goes farther, giving LCA users the ability to examine the internal workings of an actual LCA while it's operating at speed within a circuit. Design changes can be made and internal device nodes can be probed while the LCA is installed in its host system.

Figure 5.12 Data I/O's Mesa in-circuit verifier (Courtesy of Data I/O Corporation)

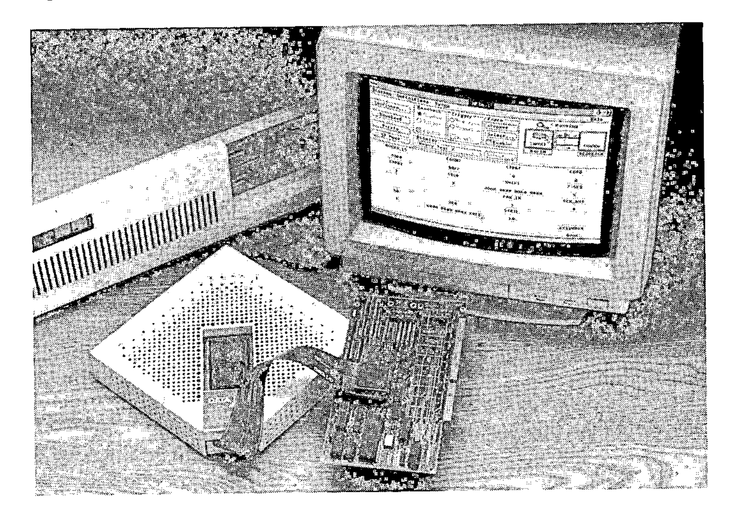

The Mesa makes use of a second LCA identical to the device under test. This *shadow device,* as it's called, operates in parallel with the in-circuit device and can be halted at any point and examined in detail. The entire process is controlled from user interface software running on an IBM PC. The Mesa unit itself, consisting of a chassis (the pod) and its attached device probe, attaches to the IBM PC via a plug-in card. The card can support up to four Mesa pods so multiple-chip systems can be examined.

Test vectors can be loaded into the Mesa system, and applied to the device under test. Alternatively, partial vectors can be combined with real-time stimulus from external sources to troubleshoot system interface problems.

PLD Tool Vendors Provide Paths to FPGAs

Many PLD tool vendors have announced that their tools can be used to design for FPGAs. Typically, the level of such support is limited, consisting of translators that will convert PLD designs to intermediate forms readable by the FPGA vendors' optimization tools. Without FPGA-specific optimization tools (such as the optimization modules produced by Exemplar Logic for Xilinx and Actel) the PLD design tools do a poor job of synthesizing designs for FPGAs.

In some cases, the PLD design tools can perform general optimizations that can enhance or replace circuit optimizations performed by the FPGA vendors' tools. Logic minimization features are available in most PLD design tools and these features can help to reduce the size of FPGA implementations. Logic minimization isn't always enough, however. FPGAs are typically more restricted in their ability to handle the wide logic gates typical of PLDs, so multilevel optimizations are often required.

FutureDesigner from Data I/O, for example, includes a factoring module that can be used to decompose large design equations into smaller pieces more appropriate for the architecture of an FPGA. FutureDesigner can output the optimized design in the form of automatically generated schematics, or as Berkeley PLA format files. These outputs can be used to interface to the FPGA vendors' tools.

While the current crop of PLD design tools aren't entirely capable of efficiently utilizing FPGAs from high-level circuit descriptions, it's likely that ongoing research and development in the areas of multilevel optimization and technology mapping will result in highly advanced design tools for these architectures within the next few years.

5.9 PLD TESTING TOOLS

As we will examine in a later chapter, PLDs are particularly prone to various types of manufacturing and production programming failures. These failures can only be detected by testing the devices by applying carefully constructed sets of test vectors (input stimulus and expected results). Developing these test vectors manually can be a monumental task, so *automatic test vector generation* (ATVG) products have been developed by various companies. These products all analyze a programmed PLD (expressed in the form of a JEDEC standard file) and report on the circuit's testability, as well as generating production test vectors. The major differences in the various products are in their level of device support, speed of operation, and ability to analyze complex sequential circuits.

Anvil ATG—Anvil Software

Anvil ATG, developed by Anvil Software of Nashua, New Hampshire, is a product that automatically generates test vectors to provide coverage of a high percentage of potential device faults. The product supports both PAL and PLA-type devices. The Anvil test generation software utilizes an event-driven timing simulator and concurrent fault simulator to model a programmed PLD for testability analysis.

Anvil ATG is capable of analyzing complex sequential circuits implemented in PLDs to determine a sequence of test vectors that will initialize the device to a known state. This means that devices do not need a preload feature to be supported in Anvil ATG. The Anvil ATG software accepts JEDEC format files as input and can therefore be used with any PLD design tool, as well as being capable of analyzing programmed devices for which there is no original source document. Anvil ATG is available for use on IBM or compatible personal computers.

PLDtest—Data I/O Corporation

PLDtest, the first ATVG tool offered for PLDs, was developed in 1985 by Data I/O. The product is still sold as a low-cost test vector generation tool and supports a limited number of PAL-type devices. PLDtest is capable of reporting on the testability (in terms of a percentage of faults detectable) for most simple registered and combinational PALs. The product will report on the test coverage of user-supplied test vectors and is capable of generating test vectors to achieve a high level of testability for combinational PALs and for registered PALs that have register preload features.

PLDtest can be used as a stand-alone product or interfaced to the ABEL design tools for more comprehensive design documentation. The product is available on the IBM PC and other MS-DOS compatible computers, as well as on Sun workstations and VAX computers (VMS or Unix operating systems).

PLDtest-Plus—Data I/O Corporation

PLDtest-Plus is a newer, more comprehensive PLD testing tool that extends the capabilities offered by PLDtest. In addition to the features offered in PLDtest,

PLDtest-Plus provides test vector generation and design analysis for nonpreloadable PLDs and for a wider variety of PLD architectures.

While PLDtest is capable of generating only single-vector test patterns, PLDtest-Plus uses advanced modeling techniques to determine a sequence of test vectors that will initialize a sequential PLD circuit to states from which potential

```
Afsim ABEL 4.00  Date Mon Aug 27 13:20:53 1990

Fuse file: 'cnt10.jed'  Vector file: 'cnt10.jed'  Part: 'P16R8'

_______________________________TESTABILITY_____________________________
Undetected :   10         Detected:   50          Total:   60

Seed Vector           : 15
Fault Coverage        : 83.33 %

______________________________Detected Faults__________________________
Vector 1
Vector 2
S-A-0 :  843   847  1291  1303
S-A-1 :  544   618
Vector 3
S-A-0 :  618  1135  1139  1322  1327
S-A-1 :  768   822   843   878
Vector 4
S-A-0 : 1099  1107
S-A-1 :  598   847   874
Vector 5
S-A-0 :  874   878  1358  1363
S-A-1 : 1024  1074  1099  1135  1174
Vector 6
S-A-1 :  534  1107  1139
Vector 7
S-A-1 : 1070
Vector 8
S-A-1 : 1066
Vector 9
S-A-0 : 1066  1070  1074
S-A-1 : 1291  1327  1363  1398  1408
Vector 10
S-A-1 :  530   590  1303  1322  1358  1394
Vector 11
S-A-0 :  822
Vector 12
Vector 13
Vector 14
Vector 15
S-A-0 : 1024
______________________________Undetected Faults__________________________
S-A-0 fuse:   530    534    544    590    598    768   1174   1394   1398
             1408
S-A-1 fuse:

AFSIM complete. Time: 11 seconds
```

Figure 5.13 PLDtest-Plus report output

PLD defects (faults) can be observed. In addition to generating test vectors, PLDtest-Plus can assist the designer in determining the suitability of various sequential circuit implementations. Figure 5.13 shows a sample fault grading and testability report produced by PLDtest-Plus. Companion products offered by Data I/O allow test vectors generated by PLDtest-Plus to be translated to formats appropriate for larger functional and in-circuit test equipment.

TESTPLA—Logical Devices, Incorporated

TESTPLA is another automatic test vector generation package sold by Logical Devices. TESTPLA is similar to Anvil ATG and PLDtest-Plus, with support for nonpreloadable PLDs. The program can be used to fault grade existing test vectors as well as being used for vector generation. TESTPLA is available for use on IBM compatible personal computers.

5.10 SUMMARY

In this chapter, we've attempted to cover the most important features of the PLD design tools available today. Design tool developers are constantly improving their offerings and the logic synthesis features of these products are becoming more powerful. In later chapters, we'll explore how these design tools can be effectively used to create PLD-based designs.

5.11 REFERENCES

Alford, Roger C. *Programmable Logic Designer's Guide,* Howard W. Sams and Company, Indianapolis, IN, 1989.

Data I/O Corporation, *ABEL User's Guide,* Data I/O, Redmond, WA, 1988.

Data I/O Corporation, *GATES Logic Synthesizer User Manual,* Data I/O, Redmond, WA, 1988.

Eurich, John. "A Tutorial Introduction to the Electronic Interchange Format." *Proceedings of the 23rd ACM/IEEE Design Automation Conference,* IEEE Press, New York, NY, (1986):327–32.

EXEL Microelectronics, Incorporated, *E^2 Data Book,* EXEL Microelectronics, San Jose, CA, 1988.

Gabay, Jon. "PLD Software Comes of Age." *Engineering Tools* (February 1988):98–102.

Goering, Richard. "PLD Tools Strengthen Design Entry, Simulation, and Test Capabilities." *Personal Engineering and Instrumentation News* (September 1989):29–38.

Lipsett, Roger and Shahdad, Moe. "VHDL—The Language." *IEEE Design and Test* (April 1986):28–41.

Milne, Bob. "PLD Designers Benefit from Better Tools." *Electronic Design* (April 27, 1989):71–77.

Monolithic Memories Incorporated, *PALASM 2 User Documentation,* Revision C, Monolithic Memories, Santa Clara, CA, 1987.

National Semiconductor Corporation, *Programmable Logic Devices Design Guide,* National Semiconductor, Santa Clara, CA, 1989.

Personal CAD Systems Incorporated, *CUPL User's Manual,* Personal CAD Systems, Los Gatos, CA, 1986.

Schreier, Paul G. "PLD Software Keeps Pace With Advanced Device Architectures." *Personal Engineering & Instrumentation News* (August 1988):31–38.

6

Getting Started

The previous chapters have provided us with an overview of the programmable logic devices and PLD-related design tools available as of this writing. In the remaining chapters of the book, we'll concentrate on the design process using these devices and tools. Before describing specific techniques useful to designers of PLD-based circuits, let's look at when these devices are best used and explore how a first-time user can get started.

6.1 CHOOSING AN IMPLEMENTATION TECHNOLOGY

There are many implementation alternatives available to the digital circuit designer. As we discussed in earlier chapters, it is possible to construct circuits out of standard logic devices or use foundry produced custom ASICs or programmable logic. For designs where speed is not an issue, another option may be to implement the function in software using a microprocessor.

There are many factors which affect this decision including cost goals, speed constraints, limitations on board size, power consumption or RF considerations, and issues of testability or servicability. Balancing the up-front design costs (the *non-recurring costs,* or NRE) against the expected manufacturing volumes and lifetime of the product is also a major factor in the decision. Let's examine in more detail the benefits and drawbacks of the more common alternatives to the basic PLD.

Standard Logic

The decision on whether to implement in standard logic devices (such as the 7400 family of TTL devices) versus PLDs is getting easier every year. The only direct advantage of standard logic over PLDs has historically been speed. With PLDs now becoming available in speeds approaching (or exceeding) the fastest TTL devices, this limitation of PLDs is quickly disappearing.

Other tradeoffs are less direct and include the requirement for somewhat different design techniques as well as the need for device programming equipment. The issue of design techniques is easy to overcome and the cost of programming equipment is relatively small when balanced against the cost savings normally realized from lower chip counts in end products.

Gate Arrays

The gate array has often been seen as a compromise between standard logic design and full-custom chip design. The gate array vendors have gone out or their way to make the design of a gate array appear similar to designing with standard logic. Gate arrays are usually designed at the level of a schematic which is drawn with basic logic gates such as NANDs and NORs or higher level macro-cells similar in function to standard logic devices. In reality, however, the gate array user often finds that most of the design time is not spent designing the logic of the circuit, but is instead spent in developing adequate test circuitry within the design and in developing test vectors to determine the correctness of the device.

Ensuring that the chip design is correct before production is critical to the success of a gate array design. This is the fundamental difference between gate arrays and PLDs. A PLD doesn't have to be right the first time (designing a gate array has been compared to developing computer software that requires three weeks to compile). The NRE for a gate array design, then, is much higher than the NRE for a PLD-based design, even if the circuit is of similar complexity. Adding to the cost is the requirement for expensive design software. Most gate array design is done on high-powered engineering workstations while the vast majority of PLD designs are accomplished through the use of personal computer-based design tools.

On the flip side, the production costs for a gate-array-based design are lower than for a PLD-based design and the overall circuit size will almost certainly be smaller. PLDs have constrained resources that limit their ability to implement extremely complex designs. A design that could be efficiently implemented in one gate array may require several dozen poorly utilized PLDs, particularly if the design requires large numbers of global signals and memory elements.

The key factors in the decision between PLD and gate array implementations are design stability and production volumes. If you assume that the NRE costs for a gate array implementation are $50,000 more than the NRE costs for a PLD implementation, and you expect to produce only a few hundred units before a major revision is needed, you will have trouble justifying the extra cost for the gate array.

FPGAs

FPGAs (described in Chapter 4) are a relative newcomer to the world of digital circuit design. These devices are an attempt to combine the benefits of gate arrays with those of PLDs. The primary advantage FPGAs have over typical PLDs is their flexibility when it comes to allocating devices resources. For example, the typical registered PLD has between one and ten flip-flops available. These flip-flops are normally associated with output pins of the device. More complex PLDs extend this number by providing a similar number of internal flip-flops. For large designs, however, this often isn't enough. Many design problems that could otherwise be solved with the addition of a few flip-flops are extremely difficult to resolve in a PLD. Even with internal flip-flops and highly configurable macrocells, the relatively inflexible architecture of a PLD places limitations on how efficiently these resources can be used.

FPGAs are designed to provide greater freedom in the selection and interconnection of devices resources such as flip-flops. This freedom comes at a price, however; the timing characteristics of an FPGA-based design are often difficult to predict in advance and the overall performance of the configured device is often unacceptable. If high speed is required, gate arrays still have the edge over FPGAs. FPGAs are still young, however, and we can expect to see their performance improve dramatically.

In terms of design effort, the FPGA has a convincing advantage over gate arrays. Although gate array vendors have reduced their turnaround times dramatically, the time spent by the designer in verifying his design can still consume months of effort. The cost of making an error in a gate array design is measured in tens of thousands of dollars, while the cost of making an error in an FPGA design ranges from almost zero (in the case of a reprogrammable device) to a few hundred dollars.

6.2 CHOOSING YOUR FIRST PLDs

Assuming that you have settled on an implementation technology and intend to design with programmable logic, how do you get started? As we saw in Chapters 3 and 4, the number of different device architectures can be overwhelming. The first-time user confronted with the list of possibilities often doesn't have a clue which devices are most appropriate for his particular circuit.

While there are nearly 300 different types of PLDs available in literally thousands of different packages and speed grades, there is no need to understand them all. There are really only a few types of devices that you need to consider for your first design efforts.

To start with, you should go with the mainstream PAL-type devices. These parts are readily available from distributors and are cheap enough that you can destroy dozens of them without putting too much of a dent in your budget. We suggest that you not attempt to use the more complex of the PLDs described in Chapter 3 until you have at least experimented with the simpler PLDs. The FPGA devices described in Chapter 4, while being extremely powerful, are also a large conceptual leap from standard logic and should also be deferred for later experimentation and use.

The most common of the simpler PAL chips in use today are the 20-pin 16L8, 16R4, 16R6, and 16R8 devices or the more general purpose 24-pin 22V10. These devices were described in detail in Chapter 3. The simple and general purpose PAL devices are appropriate for a wide variety of glue logic applications and for many state machine applications. The devices are available with registered or combinational outputs.

If you find yourself burning up more chips than you like, you can move into the electrically erasable parts, such as the GAL devices from Lattice Semiconductor, PEEL devices from ICT, or similar generic PLDs from other sources. These devices are general enough that one or two different GALs can suit virtually all of your engineering prototype requirements. Once your design has stabilized in the GAL device, it can be migrated to a cheaper fuse programmable part for production. If you intend to migrate the design in this way, however, you must be careful not to use the advanced features of these generic parts that aren't available in lower cost devices.

The trend in recent years has been for device manufacturers to generalize their device offerings, simplifying the selection of a device. For example, the 22V10 device can be configured to emulate virtually any existing 24-pin PAL device. Many experienced PLD users have found that by using slightly more expensive devices, such as the 22V10, GAL, or PEEL for most of their designs, they can actually save a considerable amount of money through simplified purchasing, part qualification, and inventory control.

Further simplification has resulted from licensing agreements between different manufacturers. The devices manufacturers realize the importance of multiple sources for devices and have responded by striking deals with competitors.

This is why seemingly identical devices are available from a number of different manufacturers. You should realize, however, that different device manufacturers often have different specifications for programming hardware and that a PLD device purchased from one manufacturer may require a different programmer or programming software than the identical device produced by a different manufacturer.

6.3 CREATING A PLD DESIGN ENVIRONMENT

There are many different ways that an engineer or engineering design team can first encounter PLDs, and each new entrant into the field has different requirements and budget constraints. For this reason, it is difficult to come up with the ideal design environment for all new users. We can, however, provide some pointers to help you put together a design environment that will allow you to experiment with programmable logic with a minimum of frustration.

To effectively utilize PLDs, you will require a means of programming your devices, and software design tools to convert your design concepts into programming patterns. It is possible to design with PLDs without either actually possessing these items, but doing so can be compared to designing with microprocessors without the use of an assembler or compiler.

Programming Your PLDs

Many beginning PLD users have attempted to save money by designing their PLD circuitry and then contracting with a distributor to do all of their device programming. Some device distributors have set up design centers that provide design support in the form of applications engineers and programming services. These centers are set up to emulate the environment that gate array and custom IC users are accustomed to.

Some of the most significant benefits of PLDs are lost, however, if you can't experiment with them as easily as you experiment with circuit breadboards—on your own workbench. Contracting out device programming to these design centers may be a cost-effective way to produce PLDs for production, but it really doesn't make sense in an engineering environment.

First and foremost, PLDs are devices that demand a hands-on approach to design. The best way to learn about these devices is by designing some very simple circuits (or copying some of the simple circuits provided later this book) and programming some devices. Even highly experienced users find themselves reworking designs and programming new chips frequently to resolve system interface problems or improve their designs.

For these reasons, we strongly suggest that do your own programming of devices. This implies, of course, that you will need a device programmer. Selecting a programmer can present more of a dilemma than selecting devices. To make an informed decision about the appropriate device programmer for your needs, you need to decide how which features are important to you and balance that against how much you are willing to spend.

If you are on a strict budget, there are many device programmers available for well under one or two thousand dollars that are quite adequate for the first-time user. The low-cost programmers are probably good enough for the devices you will want to use initially. These programmers are perfectly adequate for the simple PAL devices and their lack of advanced features is rarely a hindrance to the occasional PLD user. When you later decide that you are in the PLD design business for good, and need higher programming yields or broader device coverage, you can move up to a more full-featured programmer (perhaps donating your old low-cost programmer to one of those high school kids that you always see hanging around electronics supply stores). Whatever programmer you choose, make sure it has been certified by the device manufacturer for the devices you intend to program. Your best source of information about qualified device programmers is often the device manufacturer's field representatives. They will usually have available a list of programmers that have been tested and found to perform acceptably for their devices.

Renting a programmer may make sense if you feel you need a higher quality programmer but can't justify the expense. Renting allows you to try programmers from a number of manufacturers so you can begin to appreciate the differences in features and device support.

Programmer Features

Prices of new device programmers range from just a few hundred dollars for a simple box that programs only a few devices to many thousands of dollars for full-featured universal programmers. There are many factors that determine how much you will spend on a programmer and what features you will get for that money.

Stand-alone Versus Computer Peripheral Programmers

With the growing acceptance of personal computers, many programmer manufacturers realized that device programmers could be vastly simplified if most of the user interface (knobs, buttons, and displays) were removed from the programmer box and implemented in software. The choice was then whether this software should be implemented in the programmer itself or on a host computer system.

Some programmer manufacturers have chosen to use the IBM personal computer as the computing engine for their programmers. In addition to simplifying the user interface design of the unit, this also allows the personal computer to be used for storage of programming data and for periodic software updates. Typical of this type of programmer is BP Microsystem's PLD-1100 unit. This programmer attaches to an IBM PC through the PC's parallel port and supports some 650 different 20- and 24-pin PLDs, representing seventeen different device manufacturers.

Other programmer manufacturers determined that, in order to provide more powerful and intelligent programming options, the units would require a significant amount of computing power built-in. This approach is found in the Data I/O Unisite, the Logical Devices series of programmers, and the Stag ZL30 and PPZ. These programmers have provision for data storage and software update built-in and also feature the ability to be controlled remotely over a serial port from a host computer.

Many of these stand-alone programmers, since they are computers in themselves, are provided with built-in features that take further advantage of the available computing power. Some programmers, for example, are available with the PALASM assembler built-in. Menu interfaces lead the user of these programmers through the design entry and programming process. In some cases, the programmers have built-in CRT displays while, in others, a separate computer terminal is used for user interaction. Obviously, the stand-alone programmers are more complex and cost significantly more than those designed as simple computer peripherals. The benefits of the universal programmers are their ability to program a larger number of different devices of varying programming technologies and packages.

Building Your Own Programmer

Before commercial programmers became available, users of devices such as PROMs had to build their own programming circuits and fixtures. These users quickly learned about such things as fuse grow-back, underblow and overblow, and the difficulties of supporting ever increasing numbers of new devices.

Even with the time required to build and maintain a custom programmer, there are still many users of EPROMs who choose not to use a commercial programmer. The EPROM is relatively simple to program, and acceptable programmers can be constructed without a large effort. PLDs, however require quite different methods for programming than EPROMs, so building your own programmer for these devices presents a much greater challenge.

The major difference is that, in a device such as a PAL, the pins used to program the device are the same pins used during device operation. Whether the device is in an operating mode or a programming mode is determined by the voltages applied to the device pins. The programming voltage is called a *super-voltage*. The super-voltage required for programming varies from one manufacturer to another, so it is difficult to build a simple programmer that will work for a wide range of devices.

Not only are the programming voltages higher and of critical values, but the timing characteristics of the programming pulses are critical as well. If you intend to use more than a few devices from more than one or two sources, we strongly recommend that you not waste any time trying to build a device programmer. If you are intent on doing so, however, there is a fine article by Robert A. Freedman in the January 1987 issue of *Byte* magazine. This article describes how to build an IBM PC-based programmer that will program a small number of simple PALs. Another article, in the January 1990 issue of *Popular Electronics,* describes how to construct a rudimentary programmer that will work for ICT's 18CV8 PEEL device.

Design Tools

There is no need to spend half your engineering budget on high-priced engineering workstations and software for PLD design support. You probably already have an IBM PC, or equivalent personal computer, and design software for the simpler PLDs is readily available at a low cost.

If you are on a tight budget and don't intend to use large and complex PLDs, PALASM is probably sufficient and is available free of charge from many sources. Using a PLD assembler such as PALASM for your first design efforts has some educational value. After you have produced a few designs and manually converted them to sum-of-products equations for a particular device, you will be better able to evaluate the effectiveness of the automated design tools and will have a better understanding of how design decisions will affect the success of automated optimizations.

If you wish to experiment with higher level tools, universal PLD compilers such as ABEL and CUPL are sometimes available at reduced cost from device manufacturers. These special low-cost versions of the universal tools have usually been modified to support only a small set of devices and allow you to evaluate those devices as well as the design tools with little risk. The field representatives of the device manufacturers are your best source of information about these special versions of software.

Another low-cost approach is to use the design software produced by the device manufacturers for their devices, although doing so can limit your options should you decide to use devices produced by another manufacturer. You may have to learn the features and peculiarities of a new design tool for each different manufacturer's devices. This can become particularly frustrating if you decide to migrate existing designs from one device type to another.

If design tool cost is less of a concern, you will probably choose to use a full-featured universal PLD tool such as those described in Chapter 5. These tools all offer high-level design entry features and differ primarily in their level of device support and ability to automatically fit a design into a variety of different PLD architectures.

If you plan on using FPGAs, you should make sure that the design tool you choose for your PLD design efforts will support the migration of your PLD-based designs to FPGAs. You should be aware, however, that using FPGAs will almost certainly require an additional investment in FPGA-specific design tools.

6.4 SUMMARY

While the large number of devices and design tools presented in this book may seem overwhelming, getting your feet wet in programmable logic doesn't have to be overly stressful or expensive. A low cost development system can be assembled that will get you started. Later on, you can decide if a more powerful design environment is justified for your applications.

6.5 REFERENCES

Freedman, Robert A. "Getting Started with PALs." *Byte* (January 1987):223–30.

Monolithic Memories, Incorporated, *Programmable Logic Array Handbook,* 5th edition, Monolithic Memories, 1986.

Green, Bill. "Create Your Own ICs." *Popular Electronics* (January 1990):33–42.

7

Applying Digital Logic to PLDs

A solid understanding of digital logic fundamentals is critical to the understanding of PLDs and PLD-related design techniques. In the first sections of this chapter, we'll review these fundamentals and examine how the techniques of digital logic design can be used to help us design for PLDs. From there we'll move quickly into advanced techniques that are particularly useful for PLD designs.

The terminology of digital logic can be confusing. There are many words and terms that seem to mean the same thing but actually may not. A major goal of this chapter, therefore, is to define the language of logic design as used in this book. Even if you are a seasoned logic designer, please bear with us during the early parts of this discussion so we can make sure we are speaking a common language. This will be more important as we discuss specific optimization techniques relating to PLDs.

7.1 LOGIC CIRCUITS AND LOGIC FUNCTIONS

First, we will try at all times to distinguish between real circuits and purely functional representations of these circuits. In the following discussions, we will consider a *logic function* to be an abstract concept, while a *logic circuit* is a collection of physical circuit elements and interconnections (typically wires or metal traces). Logic circuits have real-world attributes such as size, speed, and critical timing relationships. This distinction is important to us because, when designing for PLDs, we are nearly always designing at a functional level rather than at a real implementation level.

Basic Logic Gates

Logic gates are the basic building blocks of digital logic circuits. A logic gate may be thought of as a decision-making element. A logic gate has one or more inputs (each of which can be either *true* or *false* at any given time) and a single output that produces either a true or false value based on the values of the gate inputs. To simplify later discussions, we may refer to input and output values of these gates as either logic level 1 (true) or logic level 0 (false). The numerals 0 and 1, when used in this context, are not numbers; they are logic values.

The inputs and outputs of logic gates are called *signals*. In an actual hardware implementation, a signal corresponds to a wire or metal channel that carries current from one circuit element to another. For the moment, we can assume that all signals are, at any given time, either 0 or 1, with no ambiguous signal values. When one or more logic gates are connected together and the input and output signals are given some significance (usually indicated by specific signal names), they form a logic circuit.

In describing the operation of logic gates and logic circuits, it is common to use a form of representation known as the *truth table*. A truth table simply associates combinations of inputs with resulting outputs. Truth tables can be used to fully specify a logic function (as in the following examples) or to partially specify it. We'll use fully specified truth tables to define the basic logic gates. Advanced applications of fully and partially specified truth tables will be covered later in this chapter.

The logic gates of most interest to us for Boolean logic manipulations are the AND, OR, and NOT (also called an *inverter*) gates. The most common symbolic representations of these gates is shown in Figure 7.1, along with truth tables describing their operations.

The AND, OR, and NOT gates are analogous the words "and", "or" and "not" in the english language. We could describe a logic function using the english language. For example, we might say "The function Y shall be true when the input signal A is true and the input signal B is not true or when the input signal C is true." If we wanted to build a circuit that implemented this function, it might look like the circuit shown in Figure 7.2.

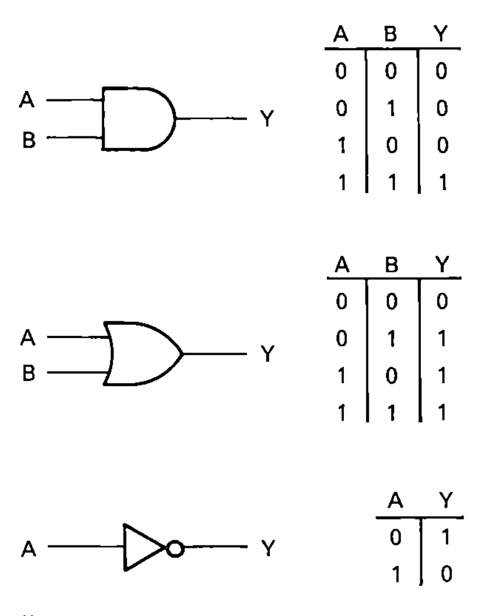

A	B	Y
0	0	0
0	1	0
1	0	0
1	1	1

A	B	Y
0	0	0
0	1	1
1	0	1
1	1	1

A	Y
0	1
1	0

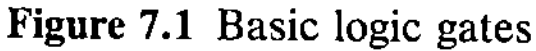
Figure 7.1 Basic logic gates

This logic circuit is what is known as a *combinational* logic circuit. This means that, for any set of input values (as shown in a fully specified truth table) there is one and only one possible circuit output value, regardless of the previous state of the circuit.

By gluing together an appropriate collection of ANDs, ORs, and NOTs, it is possible to create any combinational logic circuit. There are, of course, other types of logic gates. Designers who are used to constructing circuits out of standard TTL gates or who are used to designing circuits using *bubble logic* techniques are often more comfortable with NAND or NOR gates. These gates are widely used because they normally require less transistor circuitry for implementation than ANDs and ORs. For our purposes, however, it is more convenient to consider NAND and NOR gates to be combinations of AND and NOT gates or OR and NOT gates, respectively.

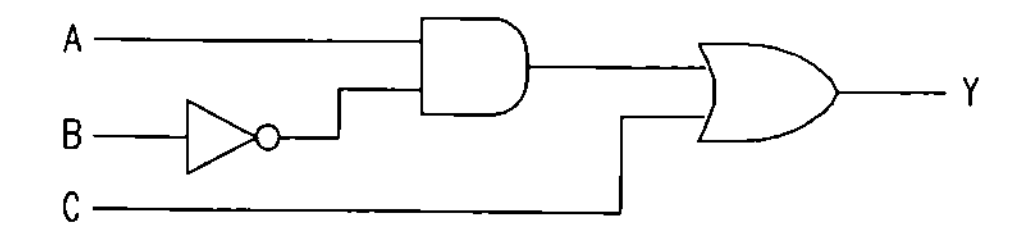

Figure 7.2 Simple logic circuit

The Exclusive-OR Gate

In addition to the AND, OR, NAND, NOR, and NOT gates, there is another type of gate that is of particular interest to users of PLDs. This is the *exclusive-OR gate,* which is often referred to as an XOR. The XOR symbol and corresponding truth table are shown in Figure 7.3.

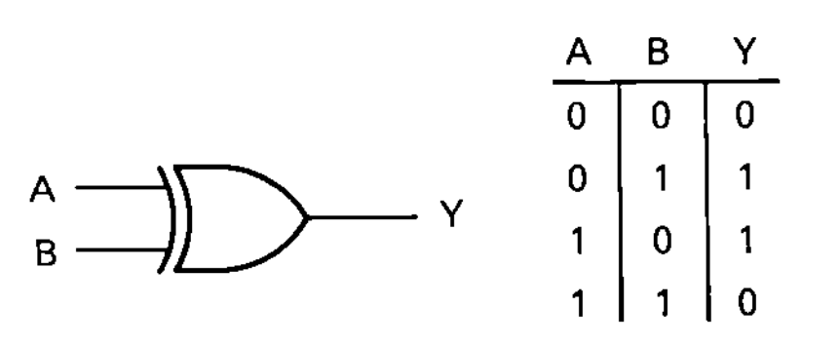

A	B	Y
0	0	0
0	1	1
1	0	1
1	1	0

Figure 7.3 Exclusive-OR gate

Like the NAND and NOR gates, the XOR can be constructed from the basic AND and OR gates, as shown in Figure 7.4. We'll discuss how the XOR gate can be used to advantage later in this chapter.

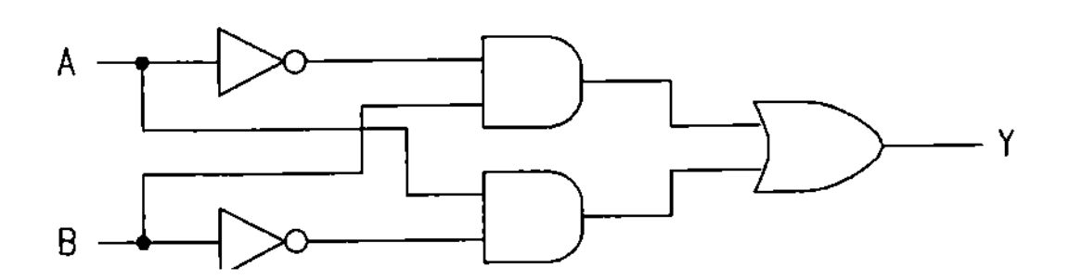

Figure 7.4 Exclusive-OR circuit

The Programmable Logic Array

The *programmable logic array,* or PLA, is a matrix of NOT, AND, and OR gates arranged as shown in Figure 7.5. This diagram is simplified by combining all of the inputs to each AND and OR gate into a single line, as we described in Chapter

Figure 7.5 Programmable Logic Array (PLA)

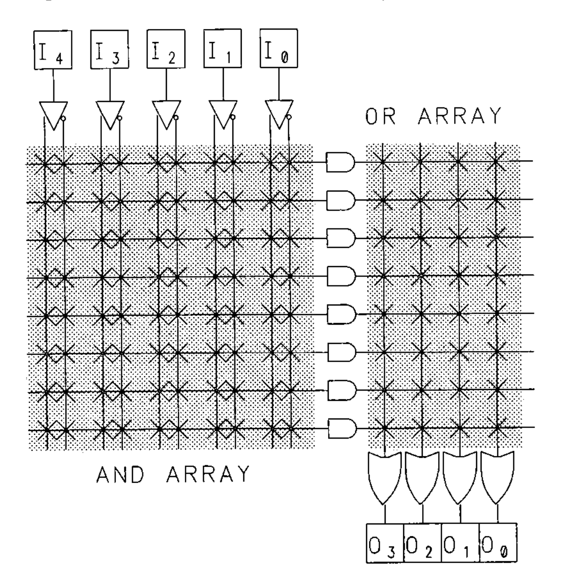

1. Each intersecting line on the diagram represents a programmable interconnect point. As you can see, there are two arrays of interconnects. The left-most array is the AND array, while the right-most is the OR array.

The simple PLA shown in the figure has four inputs, each of which is available to the AND array either directly (its *true* value) or through an inverter (its *complement*). The AND gate outputs are fed into the OR array so any AND gate output can be used as an input to any OR gate.

The PLA structure is important to us because it is the basis for nearly all of the PLDs currently in use. As we said in Chapter 1, PROMs are PLAs with fixed AND arrays, while PALs are PLAs with fixed OR arrays. The actual implementation of the PLA within a programmable device may be quite different from that shown above, but the techniques we present in this chapter are no less appropriate whether the device is constructed of ANDs and ORs, NANDs, NORs, or some other combination of logic gates.

7.2 BOOLEAN ALGEBRA FUNDAMENTALS

We have seen how logic gates can be arranged to create a simple combinational logic circuit. These logic circuits are physical implementations of *Boolean logic functions* or simply *logic functions*. Every combinational logic circuit has exactly one logic function but, for any given logic function, there are an unlimited number of logic circuits that can be constructed to implement that function. The logic design process, then, is composed of two steps: first, determining and describing the logic function and, second, implementing that function as a logic circuit.

Any logic function can be expressed in a number of ways. As we have seen, symbolic representations of logic gates can be used to describe circuits and, in fact, most simple logic functions are described in this way. In the case where discrete TTL gates are used, the logic function and logic circuit may be determined simultaneously.

For complex logic functions, however, the optimal logic circuit isn't obvious and may require a significant amount of calculation and experimentation to determine. In this situation, it is beneficial to use more conceptual representations for logic functions—representations that allow alternative logic circuits to be quickly identified and selected.

When it becomes necessary to modify a logic function, experiment with alternative circuits or manipulate the form of the circuit; these conceptual forms of representation are usually more helpful. We have seen how truth tables, and even english language statements, can be used to describe the operation of a digital logic function in a conceptual way. Another form of representation, *Boolean algebra,* has been used for many years to describe logic functions.

Boolean algebra is a system of mathematics that allows us to manipulate logic. For digital applications, where we are dealing with only two states (true and false) for any signal, Boolean algebra is really quite simple (actually, the form of Boolean algebra that we use for digital applications is more properly known as *switching algebra,* a subset of Boolean algebra). As before, we use the value 1 to indicate the true logical value and a 0 to indicate the false logical value. We'll use true/false and 1/0 values interchangeably in the remaining discussions.

Boolean Equations

At the heart of Boolean algebra are the *Boolean operators*. These operators correspond to the AND, OR, XOR, and NOT logic gates presented earlier. There are many ways of representing the AND, OR, NOT, and XOR Boolean operators and a variety of different representations are used in digital design languages and formal texts. In this book, we use the following symbols for Boolean operations:

!	NOT (complement or invert)
&	AND
#	OR
$	Exclusive OR

In most formal texts on logic design, the AND operator is omitted from Boolean equations. In this book, we will at all times include the AND operator in equations in order to remain consistent with the design languages used for PLDs.

Operator precedence (order of evaluation or binding) is the order shown in the above list. As in standard algebra, parenthesis may be used to change the order of operations. In Boolean algebra, a *variable* is a named entity (corresponding to a signal in a logic circuit) that can have one of two possible values: 1 or 0 (there is a third possible value, the *don't-care value,* that we'll consider later). A Boolean *expression* is a grammatically correct sequence of variables and Boolean operators.

A *Boolean equation* is composed of an *output variable* that is assigned a Boolean logic function through the use of the *assignment operator* (which is the equals sign). The AND, OR, XOR, and NOT gates can be represented by the following Boolean equations:

$$Y = A \mathbin{\&} B$$

$$Y = A \mathbin{\#} B$$

$$Y = A \mathbin{\$} B$$

$$Y = !A$$

To clarify these concepts, consider the following Boolean equation:

$$Y = A \mathbin{\&} B \mathbin{\#} C \mathbin{\&} !B$$

This Boolean equation describes a logic function of three input variables

($Y = f(A,B,C)$). The function can be described by the following english language statement: "The output variable Y shall be true if the input variables A and B are both true or if the input variable C is true and the input variable B is false."

Notice that, while ambiguities can exist in english language statements when the words "and" and "or" are used ("I will wear pants and shoes or sandals . . ."), no such ambiguity is possible in Boolean algebra since the OR operator always has a lower precedence than the AND operator (if this was the case in the english language, the person making the previous statement might wind up wearing no pants).

The NOT operator is a *unary operator* which means it operates on a single variable or subfunction. For Boolean algebra purposes, it is often most convenient to consider the true and complement of an input variable as two separate entities. When this is done we refer to the inputs as *literals*. In the Boolean equation we just presented, for example, the variable B appears as both a true and complemented variable. This means that the equation contains four distinct literals (A, B, $!B$, and C).

The AND, OR, and XOR operators are *binary operators* operating on two variables or subfunctions. Actually, it is sometimes more convenient to consider the AND and OR operators to be n-ary operators (operators that can operate on more than two variables) since the actual logic gates that correspond to these operators may have more than two inputs. This is allowed by the associative laws for these operators as we'll see in a moment.

As we said, Boolean algebra is a convenient method with which to describe logic functions. It is never possible, however, to completely specify a logic circuit with Boolean algebra, since there are many possible circuit interpretations of a Boolean equation. This is shown in Figure 7.6. Each of the three circuits shown will have subtly different operating characteristics due to signal propagation and gate switching delays.

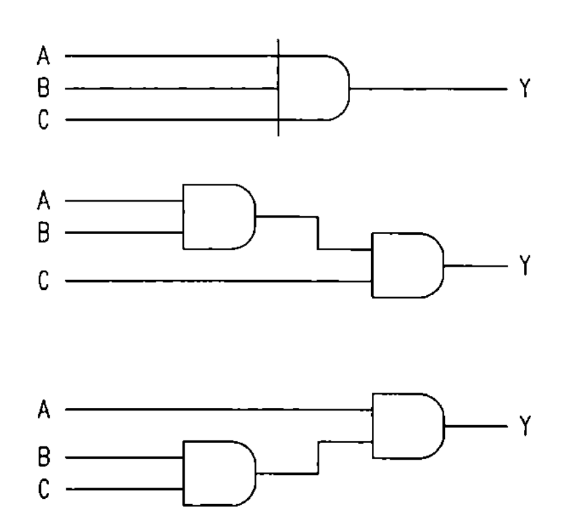

Figure 7.6 Three possible circuit implementations of a Boolean equation

Formal Rules for Boolean Operations

There are a number of operations that are used in the manipulation of Boolean equations. We will present the formal rules of the operations most useful to logic designers, although complete descriptions and proofs of these are beyond the scope of this book. The reader is referred to any one of a number of books on the subject of logic design for further study. Some of these texts are listed in the references at the end of this chapter.

The following is a list of the operations that are of greatest interest to us. We have added some exclusive-OR operations to the standard operations commonly presented. We will assume the validity of these additional operations without proof.

STANDARD OPERATIONS

1. Commutative Laws:

(a) $A \# B = B \# A$

(b) $A \& B = B \& A$

2. Identities:

(a) $A \# 0 = A$

(b) $A \# 1 = 1$

(c) $A \& 0 = 0$

(d) $A \& 1 = A$

3. Complement Identities:

(a) $A \# !A = 1$

(b) $A \& !A = 0$

4. Involution:

(c) $!!A = A$

5. Indempotence:

(a) $A \# A = A$

(b) $A \& A = A$

6. Distributive Laws:

(a) $A \# (B \& C) = (A \# B) \& (A \# C)$

(b) $A \& (B \# C) = (A \& B) \# (A \& C)$

7. Associative Laws:

(a) $A \# (B \# C) = (A \# B) \# C$

(b) $A \& (B \& C) = (A \& B) \& C$

8. Absorption:

$$(a)\ A \# (A \& B) = A$$
$$(b)\ A \& (A \# B) = A$$
$$(c)\ A \# (!A \& B) = A \# B$$
$$(d)\ A \& (!A \# B) = A \& B$$

9. Unity:

$$(a)\ A \& B \# !A \& B = B$$
$$(b)\ (A \# B) \& (!A \# B) = B$$

EXCLUSIVE-OR OPERATIONS

10. Associative Law:

$$(a)\ A \$ (B \$ C) = (A \$ B) \$ C$$

11. Distributive Law:

$$(a)\ A \& (B \$ C) = A \& B \$ A \& C$$

12. Commutative Law:

$$(a)\ A \$ B = B \$ A$$

13. XOR Identities:

$$(a)\ A \$ 1 = !A$$
$$(b)\ A \$ 0 = A$$
$$(c)\ A \$ A = 0$$
$$(d)\ A \$!A = 1$$

14. Rules of XOR Complement:

$$(a)\ !(A \$ B) = A \$!B$$
$$(b)\ !(A \$ B) = !A \$ B$$
$$(c)\ A \$ B = !A \$!B$$

In addition to these rules, there are two theorems that are widely used in Boolean equation manipulations and logic minimization.

DeMorgan's Theorem

DeMorgan's theorem is used to find the complement of a Boolean expression. This is particularly useful for programmable logic applications:

$$(a)\ !(A \# B) = !A \& !B$$
$$(b)\ !(A \& B) = !A \# !B.$$

DeMorgan's theorem can be generalized as follows: The complement of any Boolean expression can be determined by replacing each OR operator with an AND operator (while preserving the order of evaluation), replacing each AND operator with an OR operator, and replacing each literal with its complement. For example, the following two relationships are valid:

$$\text{(c) } !(A \# (B \ \& \ C)) = !A \ \& \ (!B \# !C)$$

$$\text{(d) } !((A \# !B) \ \& \ (C \ \& \ A)) = (!A \ \& \ B) \# (!C \# !A).$$

Shannon's Expansion Theorems

Shannon's theorems are used to isolate (factor out) one or more variables in a logic function. Shannon's expansion theorems may be described by the notation:

$$\text{(a) } f(A,B,C, \ldots) = A \ \& \ f(1,B,C, \ldots) \# !A \ \& \ f(0,B,C, \ldots)$$

$$\text{(b) } f(A,B,C, \ldots) = (A \# f(0,B,C, \ldots)) \ \& \ (!A \# f(1,B,C, \ldots))$$

where $f(A,B,C, \ldots)$ represents any multiple-variable Boolean logic function. Shannon's theorems are particularly useful for logic minimization and factoring purposes.

Standard Forms for Boolean Logic Functions

There are a number of general forms that may be used to express a Boolean logic function. An understanding of these forms is important for effective use of logic minimization and factoring techniques and tools.

We refer to any AND operation as a *product*. Any Boolean expression that contains one or more literals operated on by one or more AND operators is referred to as *product term*. For example, the expression

$$!A \ \& \ B \ \& \ !C \ \& \ D$$

qualifies as a product term, while the expression

$$!A \# B \ \& \ !C$$

does not, since it includes a binary operator other than AND.

A product-term expression is distinguished by the fact that there is only one set of input conditions that will result in a true evaluation of the expression.

Logical Sum

OR operators are referred to as *sums*. Any Boolean expression that contains one or more literals operated on by one or more OR operators is referred to as a *sum*.

The expression

$$!A \# B \# !C \# D$$

is an example of a sum.

Sum-of-Products

When two or more product terms are operated on by one or more OR operators, the form of the resulting expression is called *sum-of-products*. If we add an assignment operator and an output variable, we have a *sum-of-products Boolean equation,* as in the following example:

$$Y = A \,\&\, !B \,\&\, C \# A \,\&\, !B \,\&\, !C \# !A \,\&\, !B \,\&\, C.$$

Expressed as logic gates, this expression could be composed of one three-input OR gate fed by three three-input AND gates. As Figure 7.7 shows, the sum-of-products form requires two levels of logic gates (discounting the NOT gates required for the complemented input literals) for implementation. For this reason sum-of-products Boolean logic functions are often referred to as *two-level* logic functions.

The sum-of-products form is the basis for most PLD design and logic synthesis methods and maps directly into the PLA structure presented earlier. By utilizing the laws and theorems presented earlier, it is possible to express any logic function, regardless of complexity, in sum-of-products form.

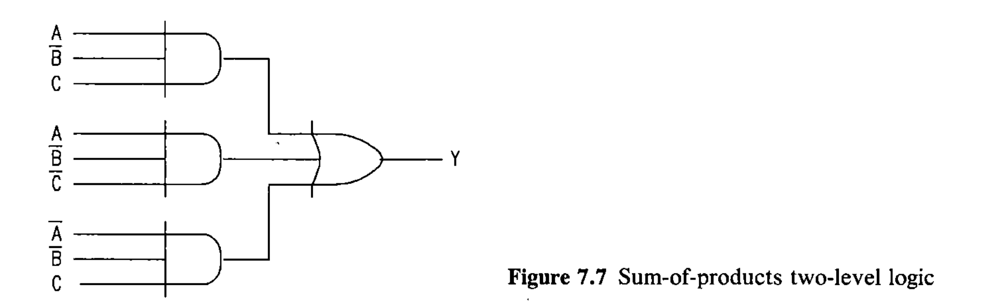

Figure 7.7 Sum-of-products two-level logic

7.3 USING BOOLEAN ALGEBRA FOR PLD APPLICATIONS

Using the basic laws and theorems presented as tools, we can now begin to experiment with various techniques that are particularly useful for PLD designs. Before starting, though, let's examine some of the reasons why these manipulations are often necessary.

The primary motivations for Boolean logic manipulations are to reduce circuit size or improve circuit speed. PLDs are, by their very nature, constrained architectures. They require that logic functions be implemented in specific forms (usually sum-of-products), and that these functions be no greater in size than the available device circuitry. PLDs are usually limited in the number of product terms that can be used for each output, so product term reduction is one of the primary goals of Boolean algebra use. If a logic function can't be reduced sufficiently for a PLD implementation, it will be necessary to perform other manipulations, such as factoring, to spread the circuitry for that function across multiple levels of logic.

Another common limitation of PLDs is the number of inputs and outputs available in each device. If a design can't be implemented into a PLD due to I/O constraints, it is necessary to either change the design or choose a different implementation. Choosing a different implementation can involve using a different type of device, using more than one device for the function, or using an entirely different implementation technology.

Speed isn't generally much of an issue when you are dealing with a single PLD, since in most cases the delay times from inputs to outputs are fixed and predictable, and PLDs are available that are fast enough for most applications. For large designs that utilize many PLDs, however, delay of signals due to multiple levels of logic can wreak havoc on a system. For this reason more advanced multilevel logic optimizations may be needed for large designs.

7.4 MINIMIZATION OF LOGIC FUNCTIONS

Logic minimization is the process of reducing the amount of circuitry required to implement a logic function. The optimal form of a logic circuit is dependent to a large extent on the architecture into which the logic function is to be implemented, as we shall see later. There are literally dozens of methods for logic minimization, and a wide variety of computer-based tools that utilize these methods. While we can't hope to describe all of these methods in this chapter, we will describe some of the more popular minimization methods, and define some concepts common to all of these methods.

Users of computer-based logic synthesis tools are often bewildered by the descriptions of algorithms and logic forms and the strange language used to describe the effectiveness of various techniques. The result, all too often, is that the users of these tools never really understand how to get the most benefit from them. In this section, we'll go over some common logic minimization techniques and, hopefully, clear up some of the confusion over terminology.

Standard Forms for Boolean Equations—Revisited

To start with, consider a Boolean logic function with three variables A, B, and C. Each variable can be expressed as either its true or complement value. If we enumerate all combinations of these three variables using the AND operator, we find that there are eight possible, as follows:

$$!A \;\&\; !B \;\&\; !C$$
$$!A \;\&\; !B \;\&\; C$$
$$!A \;\&\; B \;\&\; !C$$
$$!A \;\&\; B \;\&\; C$$
$$A \;\&\; !B \;\&\; !C$$
$$A \;\&\; !B \;\&\; C$$
$$A \;\&\; B \;\&\; !C$$
$$A \;\&\; B \;\&\; C.$$

Each of these eight expressions are referred to as a *minterm* or *canonical product*. For any logic function of n variables, there are 2^n possible minterms. These Boolean expressions are called minterms because, for each expression, there is only one set of input values that will result in the value 1. For example, the minterm $!A$ & $!B$ & C will only evaluate to 1 for one set of inputs (A, B, and C equal 0, 0, and 1, respectively). The importance of minterms lies in the fact that any logic function, regardless of complexity, can be expressed as a sum of one or more of its minterms.

Maxterms

A *maxterm* is similar to a minterm, but is related to OR operations rather than AND operations. The eight possible maxterms for three input variables are:

$$!A \;\#\; !B \;\#\; !C$$
$$!A \;\#\; !B \;\#\; C$$
$$!A \;\#\; B \;\#\; !C$$
$$!A \;\#\; B \;\#\; C$$
$$A \;\#\; !B \;\#\; !C$$
$$A \;\#\; !B \;\#\; C$$
$$A \;\#\; B \;\#\; !C$$
$$A \;\#\; B \;\#\; C.$$

These expressions are called maxterms because they will evaluate to a 1 for all but one possible set of input values.

Minsums

A sum-of-product expression that consists of a sum of minterms is referred to as a *minsum*. This form is important for many logic minimization techniques and is sometimes referred to as the *canonical sum-of-products* or *disjunctive normal form*.

For example, the three-input sum-of-products expression

$$!A \ \& \ B \ \& \ C \ \# \ A \ \& \ !B \ \& \ C \ \# \ A \ \& \ B \ \& \ !C$$

is a minsum, since all of the product terms in the expression are minterms, while the expression

$$!A \ \& \ B \ \& \ C \ \# \ A \ \& \ !B \ \& \ C \ \# \ B \ \& \ !C$$

is not a minsum, since the product term $B \ \& \ !C$ is not a minterm (the variable A is not specified in the third product term).

Realizing the Minsum for an Arbitrary Logic Function

As we said, any arbitrarily complex logic function can be realized in sum-of-products form. Further, any logic function can be realized in its minsum form. This is useful for logic minimization purposes. To demonstrate how the minsum form of a logic function can be determined, consider the following expression:

$$A \ \& \ !B \ \& \ !C \ \# \ A \ \& \ B.$$

To get this Boolean function into minsum form, we'll use the postulates and laws presented earlier. First, we know that the minsum form will require three input variables for each product term in the expression, so we use the identity postulate to get the intermediate form:

$$A \ \& \ !B \ \& \ !C \ \# \ A \ \& \ B \ \& \ 1.$$

Next, the complement postulate allows us to replace the 1 with the expression $C \ \# \ !C$ resulting in

$$A \ \& \ !B \ \& \ !C \ \# \ A \ \& \ B \ \& \ (C \ \# \ !C).$$

Finally, by utilizing the distributive law we obtain the minsum expression

$$A \ \& \ !B \ \& \ !C \ \# \ A \ \& \ B \ \& \ C \ \# \ A \ \& \ B \ \& \ !C.$$

Other Standard Forms

There are other standard forms in which Boolean logic functions can be expressed, including the product-of-sums canonical form (referred to as the *maxproduct* form), but since the vast majority of programmable logic devices are based on the sum-of-products form and most logic minimization techniques are also based on sum-of-products, we won't dwell on these other forms in this book. Again, the interested reader is referred to one of the many books written on the subject of digital logic design for further information.

Logic Minimization Methods

The various Boolean logic manipulations that we have shown can be used to minimize logic. For large Boolean expressions, however, the most efficient approach to take using these manipulations typically isn't obvious, and finding the minimal solution can be extremely time consuming.

A common method for systematically determining the minimal sum-of-products expression for a given logic function was first described by Maurice Karnaugh in 1953. This method uses a symbolic representation of the logic function. The representation, called a *Karnaugh map,* graphically depicts the function in a modified form of truth table and allows simple, organized methods to be used for minimization.

To demonstrate how a Karnaugh map can be used, we will minimize the following logic function:

$$A \mathbin{\&} B \# !A \mathbin{\&} B \# A \mathbin{\&} !B$$

A two-variable Karnaugh map for this function is shown in Figure 7.8.

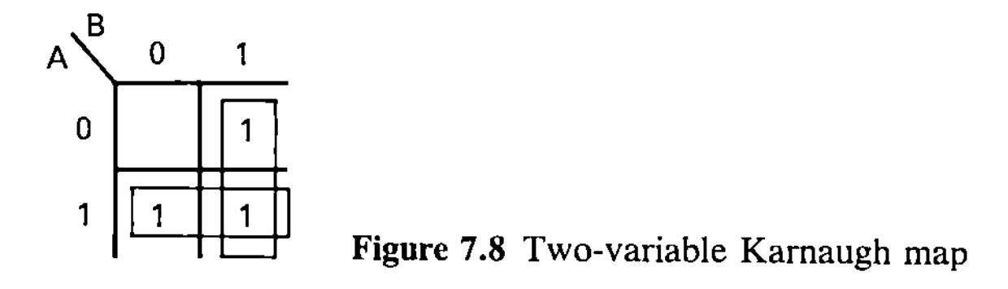

Figure 7.8 Two-variable Karnaugh map

The Karnaugh map (which we will from now on refer to as a *K-map*) is composed of a number of boxes. Each of these boxes, or *cells,* represents one possible minterm for the function. Since this function has only two variables there are 2^2 possible minterms as described earlier. Each box is identified by the values indicated on the top and sides of the box; a 1 indicates the true asserted variable, while a 0 indicates the complement of that variable. In Figure 7.9, the minterms for this Boolean function are shown in their corresponding K-map boxes.

If the minterm associated with a particular box actually exists in the minsum

A \ B	0	1
0	(!A & !B)	(!A & B)
1	(A & !B)	(A & B)

Figure 7.9 Mapping minterms onto the K-Map

form of the Boolean expression, then a 1 is written in that box. The boxes so marked are referred to as *1-cells*. Cells not marked are called *0-cells*.

To minimize a function using a K-map, we utilize the following relationship developed from our earlier Boolean identities:

$$A \mathbin{\&} f(\,) \;\#\; !A \mathbin{\&} f(\,) = f(\,).$$

The variable $f(\,)$ represents any arbitrary Boolean function. To generalize this and apply it to the K-map: if there are two product terms and one is ANDed with a variable (A) while the other is ANDed with that variable's complement ($!A$), the product terms can be combined and the variable eliminated if the product terms are otherwise identical.

This situation is identified on the K-map by pairs of horizontally or vertically adjacent 1-cells. As the figure shows, the pairs of 1-cells are grouped. Each group of two corresponding minterms can be combined into a single product term by elimination of the differing variable. For this expression, the two groupings result in the simpler sum $A \;\#\; B$.

The K-map technique is a method for determining the minimal sum-of-products representation of a logic function from its minsum form. Most logic minimization algorithms, in fact, first determine the minsum form of a sum-of-products expression before beginning the minimization process.

If you examine the K-map layout, you will notice that moving one box either vertically or horizontally always results in a single bit change; in this case, either A or B changes, but never both. This is obviously the case for the simple two-variable K-map, but what about K-maps for larger numbers of variables?

Figure 7.10 shows how a K-map for a four-variable Boolean function is written. This K-map is composed of sixteen boxes, but is still two dimensional.

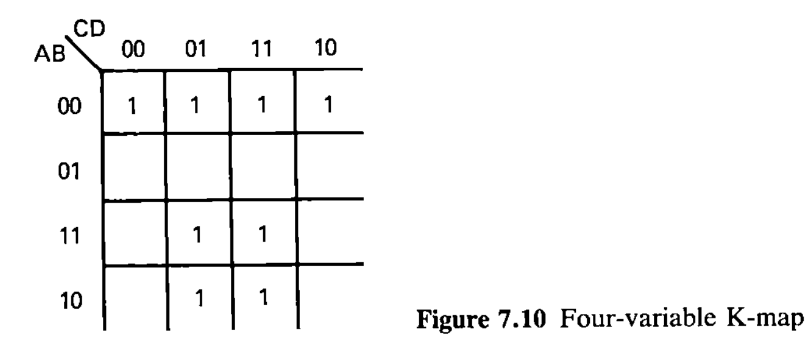

Figure 7.10 Four-variable K-map

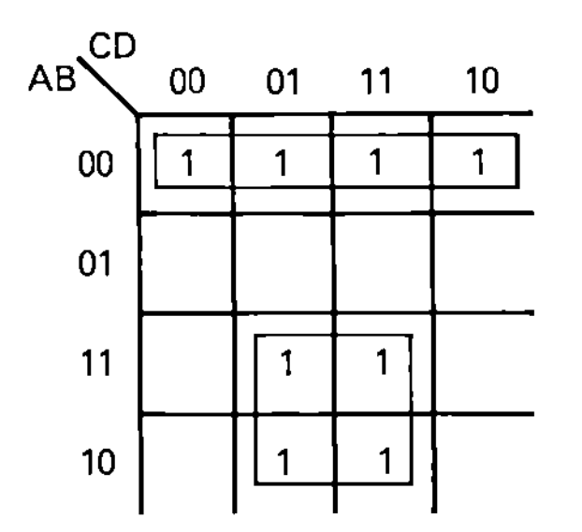

Figure 7.11 Minimizing logic with the four-variable K-map

Therefore, each box needs to represent the value of more than two variables as indicated in the numbering. In order to maintain the requirement of single bit changes between horizontally and vertically adjacent boxes, the horizontal and vertical axis values are numbered using a *gray code* (sometimes called a *reflected code*). As in the two-variable K-map, each box represents a unique minterm. The right-most box on the lower row, for example, represents the minterm *A* & !*B* & *C* & !*D*.

To reduce the four-variable minsum expression indicated in the figure, we first identify and circle all adjacent 1-cells. This is shown in Figure 7.11. Notice that, in the lower center of the K-map, there are four 1-cells grouped together in a square pattern. If these four were to be grouped as pairs, they could be combined in a number of ways.

Which combination is appropriate? Actually, this decision isn't necessary. Whenever two pairs of 1-cells are adjacent in this manner, it's an indication that all four of the minterms represented can be combined into a single product term.

Similarly, the four 1-cells that form an encircled rectangle on the upper row can be combined, resulting in a significant reduction of logic. The completely minimized expression is shown below:

!*A* & !*B* # *A* & *D*.

As you can see, the variable *C* was completely eliminated. Next, consider the expression and K-map of Figure 7.12.

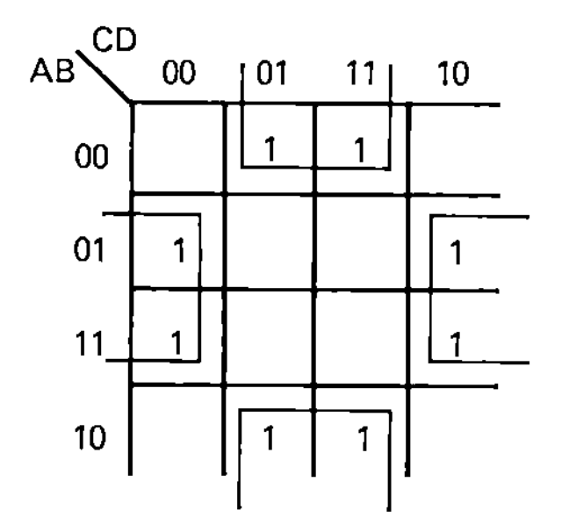

Figure 7.12 K-map for eight-minterm function

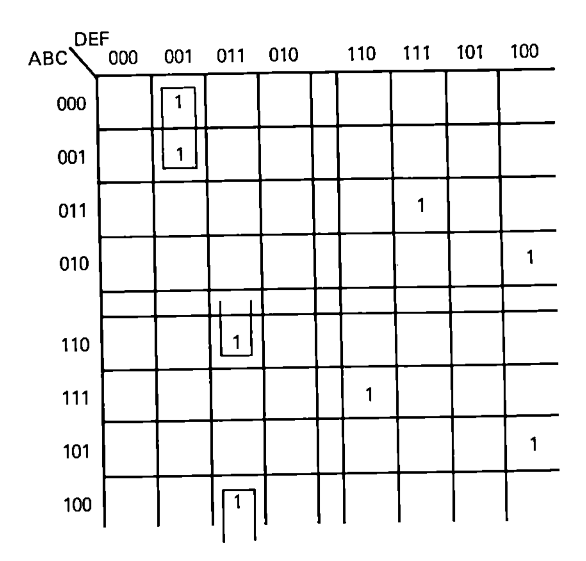

Figure 7.13 Six-variable K-map

Notice that the eight 1-cells have been grouped into two squares. This is possible because the K-map edges are logically connected; it is convenient to think of the K-map in terms of a cylinder in either the horizontal and vertical directions. The variables A and C are both eliminated and the minimized form of this expression is simply

$$B \ \& \ !D \ \# \ !B \ \& \ D.$$

Using K-maps for Larger Functions

There are many more applications of K-map techniques that are of academic interest, but are beyond the scope of this book. It is interesting to note however, that K-maps like the one shown in Figure 7.13 can be constructed to determine the minimized sum-of-products representation of expressions with more than four variables.

Since the size of the K-map increases exponentially with the increased numbers of variables, this method quickly becomes impractical for large designs. This fact, and the availability of computer-based algorithms for logic minimization, has for the most part eliminated the need for tedious manual K-map minimizations. The K-map is useful, however, for graphically demonstrating logic minimization concepts and gives important insights into the workings of computer-based logic minimization algorithms.

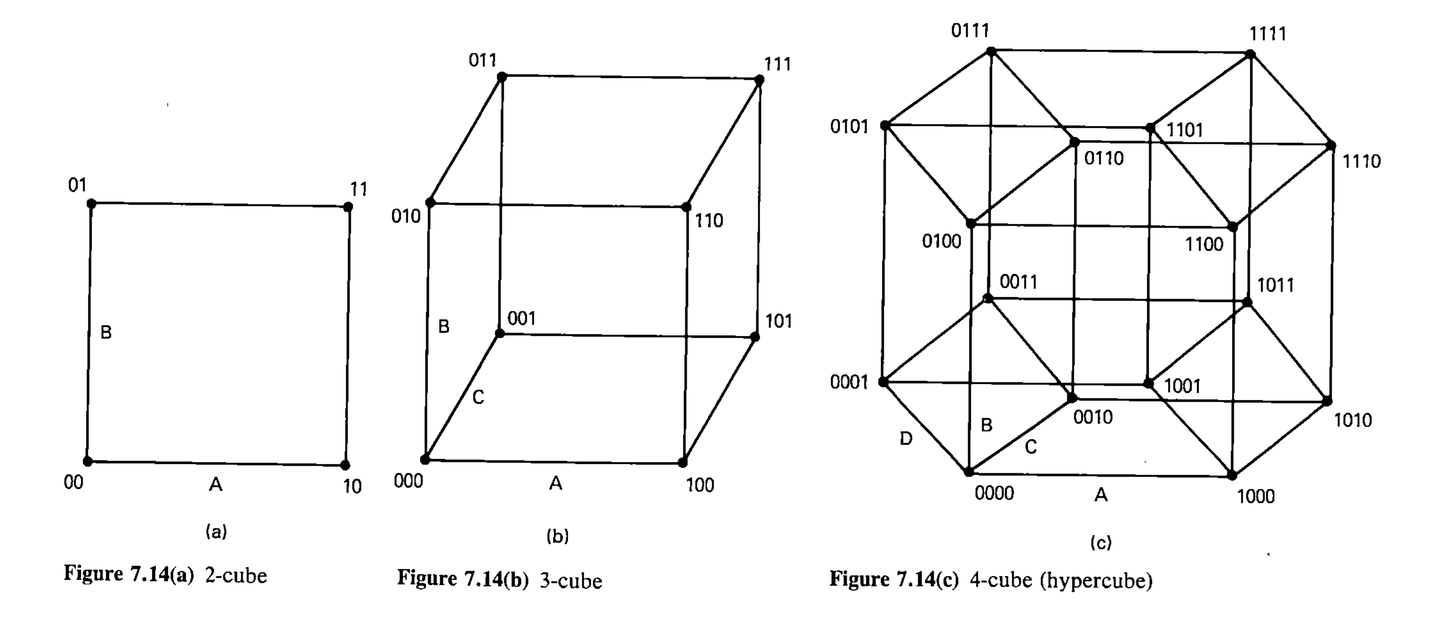

Figure 7.14(a) 2-cube

Figure 7.14(b) 3-cube

Figure 7.14(c) 4-cube (hypercube)

Cubes

Another graphic representation that is useful for describing logic minimization concepts is that of the *cube*. Consider again the case where two minterms of a function differ only in the value of one variable, as in

$$!A \ \& \ B \ \& \ C \ \# \ !A \ \& \ B \ \& \ !C.$$

As we saw in the K-map discussion, the two minterms in this minsum can be represented by sequences of ones and zeroes that correspond to true and complemented input variables. Using this representation in another way, we can assign the minterms the values 011 and 010. Now recall that a function of n variables has a total of 2^n possible minterms. If we imagine that each variable corresponds to a dimension in space, we can represent the entire set of minterms as an n-dimensional cube as shown in Figures 7.14(a), 7.14(b), and 7.14(c).

Figure 7.14(a) shows a two-dimensional cube, or *2-cube,* represented by four points connected by four line segments (it looks like a square, but for our purposes it is more convenient to think of all of the figures of various dimensions as cubes, even though they may look like lines, squares, or messy piles of toothpicks). This cube corresponds directly to its K-map representation; each point on the cube represents one box on a two-variable K-map. Movement along a horizontal line segment correspond to changes in the first variable, while movement in the vertical direction corresponds to changes in the second variable. Each point on the two dimensional cube is labeled with the corresponding value of the input variables.

Figure 7.14(b) shows a three-dimensional cube, or *3-cube*. The third dimension of this cube is drawn as diagonal line segments. Motion along these line segments corresponds to changes in the third variable. Notice that, no matter what line segment you are on, movement from one point to another always results in a change in only one of the input variables.

Figure 7.14(c) shows a four-dimensional cube (sometimes called a hypercube). This is, of course, a *4-cube*.

Determination of Cubes and Subcubes

Staying with the 3-cube for the moment, let's examine some of the ways in which the cube representation can be used. The 3-cube has, as components, a number of smaller cubes. Each face of the cube, in fact, is a 2-cube. Every 3-cube, then, has six smaller *subcubes* that are identified by the input variables represented by the line segments associated with the subcube (remember, each line segment in a cube corresponds to a change in a single input variable). Take a look at Figure 7.15. As you can see, the indicated subcube is defined by the 3-cube corners with the values 100, 101, 111, and 110.

Notice that the first variable in these four points remains unchanged while at least one of the second and third variables changes from one point to the next.

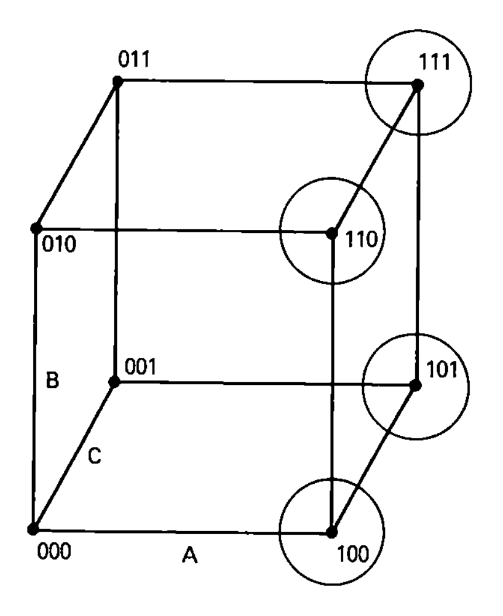

Figure 7.15 Identifying subcubes on a 3-cube

The second and third variables are, therefore, the variables that define the subcube. These variables are said to be *unspecialized* in relation to that subcube, while the first variable is *specialized.* If we specialize one of the two variables of the resulting 2-cube (or specialize two variables of the 3-cube), we end up with a *1-cube*. We can even go as far as specializing all of the variables of a cube, in which case we find ourselves with 2^n *0-cubes,* where n is the number of input variables (a 0-cube simply corresponds to one point on a cube).

How is all this academic hopscotch used? If we map the minterms of a logic function onto a cube, we can quickly recognize subcubes of the function. Consider the following three-variable minsum function:

!*A* & *B* & !*C* # !*A* & *B* & *C* # !*A* & !*B* & *C* # !*A* & !*B* & !*C* # *A* & !*B* & !*C*.

Figure 7.16 shows this function mapped onto a 3-cube. As the figure shows, four of the minterms of this function form a complete 2-cube. This 2-cube is defined by specialization of the variable *A*, so we can immediately recognize that these four minterms can be replaced by a single product term that consists of nothing but !*A*. In addition, the 1-cube that is formed by the minterms 100 and 000 can be reduced to the simpler product term !*B* & !*C*, since *A* is unspecialized for that subcube. The minimized function is then:

!*A* # !*B* & !*C*

The K-map presented earlier is simply a convenient representation of a cube, and a method for quickly identifying subcubes.

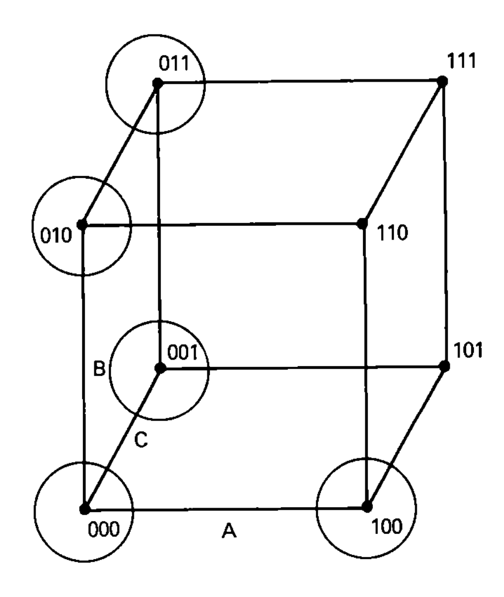

Figure 7.16 Mapping a function onto a cube

Tabular Representation of Sum-of-Products Functions

In the K-Map and cube representations, we use the values 0 and 1 to represent the complement and true values of input variables. This helps to simplify the identification and combination of cubes and subcubes within the function, but requires that the function be expressed in its minsum form.

Many logic minimization algorithms utilize another representation that doesn't require that the function be expressed in minsum form. Like the K-map representation, complement and true variables are indicated by a 0 or 1 respectively, and another symbol, the dash, is used to indicate that a variable of the function isn't used (is unspecialized) for a specific product term. To represent the four-variable function

$$A \mathbin{\&} !B \mathbin{\&} C \mathbin{\&} D \mathbin{\#} B \mathbin{\&} !C \mathbin{\#} A \mathbin{\&} D$$

in tabular form, for example, we write

1011

–10–

1––1

The dash symbol is referred to as a *don't care input condition* since, for the product term in which a dash is found, we don't care what value that particular input variable has—the product term will evaluate to 1 regardless of that input's value.

As you may have already realized, this form is quite convenient for recognizing or obtaining such things as minterms and cubes. A minterm is indicated whenever there is a product-term row that contains no dashes. A minsum form is, therefore, a table that contains no dashes whatsoever. The number of dashes in a product term indicates the dimensions of the subcube represented by that entry in the table: two dashes indicate a 2-cube, one dash indicates a 1-cube, and no dash indicates a 0-cube which is, as we said, a minterm.

To convert the above table to minsum form, you simply replace each row of the table with however many rows are required to eliminate the dashes. For example, the last row becomes four rows as follows:

```
1) 1--1 → 10-1
          11-1
2) 10-1 → 1001
          1011
   11-1 → 1101
          1111
```

Prime Implicants

In a sum-of-products expression, each of the product terms is called an *implicant* of the logic function. If any product term evaluates to a 1, it implies that the entire function will evaluate to a 1, regardless of the value of the other product terms. Consider the following minsum expression (recall that any sum-of-products expression can be expressed in minsum form):

!*A* & *B* & *C* & *D*
\# *A* & *B* & *C* & *D*
\# *A* & *B* & !*C* & *D*
\# !*A* & *B* & !*C* & *D*

Each of the four minterms in this expression is an implicant of the logic function represented by the expression since, if any of these minterms evaluates to a 1, the entire expression will also evaluate to a 1. These four minterms aren't the only implicants of the function, however.

When the expression is simplified, we find that the first two minterms can be combined into a single product term *B* & *C* & *D* by elimination of the variable *A*. Similarly, the third and fourth minterms can be combined to form the product term *B* & !*C* & *D*, also by elimination of the variable *A*. These two smaller product terms are therefore also implicants of the function, since any input condition that results in either of these product terms evaluating to 1 will result in the entire expression evaluating to 1 as well.

The implicant product terms *B & C & D* and *B & !C & D* can, of course, be simplified further by eliminating the variable *C*. The resulting product term *B & D* is what is called a *prime implicant* of this logic function. Since this logic function has only one prime implicant, that single product term represents the minimal form for the logic function.

To summarize, a prime implicant is any implicant of a logic function that is not implied by any other subfunction. A logic function may have any number of prime implicants.

Minimal Cover

The set of prime implicants for a given logic function is unique since they are determined from a unique set of minterms. It isn't always necessary, however, to use all of the prime implicants for a function to obtain the minimal sum-of-products form. This is true because one prime implicant for a function may cover a number of minterms and overlaps can exist; a single minterm can be covered by more than one prime implicant.

The goal of total sum-of-products logic minimization is to determine a minimal subset of prime implicants that will cover all of the minterms of the function. This subset of prime implicants is called the *minimal cover*. The important thing to understand about prime implicants is that the fully minimized form of a logic function will consist of nothing but prime implicants, but may not require all of the prime implicants for the function.

Essential Prime Implicant

The search for the elusive minimal cover of prime implicants can be hastened if we first attempt to determine which prime implicants are required in a fully minimized sum-of-products representation of the function. These prime implicants are known as *essential prime implicants*, and can't be eliminated from the function by any means. For example, the three-variable function

!A & !C # !A & B # A & C # A & !B

is composed entirely of prime implicants. This is not, however, the minimal sum-of-products form for this function. Two of the prime implicants, *!A & !C* and *A & !B*, are essential prime implicants and must be maintained. The remaining two prime implicants are not essential, and can be replaced by the single prime implicant *B & C*, resulting in the function

!A & !C # B & C # A & !B.

The Quine–McCluskey Procedure

The methods of logic minimization thus far presented are useful to help understand the minimization process, but are not generally applicable to large designs with more than five or six variables. In addition, these methods are difficult to implement on a computer, due to their reliance on pattern recognition.

The *Quine–McCluskey procedure* is a tabular method of logic minimization that can be performed manually or implemented on a computer. This method is composed of two algorithms. The first algorithm determines all of the prime implicants for a given logic function and the second selects from this set of prime implicants a subset that provides a minimal full cover for the function. The Quine–McCluskey procedure has been largely replaced by more efficient minimization methods, but it is nonetheless valuable for developing an understanding of computer-based logic minimization algorithms in general.

Determining Prime Implicants using the Quine–McCluskey Procedure

As in the K-Map method, the unity theorem (A & B # !A & B = B) is used. A simplistic way to describe this process is to say that the algorithm first determines the complete set of minterms for the function and then utilizes the unity theorem iteratively for all possible pairs of product terms to obtain all of the prime implicants of the function. In this process, each pair of minterms is examined to determine if the two minterms differ by exactly one position. Pairs so identified are combined into a single product term having one fewer literal than the previous two.

After all minterm pairs have been examined, the process is repeated for the resulting terms. This process is continued until we are left with the set of prime implicants. A prime implicant is identified by the fact that it can't be compared in this way with any other term.

This determination of prime implicants can be quite time consuming, particularly for functions with a large number of variables. To speed this process, the Quine–McCluskey method uses the tabular representation described earlier. To illustrate the algorithm, we'll use an example. Consider a four-input logic function

$$f(w,x,y,z)$$

whose minterms are as shown in the list of Figure 7.17. The minterms are listed in order of the number of ones in the tabular assignment. This assists in the identification of paired minterms.

Figure 7.18 shows the steps used in the process of searching for prime implicants. The second list is determined by combining those minterms that differ by one position, as described above. In the second list, the dash (don't-care) replaces the removed variable. The two terms that are combined to create the

0000	!W	&	!X	&	!Y	&	!Z
0010	!W	&	!X	&	Y	&	!Z
0100	!W	&	X	&	!Y	&	!Z
1000	W	&	!X	&	!Y	&	!Z
0101	!W	&	X	&	!Y	&	Z
1010	W	&	!X	&	Y	&	!Z
1100	W	&	X	&	!Y	&	!Z
1011	W	&	!X	&	Y	&	Z
1101	W	&	X	&	!Y	&	Z
1111	W	&	X	&	Y	&	Z

Figure 7.17 Minterm list for four-input logic function

entry in the second list are checked in the first list indicating that they can't be prime implicants.

Having checked an entry in the first list doesn't complete the comparison process for that entry; each entry in the list must be compared with every other entry to eliminate all possible nonprime implicants. This means that there are n^2 comparisons required for the first list alone, where n is the number of minterms in the list. In practice, the ordering of the list (by the number of ones in each product term) means that not all pairs need to be examined, speeding the comparison process.

When all of the entries in the first list have been so compared, the process is repeated for the second list to create the third list. Note that, during the comparisons for the second and subsequent lists, candidate pairs must have dashes

LIST 1		LIST 2		LIST 3
0000	X	00-0	X	-0-0
0010	X	0-00	X	--00
0100	X	-000	X	-10-
1000	X	-010	X	
0101	X	010-	X	
1010	X	-100	X	
1100	X	10-0	X	
1011	X	1-00	X	
1101	X	-101	X	
1111	X	101-		
		110-	X	
		1-11		
		11-1		

Figure 7.18 Determining prime implicants by tabular comparison

in the same position. It isn't possible, for example, to compare the terms 1–00 and –101.

When all the entries in the second list have been compared, there are three entries left that aren't checked. These entries are prime implicants. Attempts to compare the third list produce no combinations, so we are left with the three prime implicants from the second list plus the three in the third for a total of six prime implicants for the function. This technique for finding prime implicants has general applicability beyond the Quine–McCluskey method and is frequently used.

Determination of Minimal Cover

The second step in the Quine–McCluskey method is the determination of a minimal cover for the function, using the prime implicants found as a result of the first step. This is done by first creating a covering table as shown in Figure 7.19. The covering table shows all the minterms of the function, and which minterms are covered by which prime implicants. Obviously, it will speed the process if this covering table is constructed from the comparison information determined during the first step. The rows of the table list the prime implicants identified, while the columns represent the minterms of the function.

When the covering table has been completed, we can easily identify the essential prime implicants. These are the implicants (in this case the terms –0–0 and –10–) that are the only covers for one or more minterms. In the covering table, the essential prime implicants are flagged by an asterisk (*).

Since we know that the essential prime implicants must appear in the minimal cover, we can simplify the covering table by removing the rows corresponding to these essential prime implicants, and also removing all of the columns corresponding to minterms covered by them. This simplification leaves us with the reduced covering table shown in Figure 7.20.

This reduced table shows quite clearly that the prime implicant 1-11 covers both of the remaining minterms, so this prime implicant, combined with the two essential prime implicants previously identified, comprises the minimal cover for

	0000	0010	0100	0101	1000	1010	1011	1100	1101	1111
101-						X	X			
1-11							X			X
11-1									X	X
*-0-0	X	[X]			X	X				
--00	X		X		X			X		
*-10-			X	[X]				X	X	

Figure 7.19 Covering table

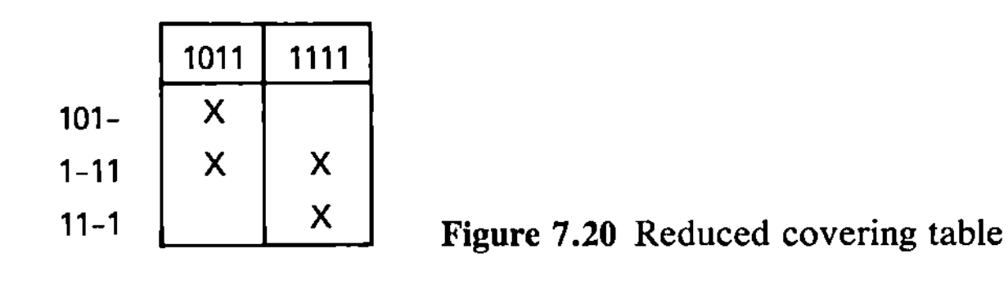

	1011	1111
101-	X	
1-11	X	X
11-1		X

Figure 7.20 Reduced covering table

the function. The minimal cover, expressed in equation form, is then

$$W \;\&\; Y \;\&\; Z \;\#\; !X \;\&\; !Z \;\#\; X \;\&\; !Y.$$

Further Applications of Quine–McCluskey

There are many extensions to the basic Quine–McCluskey method. Finding minimal covers when don't-cares are specified, and minimizing multiple output functions, can be accomplished with a modified form of the algorithm just described.

7.5 ON-SETS, OFF-SETS, AND DC-SETS

When a logic function is described, it can be expressed in one of many ways. When we describe a function using the form of Boolean equations we presented earlier, we are describing the set of input conditions that will cause the expression to evaluate true. We call this set of input conditions the *on-set*. All other input conditions will result in the expression evaluating to false. This set of implied conditions is called the *off-set*. It's also possible to use Boolean equations to explicitly specify the off-set of a function. This is done by complementing the output variable, as in

$$!Y = A \;\&\; B \;\#\; !A \;\&\; !B.$$

When we describe a function with a single on-set equation, we must assume that all conditions not covered by that equation comprise the off-set. Similarly, if the function is described by an off-set equation, it must be assumed that all unspecified conditions form the on-set.

The truth table representation is used to describe a logic function by specifying both its on-sets (indicated by ones) and off-sets (indicated by zeroes). We have used truth tables extensively in this chapter. All of these truth tables have specified the complete on-set and off-set for the corresponding logic functions.

For many (perhaps most) logic designs, there are certain input conditions that will never be encountered by the circuitry being described. There may also be input conditions that will only occur at times when the output of the circuit will be unused. This information about the design can be used to advantage when

A	B	C	Y
0	0	0	0
0	0	1	1
0	1	1	1
1	1	1	1
1	1	0	0
1	0	0	0

Figure 7.21 Incompletely specified truth table

minimizing the logic for that circuitry. If we know that, for certain input conditions, the output may be either true or false with no effect on the operation of the system, we can reduce the amount of circuitry required to implement the function. These input conditions are referred to as the don't-care set, or *dc-set.* This is best illustrated in a truth table. Figure 7.21 shows a truth table representation of a three-variable logic function.

This truth table is only partially complete; the value of the output variable *Y* is missing for the input conditions where *A, B,* and *C* are 111 or 001. This truth table is what we call an *incompletely specified* truth table.

The missing conditions are referred to as don't-care conditions for *Y.* The truth table rows that specify a 1 for *Y* are conditions belonging to the on-set, while the remaining rows are the conditions that are members of the off-set. The two unspecified conditions form the dc-set for this function. There are four possible nonredundant implementations for this function; two of these are shown in Figure 7.22 with their associated minimized Boolean equations and K-maps. As this example demonstrates, the intelligent use of don't-cares can have a dramatic affect on the size of the circuit.

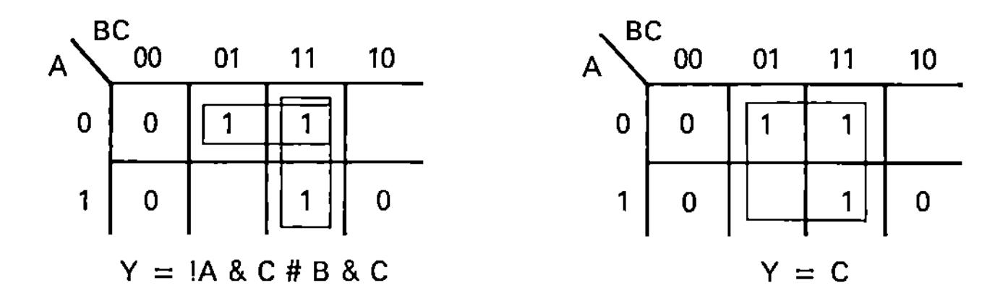

Figure 7.22 Two possible implementations of partially specified logic function

Determining the minimal implementation of a function containing don't-care conditions is simply an extension of the principles outlined earlier. To represent don't-care input conditions on a K-map, we simply place a dash (-) symbol on the cell corresponding to that input condition, as shown in Figure 7.23.

The K-map can be processed as though the dash symbol represented either a 1 or 0, whichever results in the smallest minimal cover. For this K-map, we

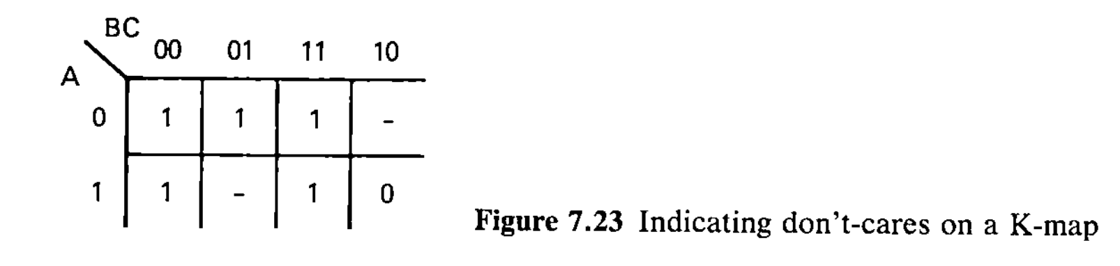

Figure 7.23 Indicating don't-cares on a K-map

can see that replacing the dash in the lower row with a 1 will result in a reduction of logic, while replacing it with a 0 will not. For many purposes, it's useful to think of the don't care value as an actual circuit condition. This is a common approach used in circuit simulators. This means that each variable now has three possible values: true, false, and don't-care.

The truth table representation allows us to specify a function by supplying the on-set and off-set, with the dc-set being implied by omission from the table. When describing a function with equations, it's usually more convenient to describe the function by supplying its on-set and dc-set. In this situation the off-set is implied. Similarly, it's possible to describe a function by supplying only the off-set and dc-set, with the on-set implied. It's also possible to describe a function by specifying the complete on-set, off-set, and dc-set, but this is usually a pointless exercise.

When you design using a combination of on-sets and off-sets, on-sets and dc-sets or off-sets and dc-sets, you must be careful not to create overlap conditions such as those shown in Figure 7.24.

A	B	C	Y
0	1	0	0
0	1	0	1

Figure 7.24 Overlapping truth table conditions

In this truth table, the same input condition is listed in both the on-set and off-set for *Y*. This is easy to detect in a truth table (or K-map), but is not so easy to detect if on-sets and off-sets are being expressed in some other form (notably Boolean equations or state transitions). This is a common area of confusion for users of computer-based logic minimization tools.

7.6 MULTIPLE-OUTPUT MINIMIZATIONS

In a typical PAL-architecture PLD, the above techniques are adequate since there is no sharing of logic from one device output to another. In PLA-type devices, however, there is opportunity for improvement. The total amount of circuitry required to implement a multiple-output function can be reduced if all outputs of the function are evaluated together.

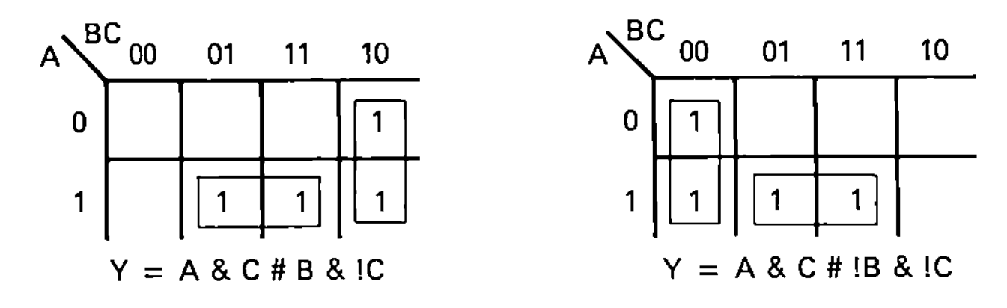

Figure 7.25 Functions with shared prime implicant

Figure 7.25 shows two three-variable functions expressed as K-maps. Each function has two prime implicants as shown. The K-maps, when overlaid, show that there is one prime implicant common to both functions. Considered together, then, it is possible to implement both of these functions using a total of three product terms.

The prime implicant that is shared by the two functions is called a *multiple-output prime implicant*. The two prime implicants that are not shared are called *single-output prime implicants*. In this example, the minimal form for the multiple-output function is the same form that results from the outputs being minimized individually. This is not, however, always the case. Consider the example functions shown in Figure 7.26.

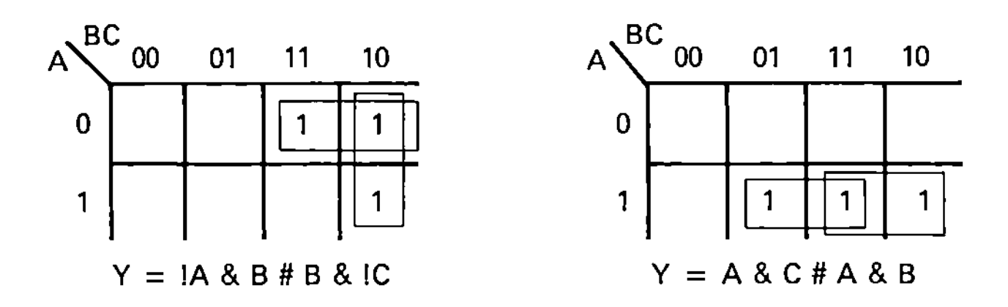

Figure 7.26 Two independent functions

If we minimize these two functions separately, we result in the groupings indicated, which correspond to the minimized equations

$$Y1 = !A \ \& \ B \ \# \ B \ \& \ !C$$

$$Y2 = A \ \& \ C \ \# \ A \ \& \ B.$$

There are no common prime implicants in these functions. If, however, we identify those 1-cells that are common to both K-maps and group in a manner that isolates those cells, we can come up with a set of implicants (not necessarily prime implicants) that will provide a minimal cover for the combined functions. Figure 7.27 shows such a grouping and the corresponding Boolean equations for the two functions.

Notice that the two functions now require a total of three product terms,

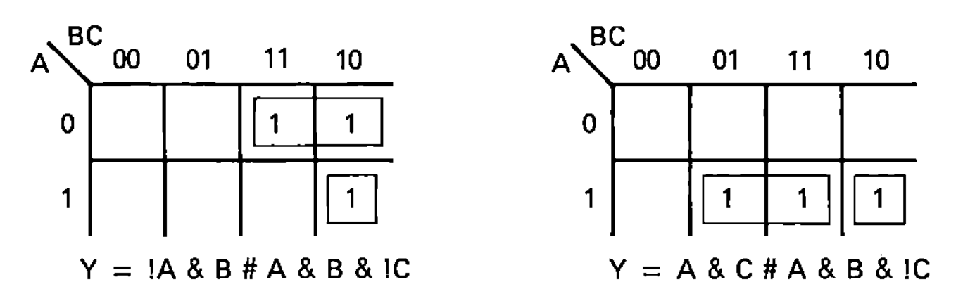

Figure 7.27 K-map groupings to isolate shared implicant

compared to the original four. Notice also that the product term A & B & !C is not a prime implicant of either function and, therefore, is not a single-output prime implicant. It is a multiple-output prime implicant for this multiple-output function, however, since there is no other multiple-output implicant for this set of functions that implies it.

7.7 MULTILEVEL OPTIMIZATIONS

All of the techniques shown above have been used to minimize sum-of-products Boolean equations. We have assumed that the sum-of-products form is the most optimal (in terms of gate requirements). In the real world, however, there are many constraints beyond the gate count that must be considered. These include the problems of fan-out, constrained numbers of inputs to gates or sum-of-product logic blocks, and problems of timing hazards. For these reasons it is often necessary to implement a design using multiple levels of logic. An example of this is found in the following sum-of-products equation:

$$\begin{aligned} Y = & \ !A \ \& \ !B \ \& \ !C \ \& \ !D \ \& \ !G \\ & \# \ !A \ \& \ !B \ \& \ F \\ & \# \ !A \ \& \ !B \ \& \ D \ \& \ !E \ \& \ !G. \end{aligned}$$

This seven-variable function might be implemented as shown in the circuit of Figure 7.28.

If, however, the constraints of the target implementation include a requirement that all gates have four or fewer input variables, some changes will be required. Figure 7.29 shows how the function can be implemented in a *multilevel* form to meet the four-input gate criteria. In a multilevel circuit, some or all of the signal paths experience a delay of more than two gates.

There are actually many possible multilevel circuits that we could have constructed to implement the function within the constraints specified. The circuit shown takes advantage of the fact that there is a term (!A & !B) that is common to all three product terms of the original sum-of-products function. This term is isolated out by a process called *factoring* and the isolated term is called a *factor*.

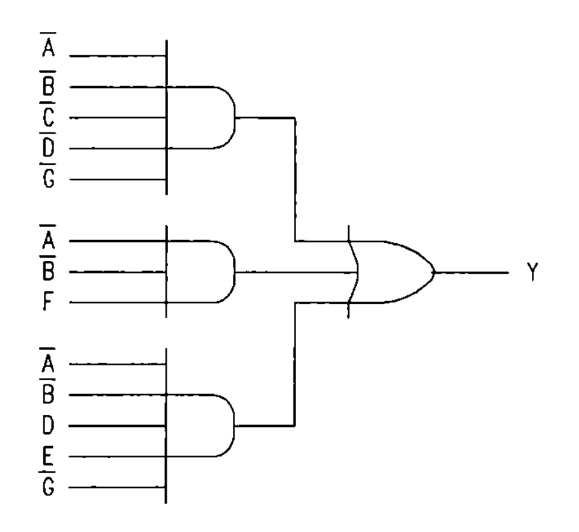

Figure 7.28 Two-level circuit implementing seven-variable function

Factoring

Factoring is the primary technique for converting two-level representations into multilevel forms. Factoring is useful for those situations where an equation is too large (too many inputs or too many product terms) to fit into the logic dedicated to one PLD output, or if other electrical constraints (such as fan-out restrictions) prevent the use of two-level logic.

In factoring, we break up a two-level sum-of-products equation into multiple levels of sum-of-products equations by using intermediate variables. For example, the sum-of-products equations

$$Y0 = A \& B \# A \& !C \# D \& B \# D \& !C$$

$$Y1 = D \& B \# D \& !C$$

$$Y2 = E \& !B \& C$$

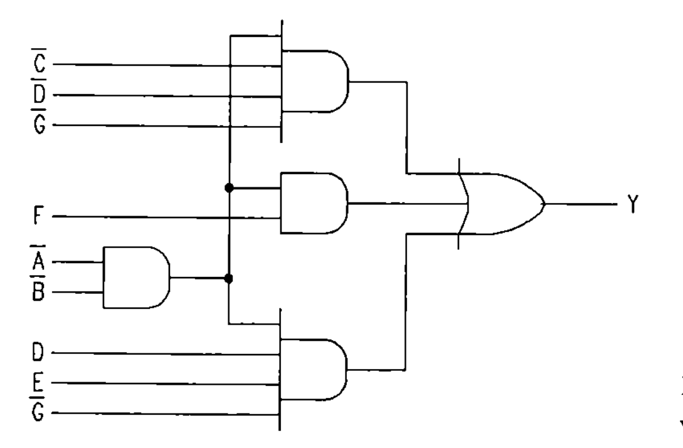

Figure 7.29 Multilevel circuit for seven-variable function

could be factored into the equations

$$Y0 = A \,\&\, f0 \,\#\, D \,\&\, f0$$

$$Y1 = D \,\&\, f0$$

$$Y2 = E \,\&\, !f0$$

$$f0 = B \,\#\, !C$$

for a savings of one product term. The variable $f0$ is the *intermediate factor variable*. In this example, a reduction in the total number of product terms was achieved, but this isn't always the case. Quite often, the constraints that drive the need for factoring result in a larger set of equations. Factoring, then, can be used as a tool for multilevel logic minimization, or as a tool for meeting other constraints that may actually result in larger circuits.

It may seem that factoring is of no use for PLDs, since they are usually inherently two-level in their design. Although it is true that most PLDs are essentially sum-of-products logic devices, the intermediate signals (factors) can often be allocated to unused combinational outputs of the PLD and fed back into the sum-of-products array as an input. Many of the newer PLDs feature folded arrays that allow multilevel logic to be mapped into the devices with no waste of I/O pins.

7.8 UTILIZING EXCLUSIVE-OR GATES

Exclusive-OR gates, or XORs, are found in an increasing number of PLDs. These gates can be used to advantage in a number of ways. When we add an XOR gate to the usual sum-of-products structure, we have a structure that is referred to as *XOR-of-sums-of-products*. This form is shown in Figure 7.30.

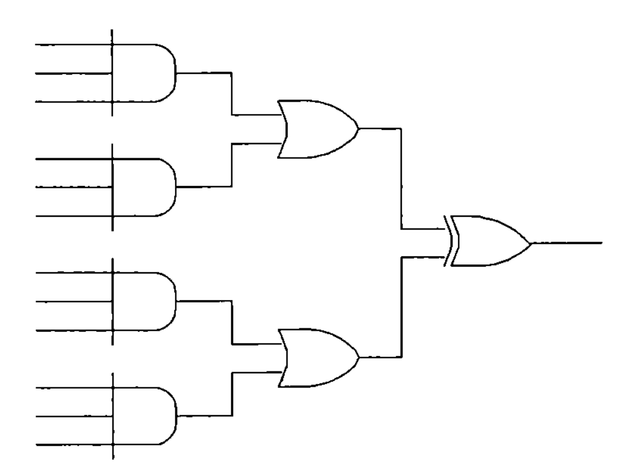

Figure 7.30 XOR-of-sums-of-products circuit

Standard Forms for XOR Boolean Equations

One form of Boolean equation that utilizes the XOR-of-sum-of-products structure is called the *Reed–Muller* form. In a Reed–Muller form, all literals of the Boolean equation are asserted true, as in the equation

$$X = (A \,\&\, B \,\#\, C \,\&\, D) \,\$\, (D \,\#\, E \,\&\, F).$$

The Reed–Muller form has important uses outside of PLDs. Like the sum-of-products form, any logic function can be expressed in Reed–Muller form. Since no input inverters are required, the Reed–Muller form requires the same number of gates (in terms of signal delays) that the sum-of-products form requires.

Another form is called the *generalized Reed–Muller,* or GRM, form. In the GRM form, there may be a mixture of true and complement literals, but no variable of the function can appear in the equation in both its true and complement form. If both the true and complement literals for a variable appear in the equation, the equation is said to be in a *mixed-GRM* form. Mixed-GRM is the form of Boolean equation that can be mapped into a XOR PLD such as the 20X8 PAL shown in Figure 7.31.

XOR Usage in PLDs

There are many different implementations of XORs in PLDs, some that are asymmetrical, and some that exploit term sharing between the OR gates. The most common application of the XOR gate in PLDs is for output polarity control. This polarity control can be either static (where one input the XOR gate is tied to ground or allowed to float high depending on the state of a device fuse) or dynamic.

The existence of XOR gates in a PLD allows some interesting applications, of which polarity control is just one. We can also exploit the behavior of XORs

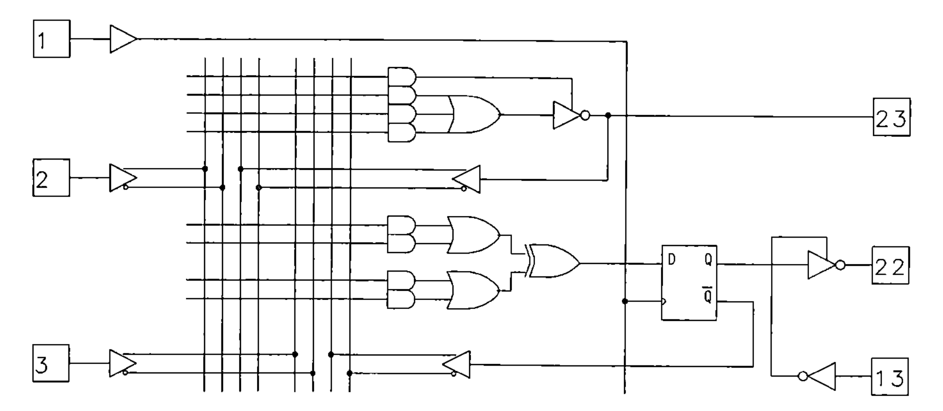

Figure 7.31 20X8 PAL output logic

for efficient polarity control of more complex equations that already utilize the single XOR gate in the device. Consider the following XOR equation:

$$X = (A \# C \,\&\, B \# C \,\&\, D) \,\$\, (E \# F).$$

If we wish to implement this equation in a device such as the 20X8 PAL, it will be necessary to make some modifications to the equation since, as the equation is written, it is inappropriate for the structure of the device.

First, it will be necessary to complement both sides of the equation since the 20X8 features inverters on its outputs. This is quite easily done with an XOR equation, if we utilize the XOR complement identities presented earlier. According to those rules, it is possible to complement the entire equation simply by complementing one sum-of-products input to the XOR. For the above equation, either of the following equations might be appropriate:

$$!X = (A \# C \,\&\, B \# C \,\&\, D) \,\$\, (!E \,\&\, !F)$$

$$!X = (!A \,\&\, !B \,\&\, !D \# !A \,\&\, !C) \,\$\, (E \# F).$$

In the first equation, the right side of the XOR operator was complemented ($E \# F$ became $!E \,\&\, !F$), while in the second equation, the left side of the XOR was complemented ($A \# C \,\&\, B \# C \,\&\, D$ became $!A \,\&\, !B \,\&\, !D \# !A \,\&\, !C$).

Complementing the inputs to the XOR gates was accomplished through the use of DeMorgan's theorem. Notice that both forms of the complemented equation use a similar amount of product terms and literals and, therefore, require the same amount of circuitry. The first equation, however, cannot be implemented in the 20X8 device. As shown in Figure 7.31, the 20X8 has only two product terms available on each side of the fixed XOR gate. The first equation listed above requires three product terms for one of the XOR inputs. For this implementation, the second form of complemented equation is the preferred form.

It is important to understand that, in the absence of a specific target device type, there is no "right answer" when making these sorts of equation optimizations. The first of the two equations shown above, while not being appropriate for the 20X8 architecture, would be in the correct form for other types of devices (the Cypress 330, for example, which has many product terms allocated to one input of the XOR gate, and only one product term allocated to the other).

The XOR complement identities can also be used to help minimize logic in situations where no equation inversion is desired. Consider the equation

$$!X = (A \# C \,\&\, B \# C \,\&\, D) \,\$\, (E \# F).$$

This is the same equation we just presented, but with an already complemented output. As we already noted, this equation cannot be directly mapped into the 20X8 device because there are too many product terms on the left side of the equation. We can modify this equation, however, so that it will fit. All that is required is that we complement both sides of the XOR operator in accordance with the XOR complement identities. The resulting equation

$$!X = (!A \ \& \ !B \ \& \ !D \ \# \ !A \ \& \ !C) \ \$ \ (!E \ \& \ !F)$$

will fit into the 20X8 device.

Further applications of output polarity control, including dynamic control, are presented in Chapter 8. Now let's look at how XOR gates can be used to help optimize logic that may or may not have been originally expressed with XORs.

Using XOR Factoring

A lesser-known application of XOR gates is a technique called *XOR factoring*. XOR factoring is a method that can be used to reduce the number of product terms needed to implement a design in an XOR device. Consider the following sum-of-products equation:

$$\begin{aligned} Y = {} & A \\ & \# \ X0 \ \& \ X1 \ \& \ X3 \ \& \ X4 \ \& \ X5 \ \& \ X6 \ \& \ X7 \\ & \# \ X2 \ \& \ X3 \ \& \ X4 \ \& \ X5 \ \& \ X6 \ \& \ X7 \\ & \# \ !X0 \ \& \ !X7 \\ & \# \ !X1 \ \& \ !X7 \\ & \# \ !X2 \ \& \ !X7 \\ & \# \ !X3 \ \& \ !X7 \\ & \# \ !X4 \ \& \ !X7 \\ & \# \ !X5 \ \& \ !X7 \\ & \# \ !X6 \ \& \ !X7. \end{aligned}$$

As you can see, this equation consumes ten product terms. Because of this, it is not possible to implement this equation in a PLD with only eight product terms per output. It is possible, however, to implement this equation in just four product terms, using an XOR type of device and XOR factoring.

The technique of XOR factoring is based on the XOR identities presented earlier. These identities allow us to XOR any Boolean expression with a second expression, and then XOR the resulting larger expression with the second expression again resulting, finally, with the original expression. For example, the equation

$$X = A \ \& \ B;$$

is functionally equivalent to the equation

$$X = B \$ (B \$ (A \& B)).$$

In effect, the two XORs cancel each other out.

By identifying certain commonly used subexpressions in an equation, and factoring these subexpressions out with an XOR, we can reduce the number of product terms required for the primary expression in the equation. In many cases, this can dramatically reduce the total number of product terms required.

Using the previous equation for Y as an example, we first modify the sum-of-products equation as follows:

$$
\begin{aligned}
Y = {} & (!A \;\&\; X7) \\
& \$ \; ((!A \;\&\; X7) \\
& \$ \; (A \\
& \# \; X0 \;\&\; X1 \;\&\; X3 \;\&\; X4 \;\&\; X5 \;\&\; X6 \;\&\; X7 \\
& \# \; X2 \;\&\; X3 \;\&\; X4 \;\&\; X5 \;\&\; X6 \;\&\; X7 \\
& \# \; !X0 \;\&\; !X7 \\
& \# \; !X1 \;\&\; !X7 \\
& \# \; !X2 \;\&\; !X7 \\
& \# \; !X3 \;\&\; !X7 \\
& \# \; !X4 \;\&\; !X7 \\
& \# \; !X5 \;\&\; !X7 \\
& \# \; !X6 \;\&\; !X7).
\end{aligned}
$$

For this equation, we have chosen the term $!A \;\&\; X7$ for a factor. When we convert the equation to mixed GRM form (preserving $!A \;\&\; X7$ as one operand of the XOR) and minimize the two sum-of-products XOR inputs, we get an equation of the form

$$
\begin{aligned}
Y = {} & (!A \;\&\; X7) \\
& \$ \; (!A \;\&\; X0 \;\&\; X1 \;\&\; X3 \;\&\; X4 \;\&\; X5 \;\&\; X6 \;\&\; X7 \\
& \# \; !A \;\&\; X2 \;\&\; X3 \;\&\; X4 \;\&\; X5 \;\&\; X6 \;\&\; X7 \\
& \# \; !A \;\&\; X0 \;\&\; X1 \;\&\; X2 \;\&\; X3 \;\&\; X4 \;\&\; X5 \;\&\; X6).
\end{aligned}
$$

This equation requires only four product terms, and will fit quite efficiently in a PLD with fixed XOR gates.

How did we decide on the XOR factor $!A$ & $X7$? Partly through trial and error: we chose a variety of different candidate factor terms based on which literals were observed most frequently in the equation and experimented to determine which would result in the most savings of logic. We deliberately chose a single-term XOR factor that was relatively simple. There is no reason, however, why an XOR factor can't be a complex expression that results in an even split of the equation into two similarly sized components. As you might imagine, then, the number of possible XOR factors is enormous, including subexpressions of all sizes.

There are algorithmic methods for determining lists of candidate XOR factors, but no known algorithm will determine the optimal factor for a restrictive XOR configuration (one with limited or asymmetrical sum-of-products XOR inputs). The choice of XOR factor is affected to a great extent by the architecture of the device into which the design is being implemented. Clearly, if one of the OR gates that feeds the XOR has more product terms available than the other, this needs to be taken into consideration when selecting a factor term.

Choosing an XOR Factor

The trial and error approach to finding an appropriate XOR factor can be time consuming if it isn't done in a methodical way. One of the best methods we are aware of is to choose factors that are composed of one or more subexpressions known as *level-0 kernels*. A factor is a subexpression that results from factoring a larger equation. Level-0 kernels are those factors that cannot be factored any further. In this sense, the level-0 kernel can be thought of as being analogous to a prime number. For example, the sum-of-products equation

$$Y = A \,\&\, B \,\#\, C \,\&\, B$$

can be factored into the equations

$$Y = F1 \,\&\, F2$$
$$F1 = A \,\#\, C$$
$$F2 = B.$$

The two factors represented by $F1$ (A # B) and $F2$ (B) are level-0 kernels of Y because they cannot be factored any further.

7.9 SEQUENTIAL APPLICATIONS IN PLDS

All of the discussions up to this point have been oriented toward the design and optimization of purely combinational logic. Combinational logic is distin-

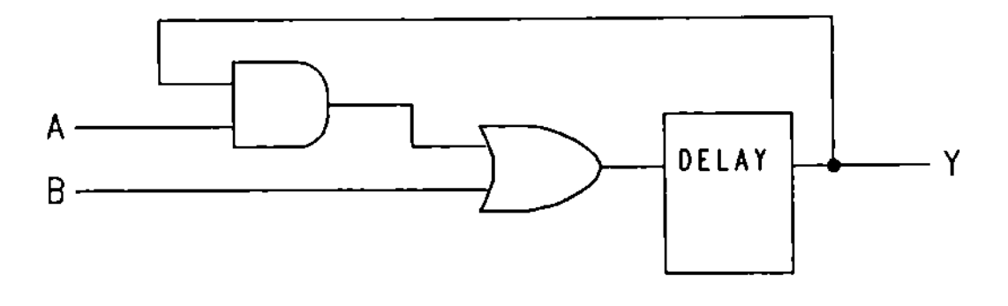

Figure 7.32 Typical model of a sequential circuit

guished by the fact that it contains no memory elements and has no feedback loops. A combinational circuit is so named because, for any given input combination, the output (or outputs) will produce a known logical result regardless of the circuits previous condition. *Sequential logic circuits* are those circuits whose behavior is dependent not only on the inputs to the circuit, but also on previous output conditions. The previous output condition is provided to the circuit by means of *feedback.*

A typical model of a sequential circuit is shown in Figure 7.32. There are two fundamental parts to a sequential circuit such as this. The first is the combinational logic that decodes the current state of the circuit as observed on the output. The second is the feedback loop and associated delay element. The delay element may be nothing more than the propagation delay of the gates in the combinational logic, or as complex as a clocked memory element (such as a flip-flop).

Synchronous Sequential Circuits

A *synchronous sequential circuit* is one that features some form of periodically clocked memory element to ensure proper synchronization of circuit outputs. Such a circuit operates in regular cycles of sufficient duration that the propagation delays within the combinational logic portion do not affect the operation of the circuit as a whole. In each cycle of operation there exists, simultaneously, information about the *current state* of the circuit and the *next state*. For this reason, these circuits are often called *state machines*.

A simple state machine is illustrated in Figure 7.33. This circuit is composed of combinational logic that decodes the previous state of the machine (as stored in the clocked memory element) and the state machine inputs to determine the

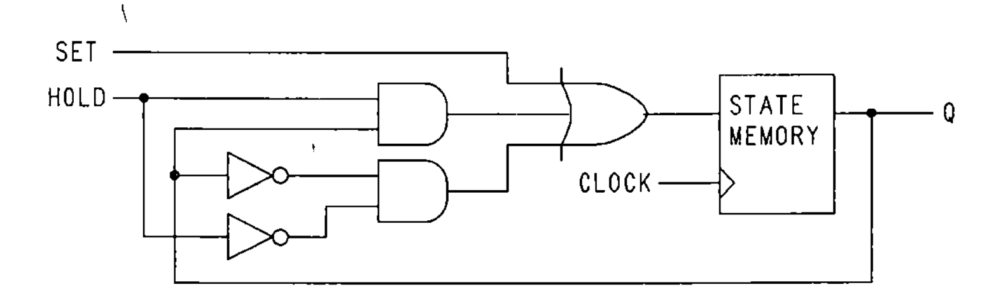

Figure 7.33 Simple state machine circuit

next state. The next state is fed to the memory element and, when the memory element is clocked, the next state becomes the current state.

This simple state machine has only one fedback signal and one corresponding memory element. Since there are only two possible values for the signal and memory element, this state machine has only two possible states. The number of states possible in a state machine is 2^n, where n is the number of memory elements, or *state bits,* in the machine.

Synchronous Memory Elements

There are a variety of different types of memory elements used in PLDs. The most common of these is the *edge-triggered D-type flip-flop* shown in Figure 7.34. As the timing diagram indicates, the D-type flip-flop stores the value applied to its *D* (data) input whenever the *Clk* input transitions from a low to a high. The amount of time required to load the *D* input, during which that input must be stable, is called the *setup and hold* time. Once a value has been loaded into the D-type flip-flop, the *Q* output of the flip-flop will remain at that value until the next rising clock transition is observed on the *Clk* input.

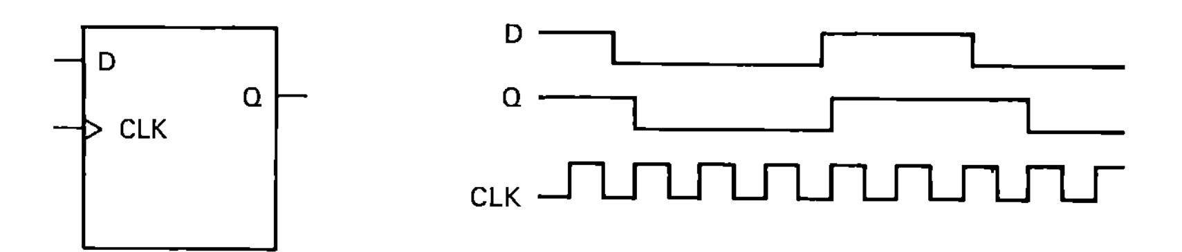

Figure 7.34 Edge-triggered D-type flip-flop

A distinguishing characteristic of a clocked D-type flip-flop is its inherent default state; when no information is presented to the flip-flop, it will always return to the low state. This fact can be used to advantage when designing state machines, with some caveats that we'll explore later.

The *clocked T-type flip-flop,* shown in Figure 7.35, differs from the D-type flip-flop in that the *Q* output of the flip-flop doesn't reflect the value observed on the flip-flop input. Instead, a true (high) input to the *T* input during the rising clock transition will cause the *Q* output to toggle (reverse its value). This behavior is useful for counter applications.

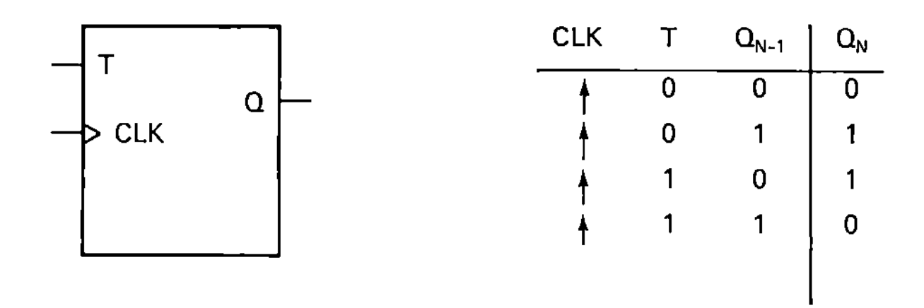

CLK	T	Q_{N-1}	Q_N
↑	0	0	0
↑	0	1	1
↑	1	0	1
↑	1	1	0

Figure 7.35 Clocked T-type flip-flop

The *clocked SR-type flip-flop* features two inputs, labeled S and R. This flip-flop is illustrated in Figure 7.36. The truth table shown in the figure defines the operation of the flip-flop. The S (set) input is used to load a true (high) value into the flip-flop during rising clock transitions, while the R (reset) is used to clear the flip-flop during rising clock transitions. If both the S and R inputs are asserted true during the rising clock transition, the result observed on the Q output is indeterminate. The SR-type flip-flop does not have an inherent default state. Unspecified circuit conditions will result in the flip-flop remaining in the previous state.

S
CLK
R
Q

CLK	S	R	Q_{N-1}	Q_N
↑	0	0	0	0
↑	0	0	1	1
↑	1	0	X	1
↑	0	1	X	0
↑	1	1	X	?

Figure 7.36 Clocked SR-type flip-flop

The *clocked JK-type flip-flop,* shown in Figure 7.37, is appearing in an increasing number of PLDs. This versatile flip-flop has two data inputs, labeled J and K. The JK-type flip-flop operates the same as an SR-type flip-flop, with the exception of the case when both inputs are true. In this case, the flip-flop toggles. This is shown in the truth table.

J
CLK
K
Q

CLK	J	K	Q_{N-1}	Q_N
↑	0	0	0	0
↑	0	0	1	1
↑	1	0	X	1
↑	0	1	X	0
↑	1	1	1	0
↑	1	1	0	1

Figure 7.37 Clocked JK-type flip-flop

The value of the JK-type flip lies in the fact that, by adding a simple interconnection or inverter gate between the J and K inputs, it can be made to function as a D-type or T-type flip-flop. This is shown in Figure 7.38.

Flip-flop Emulation

One technique that can be used to advantage when designing sequential circuits in PLDs is that of *flip-flop emulation*. Flip-flop emulation will frequently allow a

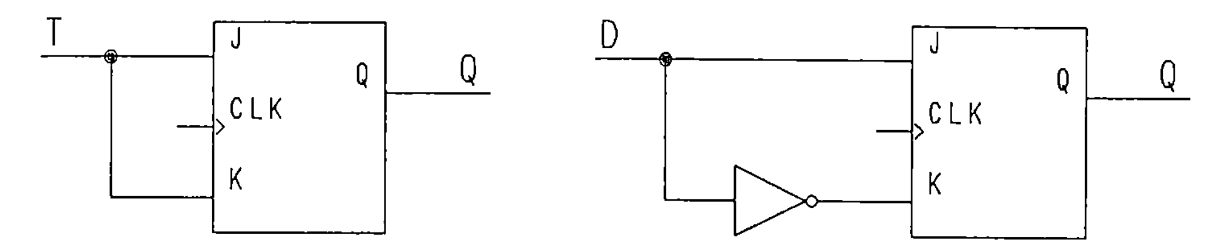

Figure 7.38 Emulating D-type and T-type flip-flops with JK-type

sequential design to be implemented in a device that features flip-flops that are not of the type required for a particular function. PLDs that feature XOR gates are particularly appropriate for these emulations.

The most common PLDs in use, the PAL-type devices, utilize D-type flip-flops as memory elements. This means that certain classes of sequential designs, most notably counters, are difficult to efficiently implement in PALs.

Quite often, the type of flip-flop that is most appropriate for a design isn't available in common PLDs. PLDs that feature configurable flip-flops (D/JK or D/T, for example) can simplify the job of designing state machines, at the expense of device complexity. Even without configurable flip-flops, the use of flip-flop emulation techniques can help simplify the determination of sequential logic for PLDs.

It can be shown that any of the flip-flop types described above can be used to emulate any of the other three flip-flops, if appropriate input forming logic and/or output feedback is provided. We have seen already how a JK-type flip-flop can be used to emulate D-type and T-type flip-flops. Another example of flip-flop emulation is shown in Figure 7.39. As shown in the figure, it is possible to emulate

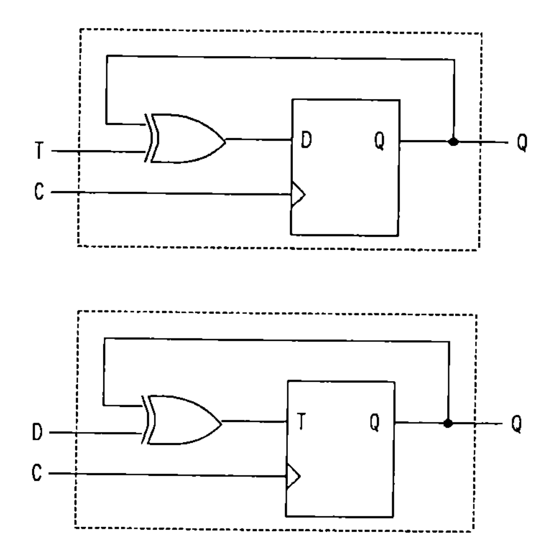

Figure 7.39 D-type and T-type flip-flop emulation using XOR gate

a clocked T-type flip-flop with a clocked D-type flip-flop (the reverse is also true), if an XOR gate is utilized. This XOR gate could be one that is implemented directly in the PLD or expanded into the sum-of-products form for implementation in non-XOR devices. Specific examples of flip-flop emulation are explored in Chapter 8.

Determining the Logic for a Finite State Machine (FSM)

A four-state machine is shown as a block diagram in Figure 7.40. We haven't shown the circuitry for this machine. Instead, we show in Figure 7.41 two representations of the machine's operation. The first, Figure 7.41(a), is a representation called a *state graph* (sometimes called a *bubble diagram*). The state graph contains circles corresponding to the four states of the machine. Each state is identified by a binary value which is the *state value* for that state. Each arrow corresponds to a *transition* from one state to the next and is annotated by the set of input conditions that will cause that transition.

Figure 7.41(b) shows another representation of the state machine—the *state table*. A state table is nothing more than a truth table which is segmented horizontally into three (or four) sections. The first two sections correspond to the current state of the machine and the state machine inputs, respectively. These two sections are the truth table inputs. The third section corresponds to the output of the truth table, and represents the next state of the machine. The fourth section (which we will utilize later) represents circuit outputs that are not state bits.

Each horizontal row of the state table represents one state transition for the

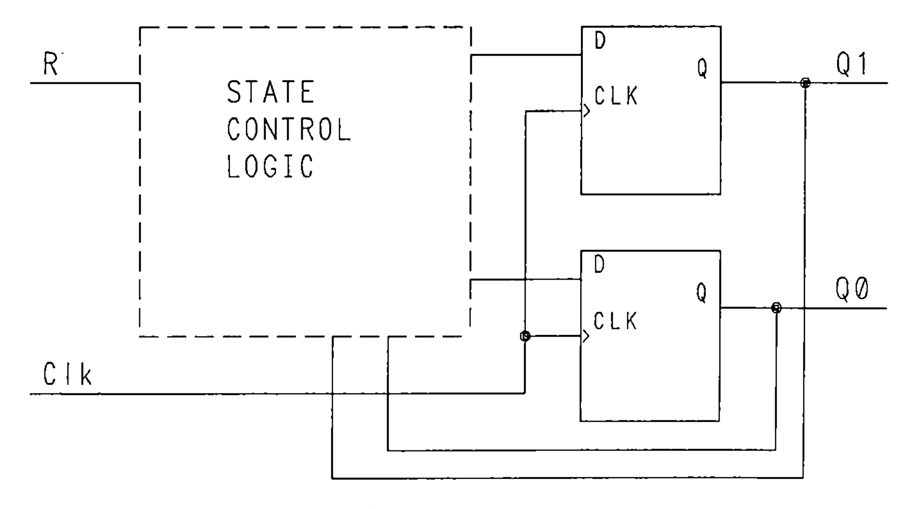

Figure 7.40 Four-state FSM block diagram

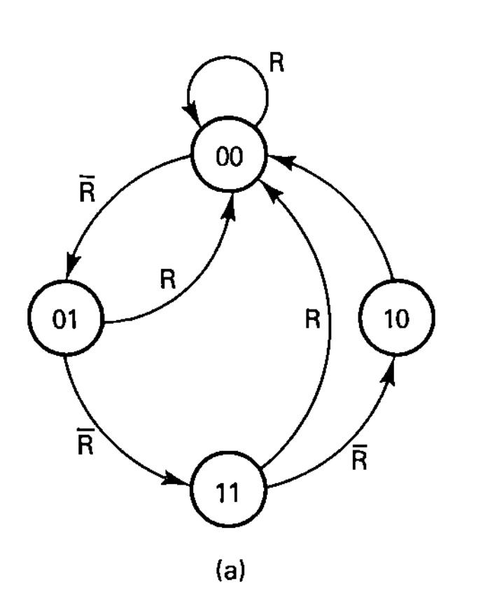

(a)

CURRENT STATE		INPUTS	NEXT STATE	
Q1	Q0	R	Q1	Q0
0	0	1	0	0
0	0	0	0	1
0	1	1	0	0
0	1	0	1	1
1	1	1	0	0
1	1	0	1	0
1	0	X	0	0

(b)

Figure 7.41(a) State graph of four-state FSM

Figure 7.41(b) State table representation

state machine. If you compare the state graph with the state table, you will see that there is one state table row entry for each transition arrow in the graph.

The state table is a convenient form for evaluating state transitions and determining transition logic. The determination of transition logic is done independently for each bit of the state register. To determine the logic for each state bit, you analyze each transition, and convert the transition data to a set of on-set and off-set equations that are appropriate for the type of flip-flop being used.

We begin by converting each row of the state table into one product term that represents the current state and state machine inputs for that row. These product terms are called the *condition product terms*. The condition product terms for the state table of Figure 7.41(b) are shown below.

$!Q1 \ \& \ !Q0 \ \& \ R$

$!Q1 \ \& \ !Q1 \ \& \ !R$

$!Q1 \ \& \ Q0 \ \& \ R$

$!Q1 \ \& \ Q0 \ \& \ !R$

$Q1 \ \& \ Q0 \ \& \ R$

$Q1 \ \& \ Q0 \ \& \ !R$

$Q1 \ \& \ !Q0$

Transition logic determination for a particular flip-flop type begins with an appropriate *flip-flop transition table* (sometimes referred to as an *excitation table*) that indicates the type of equation required for each possible transition. There is one flip-flop transition table for each type of flip-flop. Flip-flop transition tables

Transition	D Input
0 → 0	Off
0 → 1	On
1 → 0	Off
1 → 1	On

(a)

Transition	S Input	R Input
0 → 0	Off	
0 → 1	On	Off
1 → 0	Off	On
1 → 1		Off

(b)

Transition	J Input	K Input
0 → 0	Off	
0 → 1	On	
1 → 0		On
1 → 1		Off

(c)

Transition	T Input
0 → 0	Off
0 → 1	On
1 → 0	On
1 → 1	Off

(d)

Figure 7.42(a) D-type flip-flop transition table

Figure 7.42(b) SR-type flip-flop transition table

Figure 7.42(c) JK-type flip-flop transition table

Figure 7.42(d) T-type flip-flop transition table

for the D-type, SR-type, JK-type, and T-type flip-flops are shown in Figure 7.42(a) through 7.42(d).

Armed with our flip-flop transition tables, we can now generate logic for each bit of the state machine. Let's begin by mapping the condition product terms of Figure 7.43 into on-set and off-set equations for D-type flip-flops. To do this, we refer to the flip-flop transition table for D-type flip-flops (Figure 7.42(a)). This table shows us which transitions (in terms of bit changes) will require on-sets or off-sets. Beginning with $Q1$, we can see from the table that a change from state 0 to 1, or from 1 to 1, will require an on-set equation, while a change from state

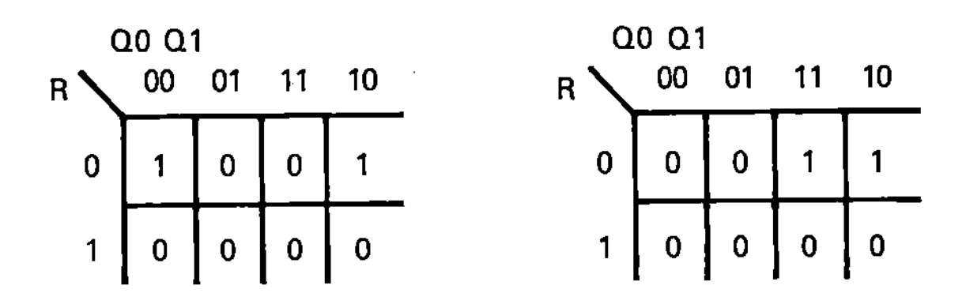

R \ Q0 Q1	00	01	11	10
0	1	0	0	1
1	0	0	0	0

R \ Q0 Q1	00	01	11	10
0	0	0	1	1
1	0	0	0	0

Figure 7.43 K-map representation of four-state FSM logic

0 to 0, or 1 to 0, will require an off-set equation. Using this information, we can now map the condition product terms for each transition into on-sets and off-sets. The resulting on-sets and off-sets can be mapped directly into a K-map (with on-sets corresponding to 1-cells, and off-sets corresponding to 0-cells) or written as equations. Both forms are shown in Figure 7-43.

The K-map can be used to minimize the logic, using the methods presented earlier. As you can see from the K-map representation, this state machine is completely specified; there are no undefined states, and all transitions are accounted for. Now let's generate the logic for a more complex state machine—one that has undefined states and unspecified transitions. A state graph, state table, and corresponding condition product terms for this state machine are shown in Figure 7.44.

To implement this state machine in JK-type flip-flops, we apply the flip-flop transition table for JK-type flip-flops to the condition product terms. Once again, we use the bit value change (from current states to next states) to map into the flip-flop transition table. The transition table for JK-type flip-flops has entries for both the J and K inputs to the flip-flops. As you can see from the table, the two inputs of a JK-type flip-flop require the same number of condition product terms as a D-type. This is because the JK-type flip-flop retains its data until cleared and can exploit don't-care conditions during toggle operations (transitions from 0 to 1, or 1 to 0).

The on-sets and off-sets for this state machine are shown in K-map form in Figure 7.45. The indicated groupings, utilizing don't-care minimizations, result in

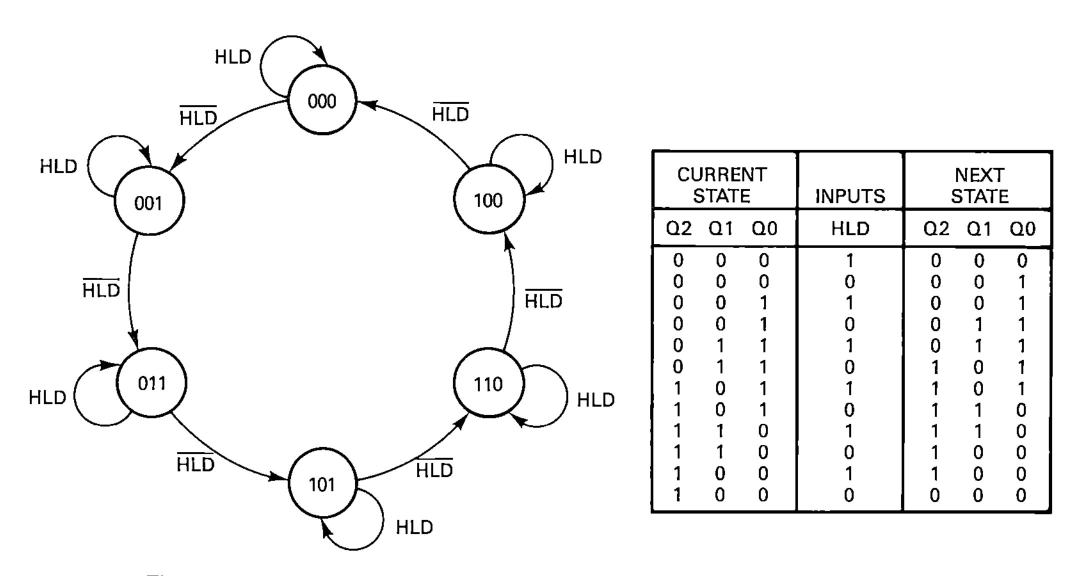

CURRENT STATE			INPUTS	NEXT STATE		
Q2	Q1	Q0	HLD	Q2	Q1	Q0
0	0	0	1	0	0	0
0	0	0	0	0	0	1
0	0	1	1	0	0	1
0	0	1	0	0	1	1
0	1	1	1	0	1	1
0	1	1	0	1	0	1
1	0	1	1	1	0	1
1	0	1	0	1	1	0
1	1	0	1	1	1	0
1	1	0	0	1	0	0
1	0	0	1	1	0	0
1	0	0	0	0	0	0

Figure 7.44 State table for incompletely specified FSM

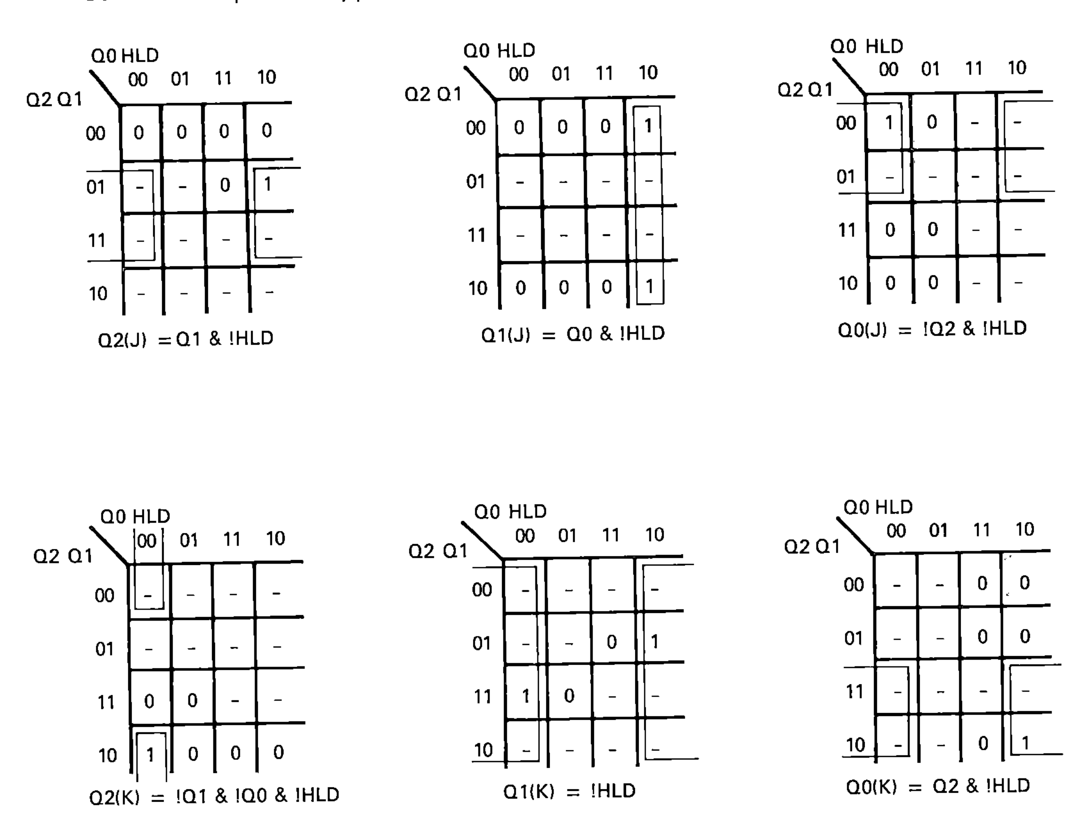

Figure 7.45 K-maps for J and K inputs to JK-type flip-flops

the following optimized equations for the J and K inputs to the state register flip-flops:

$$Q2(J) = Q1 \ \& \ !HLD$$

$$Q2(K) = !Q1 \ \& \ !Q0 \ \& \ !HLD$$

$$Q1(J) = Q0 \ \& \ !HLD$$

$$Q1(K) = Q1 \ \& \ !HLD$$

$$Q0(J) = !Q2 \ \& \ !HLD$$

$$Q0(K) = Q2 \ \& \ !HLD.$$

Mealy and Moore Machines

The previous state machines are examples of *Moore model* state machines. A Moore model machine is one in which the outputs of the machine are functions

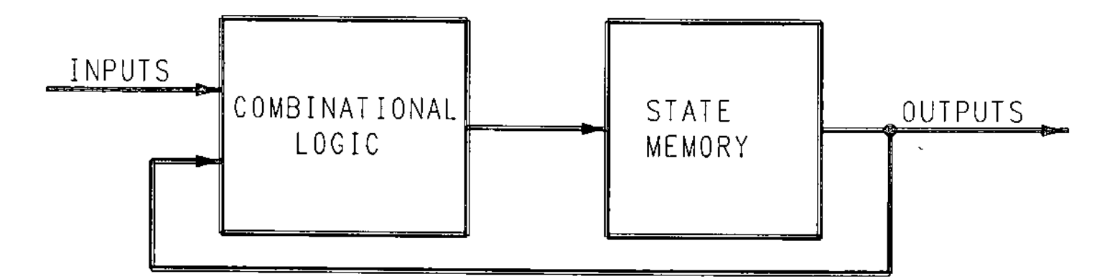

Figure 7.46 Moore model state machine

only of the current state of the machine. The typical example of a Moore model machine is a synchronous counter circuit.

In a *Mealy model* machine, the outputs of the machine are determined not only by the current state of the machine, but also by decoding of state machine inputs. In a Mealy model machine, some or all of the state machine outputs will change state asynchronously as the inputs to the machine change. In a Moore machine, all outputs are synchronized with the state registers. These two types of state machines are diagrammed in Figures 7.46 and 7.47.

The model of machine you are constructing can have an impact on the types of PLDs in which the design can be implemented. For example, if you are designing a Mealy model state machine, you probably aren't concerned with the actual state values used for each state of the machine and may not even require that the state bits be accessible outside of the PLD.

Register Limitations in PLDs

The major limitation when designing large state machines in simple PLDs is the limited number of registered outputs. This limitation forces the designer to carefully construct circuits such as state machines to minimize the number of memory elements required. It may be necessary to combine functions (state memory bits and outputs, for example) to fit the design into the constraints of the PLD architecture.

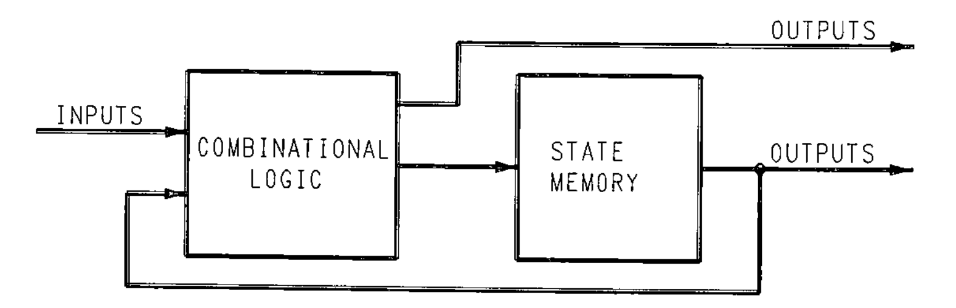

Figure 7.47 Mealy model state machine

Minimization of Finite State Machines

The amount of circuitry required to implement a state machine can be greatly reduced by the careful choice of bit values for the machine's states. How do you determine the optimal state assignment? In general, you should number the states for as few bit changes as possible for the transitions defined. If you are using D-type flip-flops, you should also attempt to minimize the number of transitions to states with many values of 1 in the state registers (state 011 costs more than state 100, in terms of transition logic required). This is not, however, always the optimal assignment, particularly for PLD implementations. In a PLD, it is often necessary to optimize a single bit of a state machine at the expense of other state bits if the logic for that particular bit is too large for the device.

Asynchronous Circuits

When no globally synchronized memory element is used in a sequential circuit, the circuit is said to be *asynchronous*. Asynchronous circuits must be very carefully designed to avoid conditions in which the circuit oscillates uncontrollably, as in the circuit of Figure 7.48.

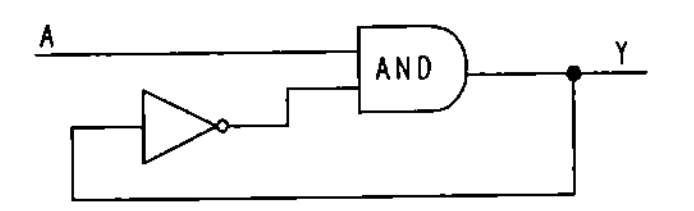

Figure 7.48 Oscillating asynchronous circuit

This circuit will oscillate when the input A is asserted true. In addition to problems of unpredictable oscillations, asynchronous circuits are also frequent victims of circuit hazards which are dealt with in detail in Chapter 10.

PLDs can be used for asynchronous state machine applications, but you must be careful to avoid hazard and testability problems. The cardinal rule to observe for asynchronous circuits in PLDs is this: an asynchronous circuit implemented in a PLD must always stabilize to a known value with no possibility of critical race conditions. You should never design a circuit in a PLD that depends on critical timing paths to stabilize properly.

7.10 SUMMARY

In this chapter, we have reviewed many digital logic fundamentals, and applied them manually to the problems of circuit design and optimization for PLDs. A complete discourse on digital logic and logic minimization techniques is beyond the scope of this book. The reader is referred to the many books and papers on the subject, some of which are listed below.

7.11 REFERENCES

Brayton, Robert K., Hachtel, Gary D., McMullen, Curtis T., Sangiovanni-Vincentelli, Alberto L., *Logic Minimization Algorithms for VLSI Synthesis,* Kluwer Academic Publishers, Hingham, MA, 1984.

Breeding, Kenneth J. *Digital Design Fundamentals,* Prentice Hall, Englewood Cliffs, NJ, 1989.

Chirlian, Paul M. *Digital Circuits with Microprocessor Applications,* Matrix Publishers, Beaverton, OR, 1982.

Hollaar, Lee A. "Direct Implementation of Asynchronous Control Units." *IEEE Transactions on Computers* (December 1982): 1133–41.

Mano, M. Morris. *Digital Design,* Prentice Hall, Englewood Cliffs, NJ, 1984.

Unger, Steven J. *The Essence of Logic Circuits,* Prentice Hall, Englewood Cliffs, NJ, 1989.

8

High-level Design Techniques

In this chapter, we'll examine in detail how a variety of common circuits can be implemented in PLDs using high-level design techniques. The basic circuits presented here are commonly used in much larger designs and help to illustrate a number of important design considerations.

As we have pointed out in previous chapters, PLDs are devices with constrained architectures. This leads to limitations in the size and configuration of circuits that are implemented in them. While automated design tools can help in the conversion of design concepts into working circuits, the designer must still understand the limitations of the target technology. By showing how some relatively simple circuits can be implemented in PLDs, we'll be able to demonstrate some of the many design tradeoffs that are required when using PLDs for designs of all sizes.

8.1 DESIGN METHODS

When a high-level concept needs to be implemented as an actual circuit, the designer must choose a design method (or combination of methods) that will allow that circuit to be created in a reasonable amount of time and that will allow future design changes with a minimum of effort. In Chapter 5, we described a number of possible design methods ranging from structural entry to various forms of behavioral description. PLD design tools generally favor the use of behavioral description methods and it's those forms of design entry that we will concentrate on here.

When using behavioral description methods, there are many possible formats to choose from, even if you are using only a single design tool. The choice of description format is based on many factors, including the type of circuit to be created, the designer's experience, and the availability of design tools that support that method. The design examples in this chapter utilize the ABEL-HDL design language, but most of the concepts covered are readily transferable to other design tools.

8.2 THE ABEL HARDWARE DESCRIPTION LANGUAGE

Describing designs with the ABEL *hardware description language* (HDL) is fairly straightforward. The syntax of the language was designed to be flexible and device independent in order to make the translation of design concepts into real circuits as natural as possible. Hardware programming in ABEL-HDL is similar in many respects to software programming in languages such as Pascal or C. ABEL designs are entered with a text editor and are then compiled into an internal form that may be merged with other design elements, optimized, and finally converted into a form that can be implemented as a PLD-based circuit. Completed PLD designs can be simulated by the ABEL simulator, which models the target PLD for design verification purposes.

HDLs such as ABEL-HDL differ from software programming languages in that they are used to describe functions that are inherently parallel. All statements in an ABEL-HDL design may be thought of as being executed at the same time. This is particularly important to realize when describing sequential circuits; the sequential operation of a circuit is never a function of the order in which ABEL-HDL language statements describing that circuit are entered.

ABEL-HDL, like many PLD design languages, provides different textual entry formats which may be combined as needed to meet the specific requirements of the design. The methods available in ABEL-HDL are: high-level equations, truth tables, and state descriptions.

To demonstrate how these design description methods can be used, we'll present a number of simple examples. Many of these examples demonstrate design concepts that are applicable to larger designs, as well as demonstrating the use of high-level design description methods.

8.3 USING HIGH-LEVEL EQUATIONS

Before launching into more interesting circuits, we'll first describe how high-level equations are used to describe designs. In ABEL-HDL, the complex expressions available for high-level equations are also available for use other design description methods (such as state descriptions). For this reason it's important to understand how high-level equations relate to lower-level Boolean equations.

First, consider the design of a simple twelve-input, four-output multiplexer. A digital multiplexer such as this is used to select one or more inputs from a larger set of inputs, and route these signals to a corresponding number of outputs. The selection of inputs is made by providing information to the multiplexer's data selection inputs. At the least, a multiplexer requires n data selection inputs in order to select between 2^n possible signal routings.

Our simple multiplexer, shown in Figure 8.1, selects one of three sets of four inputs ($a0$-$a3$, $b0$-$b3$, $c0$-$c3$) and routes the signals to the outputs ($y0$-$y3$) as

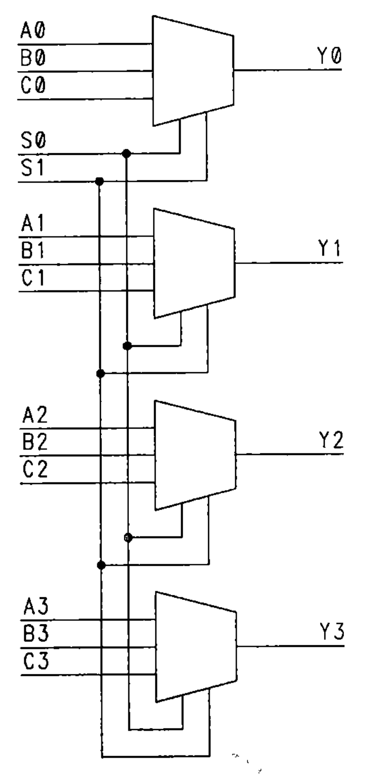

Figure 8.1 Block diagram of twelve to four multiplexer

indicated by the values appearing on the select input lines ($s0$ and $s1$). The possible values for $s1$ and $s0$ are shown below, along with the resulting signal routings:

$$
\begin{array}{ll}
s1 = \text{low},\ s0 = \text{high} & y3 - y0 = a3 - a0 \\
s1 = \text{high},\ s0 = \text{low} & y3 - y0 = b3 - b0 \\
s1 = \text{high},\ s0 = \text{high} & y3 - y0 = c3 - c0 \\
s1 = \text{low},\ s0 = \text{low} & y3 - y0 = \text{all low.}
\end{array}
$$

To simplify the design of the multiplexer, we have grouped the various inputs and outputs into sets, as indicated in the block diagram. To describe this design, Boolean equations are utilized in the ABEL description file that is shown in Figure 8.2. The file begins with a design name and description, followed by signal and set declarations.

In the design file, all of the input and outputs are declared (in this design the outputs are active high; we'll discuss later how to specify active low signals). The declared signals are grouped into sets through the use of constant set declarations. This simplifies the description of the circuit. The device declaration, pin declarations and constant set declarations comprise the declarations section of the ABEL design description. The device declaration is optional.

Following the declarations section, the equations section appears. This section contains the actual description of the multiplexer circuit in the form of a high-level equation. As you can see, the equation is simple and easy to understand. The relational operator == ("is equal to") is used to provide a comparator function for the select input set. Each line of the equation corresponds to one of the possible multiplexer selections.

When processed by the ABEL language processor, the multiplexer equation is converted into four separate sum-of-products Boolean equations. These equations can then be mapped into the product terms of the selected PLD directly, or can be further processed by logic minimization routines to reduce the number of product terms used (if required to fit into the PLD) and to improve testability.

For a better understanding of design techniques to be described later, it's important to understand how a high-level equation such as this is converted into sum-of-products Boolean equations. We'll show how ABEL does this by going through the conversion process manually for this simple multiplexer equation.

First, the declarations for *A, B, C, Select,* and *Y* are substituted into the equation to form

$$
\begin{aligned}
[y3, y2, y1, y0] = {} & ([s1, s0] == 1)\ \&\ [a3, a2, a1, a0] \\
& \#\ ([s1, s0] == 2)\ \&\ [b3, b2, b1, b0] \\
& \#\ ([s1, s0] == 3)\ \&\ [c3, c2, c1, c0];
\end{aligned}
$$

Next, the numeric constants are expanded to match the set widths of the

```
module MUX12T4
title '12 to 4 multiplexer
Michael Holley and Dave Pellerin'

        mux12t4          device   'P16V8S';

        a3,a2,a1,a0      pin      1,2,3,4;
        b3,b2,b1,b0      pin      5,6,7,8;
        c3,c2,c1,c0      pin      9,11,12,13;
        y3,y2,y1,y0      pin      14,15,16,17;
        s1,s0            pin      18,19;

        select  = [s1..s0];
        Y       = [y3..y0];
        A       = [a3..a0];
        B       = [b3..b0];
        C       = [c3..c0];

equations
        Y = (select == 1) & A
          # (select == 2) & B
          # (select == 3) & C;


declarations

        X,H,L   = .x.,!0,0;

test_vectors ([select, A, B, C] -> Y)
               [  1   , 1, X, X] -> 1; "select = 1, lines A to output
               [  1   ,10, H, L] -> 10;
               [  1   , 5, H, L] -> 5;

               [  2   , H, 3, H] -> 3; "select = 2, lines B to output
               [  2   ,10, 7, H] -> 7;
               [  2   , L,15, L] -> 15;

               [  3   , L, L, 8] -> 8; "select = 3, lines C to output
               [  3   , H, H, 9] -> 9;
               [  3   , L, L, 1] -> 1;

end
```

Figure 8.2 Twelve to four multiplexer design file

expressions in which they are found. For this equation, the constant expansion results in the equation

$$\begin{aligned}[y3, y2, y1, y0] = ([s1, s0] &== [0, 1]) \& [a3, a2, a1, a0] \\ \# ([s1, s0] &== [1, 0]) \& [b3, b2, b1, b0] \\ \# ([s1, s0] &== [1, 1]) \& [c3, c2, c1, c0];\end{aligned}$$

Now that the equation has been normalized in terms of set widths, the relational operators can be converted to Boolean operators. The rule for conversion of the '= =' operator results in the equation

```
[y3, y2, y1, y0] = (!s1 & s0) & [a3, a2, a1, a0]
                 # (s1 & !s0) & [b3, b2, b1, b0]
                 # (s1 & s0) & [c3, c2, c1, c0];
```

The next phase of the conversion is distribution of the AND operator into the set operands. This conversion results in

```
[y3, y2, y1, y0] = [a3 & !s1 & s0,
                    a2 & !s1 & s0,
                    a1 & !s1 & s0,
                    a0 & !s1 & s0]
                 # [b3 & s1 & !s0,
                    b2 & s1 & !s0,
                    b1 & s1 & !s0,
                    b0 & s1 & !s0]
                 # [c3 & s1 & s0,
                    c2 & s1 & s0,
                    c1 & s1 & s0,
                    c0 & s1 & s0];
```

The equation is now in a form that can be separated easily into individual equations for each of the four outputs:

```
y3 = a3 & !s1 & s0
   # b3 & s1 & !s0
   # c3 & s1 & s0;
y2 = a2 & !s1 & s0
   # b2 & s1 & !s0
   # c2 & s1 & s0;
y1 = a1 & !s1 & s0
   # b1 & s1 & !s0
   # c1 & s1 & s0;
```

$$y0 = a0 \ \& \ !s1 \ \& \ s0$$
$$\# \ b0 \ \& \ s1 \ \& \ !s0$$
$$\# \ c0 \ \& \ s1 \ \& \ s0;$$

The equations, now in a sum-of-products form, can be mapped directly into

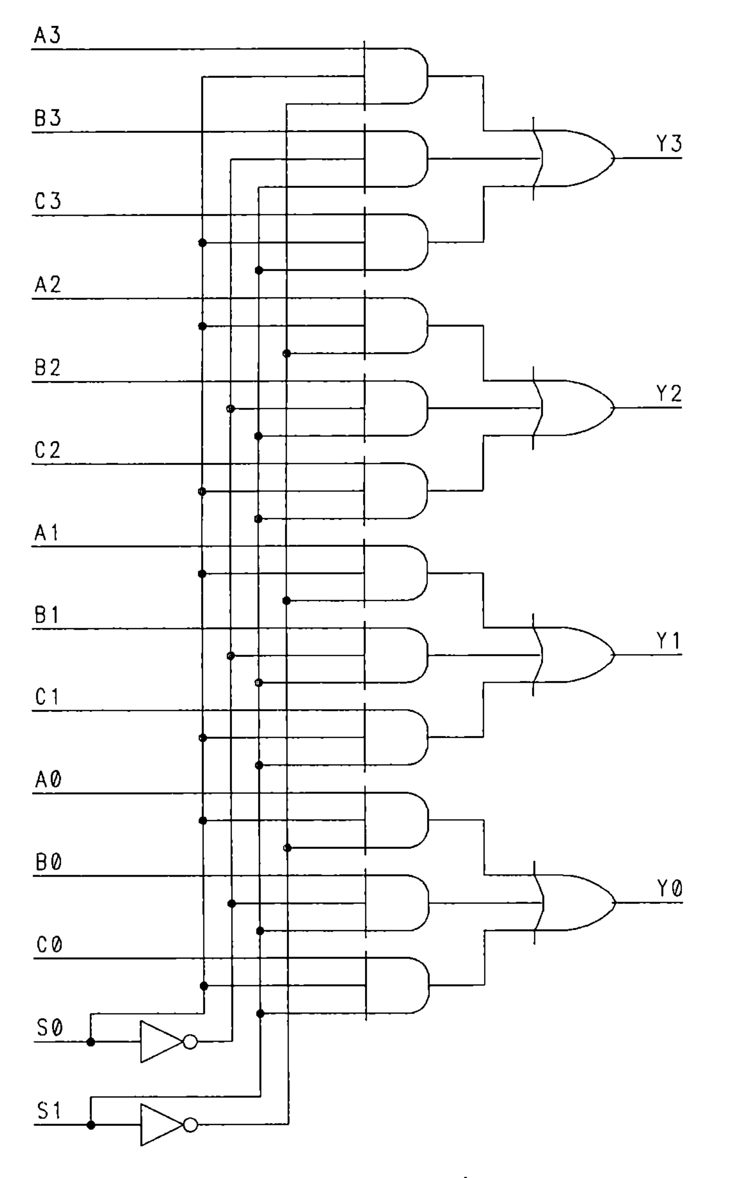

Figure 8.3 Multiplexer schematic representation

a PLD. If necessary, the Boolean equations can be processed by logic minimization modules. For this design, further minimization is unnecessary.

As you can see, a number of complex operations are performed to convert a high-level equation into sum-of-products equations appropriate for the architecture of a PLD. For comparison purposes, the multiplexer design is shown in schematic form in Figure 8.3.

8.4 DESCRIBING ADDRESS DECODERS WITH EQUATIONS

Address decoders are perhaps the most common application for PLDs. A decoder is a circuit that translates n binary inputs into one of up to 2^n outputs (code converter circuits that translate binary input data into a pattern of multiple outputs are also sometimes referred to as decoders). If the decoding function results in the same output for multiple input combinations (either by design or due to don't-care input conditions) then the number of outputs required will be less than 2^n.

An address decoder is a specialized decoder that performs a magnitude comparison function to determine the proper decoding of its inputs into a relatively small number of outputs. The typical use for such a circuit is in enabling different sections of memory, based on the memory address observed on an address bus.

The ABEL design shown in Figure 8.4 describes an address decoder that monitors the most significant seven bits of a 16-bit address bus and enables one of four different blocks of memory based on that value. Written in the ABEL language, this design is simple enough that little explanation is required. There are some important points that can be made about this design, however, that have application for more complex circuits.

First, notice that the declarations for the outputs (*SRAM, PORT, UART,* and *PROM*) are made with an active-low indication (the ! operator applied to the signal name). The active-low declarations specify that these signals will be low under the conditions indicated by the design equations, rather than high. The reason for this is that the memory chips and I/O circuitry that are being controlled with this decoder have active-low enable inputs. We have used a 16L8 PAL device for the design, as specified in the declarations section of the design file. It would also be possible to use a 16H8 PAL (or a simple PROM, for that matter) since the minimized Boolean equations that result from the description will fit into devices that have inverting or non-inverting outputs. Although this active-low design uses fewer product terms in a PLD with inverting outputs, it's important to realize that an active-low circuit doesn't necessarily require a PLD with inverting outputs. We'll discuss why a little later.

Another thing of interest about this ABEL description is the use of a 16-bit set declaration (the seven address line inputs padded to 16 bits with no-connects). The use of a 16-bit set allows the decoder to be specified in terms of actual memory address boundaries. Using no-connects in this way can vastly simplify the description of more complex circuits. You must be careful when doing this, however.

```
module DECODE
title 'Address Decoder Example'

        decode device 'P16L8';

        !SRAM,!PORT,!UART,!PROM         pin 14,15,16,17;
        A15,A14,A13,A12,A11,A10,A9      pin 3,4,5,6,7,8,9;

        H,L,X,Z,C = 1,0,.X.,.Z.,.C.;

        Address = [A15..A9,X, X,X,X,X, X,X,X,X];

Equations
        SRAM = (Address <  ^h8000);
        PORT = (Address >= ^h8000) & (Address <= ^h81FF);
        UART = (Address >= ^h8200) & (Address <= ^h83FF);
        PROM = (Address >= ^h8400);

Test_Vectors
       (Address -> [!SRAM,!PORT,!UART,!PROM])
        ^h0000  -> [  L ,   H ,   H ,  H  ];
        ^h1000  -> [  L ,   H ,   H ,  H  ];
        ^h4000  -> [  L ,   H ,   H ,  H  ];
        ^h7000  -> [  L ,   H ,   H ,  H  ];
        ^h7FFF  -> [  L ,   H ,   H ,  H  ];
        ^h8000  -> [  H ,   L ,   H ,  H  ];
        ^h81FF  -> [  H ,   L ,   H ,  H  ];
        ^h8200  -> [  H ,   H ,   L ,  H  ];
        ^h83FF  -> [  H ,   H ,   L ,  H  ];
        ^h8400  -> [  H ,   H ,   H ,  L  ];
        ^hC000  -> [  H ,   H ,   H ,  L  ];
        ^hFFFF  -> [  H ,   H ,   H ,  L  ];
end
```

Figure 8.4 Address decoder design file

In particular, you should exercise care when using no-connects in the high-order bits of an address word. Notice also that test vectors have been provided in the source document and are written using the same high-level set notation as was used in the description of the circuit. These test vectors are used by the ABEL simulator to verify that the design functions as we intend it to.

8.5 COMPARATOR CIRCUITS

The address decoder described in Figure 8.4 utilizes a comparator function to decode the address inputs. Comparators such as this are used in many complex logic designs, so it's important to fully understand how they are constructed. Comparators come in two basic flavors—identity comparators and magnitude comparators.

Identity Comparators

An identity comparator is a circuit that compares the values observed on a set of inputs against either a fixed pattern or another set of inputs. When the comparison is made against a fixed set of values, the circuit is a decoder and is a simple AND operation, as shown in Figure 8.5.

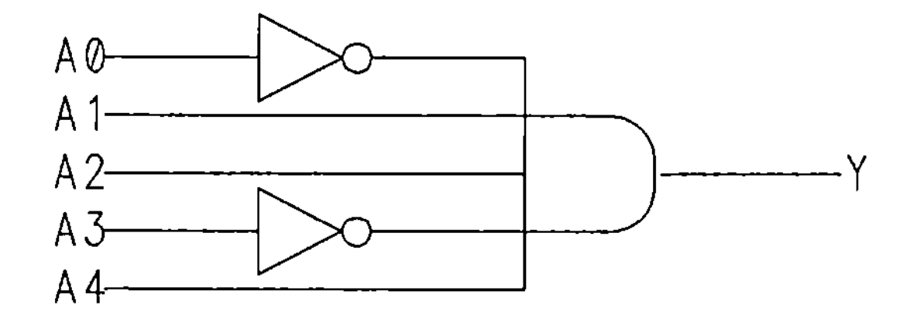

Figure 8.5 Fixed identity comparator circuit

When two sets of inputs are compared, the logic is somewhat more complex, but is easily derived if you consider that such an "equal to" comparison function is equivalent to an XNOR operation between each bit of the input sets and, correspondingly, a "not equal to" function is equivalent to an XOR. This is shown in Figure 8.6.

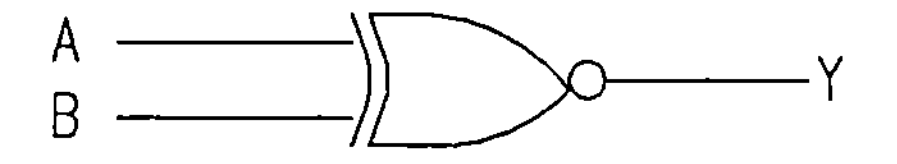

Figure 8.6 Equality comparison of two input values

An important fact about equality (or inequality) comparators is that the amount of circuitry required to implement the comparator increases in a linear fashion with an increase in the size of the input sets, if XNOR (or XOR) gates are utilized. If the comparator is implemented in sum-of-products, however, the increase will be exponential. Figure 8.7 shows a circuit that implements an 8-bit equality comparator using XNOR gates.

Figure 8.7 8-bit equality comparator circuit

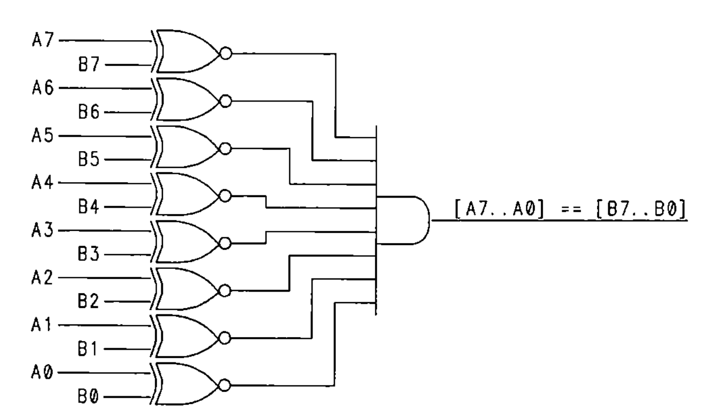

```
module EQUAL8
title '8-Bit Identity Comparator'

        equal8  device  'P22V10';

        A7,A6,A5,A4,A3,A2,A1,A0 pin 2 ,3 ,4 ,5 ,6 ,7 ,8 ,9;
        B7,B6,B5,B4,B3,B2,B1,B0 pin 10,11,13,14,15,21,22,23;

        A_EQ_B  pin 18;

        A = [A7..A0];
        B = [B7..B0];

Equations
        A_EQ_B  = ([A7..A0] == [B7..B0]);

test_vectors
       ([  A   , B ] -> A_EQ_B)
        [^h00,^h00] ->    1  ;
        [^h0F,^h00] ->    0  ;
        [^h10,^h0F] ->    0  ;
        [^h11,^h10] ->    0  ;
        [^h0F,^h10] ->    0  ;
        [^h10,^h11] ->    0  ;
        [^hFF,^hFF] ->    1  ;
end
```

Figure 8.8 8-bit equality comparator design file

The eight XNOR gates shown in the circuit each provide a comparison function for one pair of bits for the two input sets. The ABEL design shown in Figure 8.8 implements this same circuit using a high-level equation. The == operator used in the equation translates into the same circuitry as shown in Figure 8.7.

To implement this design in the PAL-type device indicated in the design file, ABEL will have to flatten the == operator into sum-of-products Boolean equations. The size of the resulting Boolean equations (shown in Figure 8.9) are too large for one output of a simple PLD such as the 16L8, so a more complex PLD

```
!A_EQ_B =  A0 & !B0
        # !A0 &  B0
        #  A1 & !B1
        # !A1 &  B1
        #  A2 & !B2
        # !A2 &  B2
        #  A3 & !B3
        # !A3 &  B3
        #  A4 & !B4
        # !A4 &  B4
        #  A5 & !B5
        # !A5 &  B5
        #  A6 & !B6
        # !A6 &  B6
        #  A7 & !B7
        # !A7 &  B7;
```

Figure 8.9 Minimized Boolean equations for 8-bit equality comparator

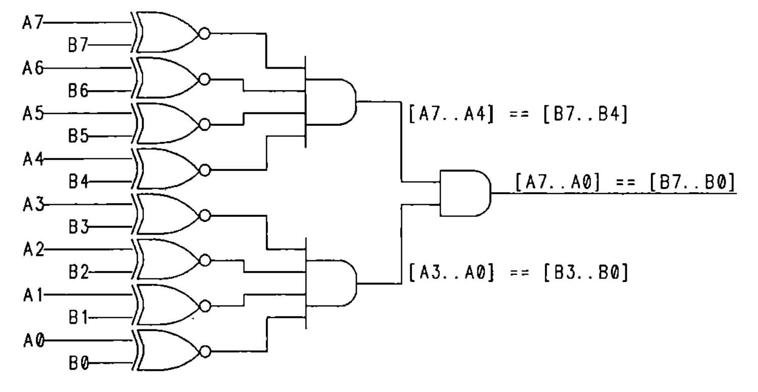

Figure 8.10 8-bit comparator implemented as two 4-bit comparators

is used. An equality comparator circuit that compares two input sets will always require at least $2n$ product terms, where n is the number of bits in each input set. This assumes that the circuit is being implemented in a PLD with either negative or programmable output polarity. If the circuit is being implemented in a positive polarity device (one with no output inversion) the number of product terms required is 2^n. This is a general rule when designing identity comparators: equality comparators are most efficiently implemented with negative polarity, while inequality comparators are better implemented with positive polarity. The 22V10 device selected has sixteen product terms available on one of its output pins, as well as programmable output polarity and can, therefore, implement the comparator function.

In order to make this circuit fit into the limited resources of a simpler device, we need to make some changes. The simplest way to make the design fit is to utilize additional PLD outputs and implement the design in multilevel logic. Figure 8.10 shows the 8-bit identity comparator implemented as two 4-bit comparators ANDed together. The ABEL design is shown in Figure 8.11.

The two 4-bit comparators are assigned to outputs named *TEMP1* and *TEMP2*. These outputs are then fedback and ANDed together to achieve the desired function for *A_EQ_B*.

Magnitude Comparators

When designing a *magnitude comparator* (such as an address decoder) that compares a set of inputs against one or more fixed values to determine a relative result (greater than or less than), the amount of circuitry required is less obvious. This is because the binary pattern of values to be compared determines the amount of circuitry. If you need to estimate the size of such a comparator, simply count the number of ones and zeroes in the binary form of the fixed value; at most, each one will require one product term if implemented with positive polarity, while the negative polarity form may require a product term for each zero. The maximum

```
module EQUAL8A
title '8-Bit Identity Comparator'

        equal8a device  'P22V10';

        A7,A6,A5,A4,A3,A2,A1,A0 pin  1, 2, 3, 4, 5, 6, 7, 8;
        B7,B6,B5,B4,B3,B2,B1,B0 pin  9,10,11,13,14,15,22,23;

        A_EQ_B  pin 18;
        TEMP1   pin 20;
        TEMP2   pin 21;

        A = [A7..A0];
        B = [B7..B0];

Equations
        TEMP1    = ([A7..A4] == [B7..B4]);
        TEMP2    = ([A3..A0] == [B3..B0]);

        A_EQ_B   = TEMP1 & TEMP2;

test_vectors
       ([  A    , B ] -> A_EQ_B)
        [^h00,^h00] ->    1  ;
        [^h0F,^h00] ->    0  ;
        [^h10,^h0F] ->    0  ;
        [^h11,^h10] ->    0  ;
        [^h0F,^h10] ->    0  ;
        [^h10,^h11] ->    0  ;
        [^hFF,^hFF] ->    1  ;
end
```

Figure 8.11 Multilevel implementation of comparator using ABEL

number of product terms required when you have a choice of output polarity is, therefore, $n/2$, where n is the number of bits in the input set.

Even more troublesome (although more predictable) are comparator circuits that compare two arbitrary sets of inputs to determine which is greater. A magnitude comparator such as this, if implemented in two-level logic, will always require at least 2^n product terms to implement where n is the number of bits in each input set. Consider, for example, the following ABEL equation:

$$A\_GT\_B = [a7..a0] > [b7..b0];$$

This harmless looking equation will generate 255 product terms (383 product terms if implemented in a PLD with inverting outputs) and couldn't possibly be implemented in a simple PLD. With an understanding of how comparators work, we can determine a more efficient solution.

A comparator circuit of this sort is quite regular in its construction, and this fact allows us to design a circuit that operates in a more procedural manner. The algorithm for comparing two sets of inputs is straightforward. We'll show how

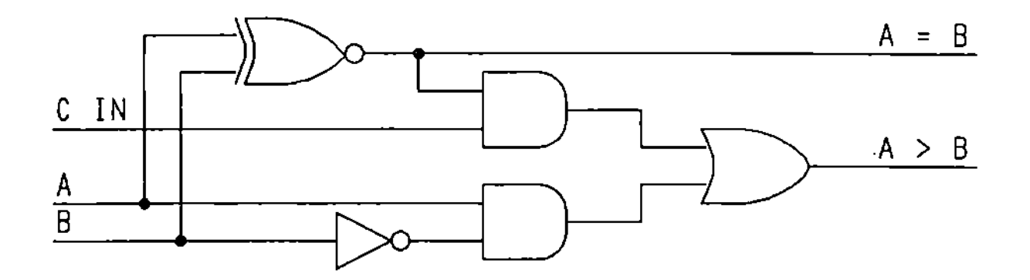

Figure 8.12 Single-bit magnitude comparator with carry-in

the algorithm works by demonstrating how it's used to compare two sets of four inputs each. First, the bit positions of each number are labeled as follows:

$$A3\ A2\ A1\ A0$$

$$B3\ B2\ B1\ B0.$$

To compare these two inputs sets, we scan the two sets, beginning with the least significant pair of bits (*A*0 and *B*0) and perform a comparison operation on each bit pair in sequence. For each bit pair, if the *An* bit is greater than the *Bn* bit, then we know that the value of set *A* is greater than the value of set *B* up to that point. If the two bits are equal, then previous bit pair comparisons must be used to determine the results. The algorithm for each bit can be expressed in the form of the single-bit comparator circuit with carry-in shown in Figure 8.12. The information about the previous bit in a multiple-bit comparison is provided to the comparator circuit through the *C_IN* signal.

Any number of these single-bit comparators can be cascaded to create a multiple-bit ripple comparator. Figure 8.13 shows an 8-bit magnitude comparator circuit implemented in this way. This circuit could be implemented in a PLD, if a device was used that had sufficient input and output pins available or had the capability for multilevel logic. (This representation would be ideal for implementation in an LCA device, for example, since an LCA is composed of many internal logic modules of limited size.)

It's also possible to construct an 8-bit magnitude comparator out of two cascading 4-bit magnitude comparators. With a properly designed 4-bit comparator, it's possible to cascade as many comparators as needed to compare large input sets. A 4-bit comparator circuit with carry-in that provides this capability is shown in Figure 8.14.

Like the cascaded 1-bit comparators, each succeeding comparator stage will add a fixed amount of delay time to the comparator circuit as a whole. The block diagram in Figure 8.15 shows two 4-bit magnitude comparators cascaded to create an 8-bit magnitude comparator. This circuit will operate significantly faster than the ripple comparator shown in Figure 8.13, and is more appropriately sized for implementation in PLDs. The corresponding ABEL design file is shown in Figure 8.16.

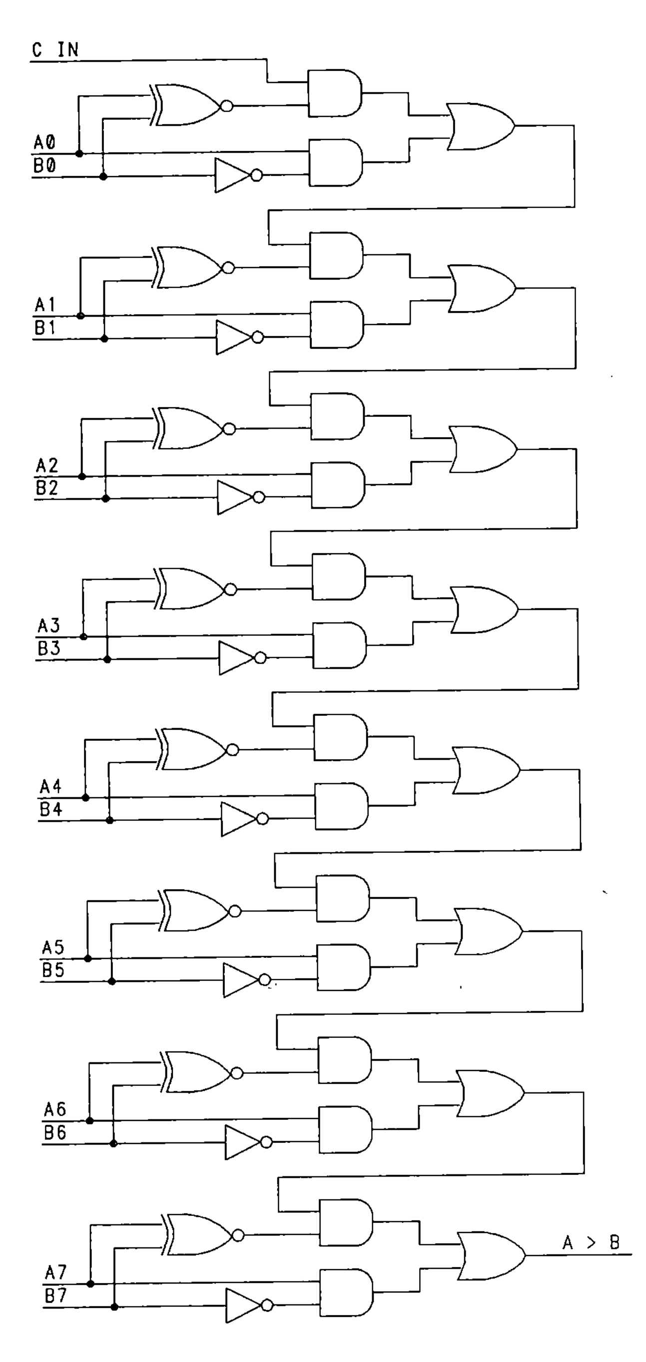

Figure 8.13 8-bit ripple magnitude comparator circuit

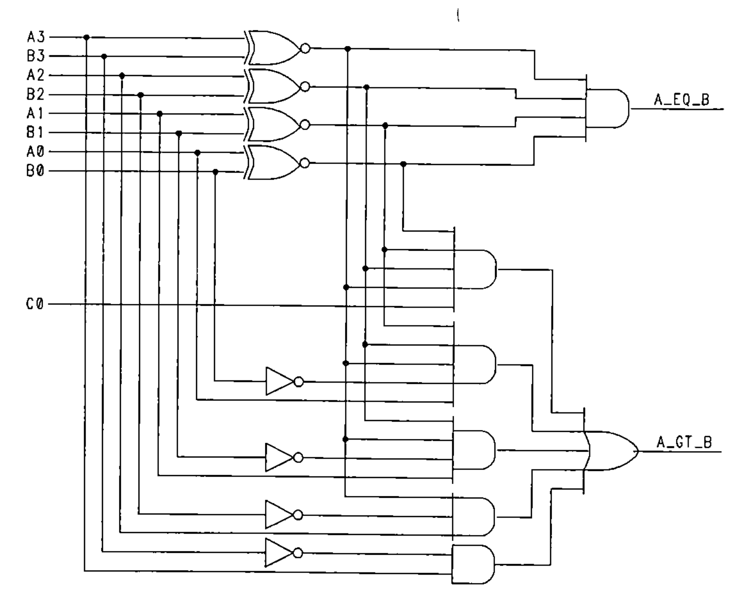

Figure 8.14 4-bit magnitude comparator with carry-in

Figure 8.15 Cascading two four-bit magnitude comparators

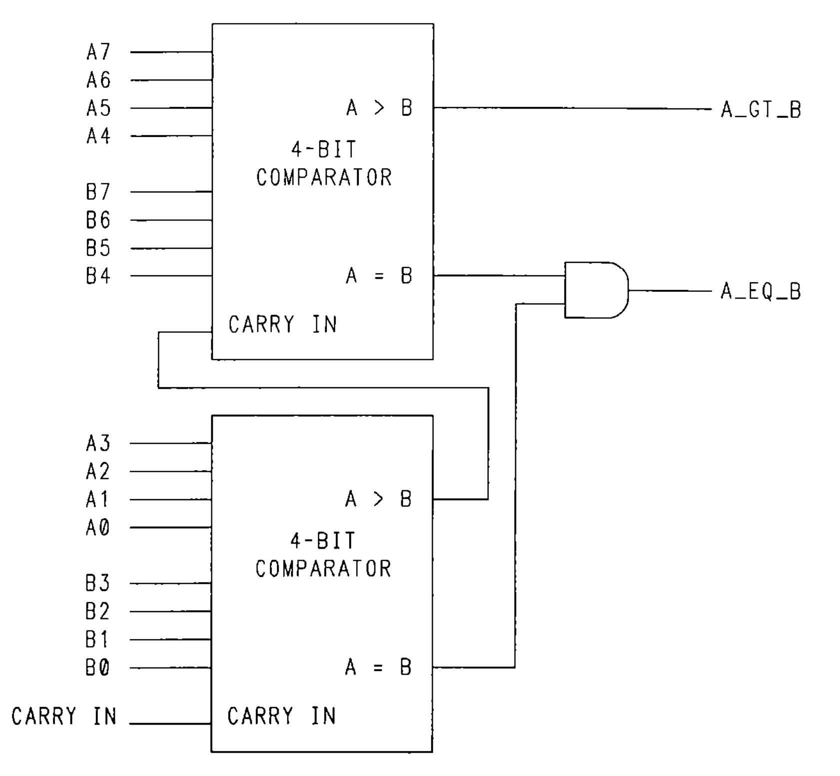

```
module COMP8A
title '8-Bit Magnitude and Identity Comparator'

        comp8a  device  'P22V10';

        A7,A6,A5,A4,A3,A2,A1,A0 pin 1 ,2 ,3 ,4 ,5 ,6 ,7 ,8;
        B7,B6,B5,B4,B3,B2,B1,B0 pin 9 ,10,11,13,14,15,22,23;

        A_GT_B  pin 19;
        GT_0    pin 17;
        GT_1    pin 18;
        GT_2    = A_GT_B;

        A_EQ_B  pin 16;
        EQ_1    pin 20;
        EQ_2    pin 21;

        A = [A7..A0];
        B = [B7..B0];

Equations
        A_EQ_B  = EQ_2 & EQ_1;

        EQ_2    = ([A7..A4] == [B7..B4]);
        EQ_1    = ([A3..A0] == [B3..B0]);

        GT_2    = ([A7..A4] >  [B7..B4]) # EQ_2 & GT_1;
        GT_1    = ([A3..A0] >  [B3..B0]) # EQ_1 & GT_0;

test_vectors
       ([ A   , B ,GT_0] -> [A_GT_B,A_EQ_B])
        [^h00,^h00,  0 ] -> [  0   ,   1  ];
        [^h0F,^h00,  0 ] -> [  1   ,   0  ];
        [^h10,^h0F,  0 ] -> [  1   ,   0  ];
        [^h11,^h10,  0 ] -> [  1   ,   0  ];
        [^h0F,^h10,  0 ] -> [  0   ,   0  ];
        [^h10,^h11,  0 ] -> [  0   ,   0  ];
        [^hFF,^hFF,  0 ] -> [  0   ,   1  ];
        [^h55,^h55,  0 ] -> [  0   ,   1  ];
        [^h55,^h55,  1 ] -> [  1   ,   1  ];
end
```

Figure 8.16 Cascading 4-bit magnitude comparators described in ABEL

8.6 PITFALLS OF USING HIGH-LEVEL DESIGN METHODS

As the comparator circuit demonstrated, circuits that are simple to describe at a high-level can require large amounts of logic to implement. Because high-level equations are so easy to write, we tend to forget or misunderstand how much logic can be created from a single, seemingly simple equation.

Not only do simple design descriptions (such as the equation $F = A > B$) often produce unmanageable amounts of logic, they also may not result in the optimum implementation of a circuit, due to design tradeoffs that must be made to adapt a design to the restrictive requirements of a particular PLD architecture.

The implementation of the eight-bit magnitude comparator as a ripple comparator and as a four-level logic circuit (two levels of sum-of-products equations) demonstrates how the constraints of the target architecture can have an impact on the design.

8.7 POLARITY, ACTIVE LEVEL, ON-SETS, AND OFF-SETS

Another area where architecture constraints can affect the use of high-level design descriptions is in the area of *device polarity*. Device polarity, circuit active levels, on-sets, off-sets, and the inter-relationships between these concepts are perhaps the most confusing aspects of PLD design and PLD design tools. While we can't expect to describe how every PLD design tool handles device polarity and circuit active levels, we can at least describe the problem, so you will have less trouble understanding why various design tools function as they do. First, we'll define some frequently misunderstood terms.

Active Level

With the exception of the address decoder design, all of the previous examples have described circuits that are active-high. An *active-high circuit* is one that produces a high voltage on its output to represent a true assertion. When a true assertion is represented by a low voltage, the circuit is said to be *active-low*. When using high-level design tools, the active level of a circuit can (if desired) be thought of as an implementation detail that doesn't affect the description of the logic function and isn't affected by the architecture of the target PLD.

Device Polarity

The meaning of the term *polarity* has been less well defined. To clear things up somewhat, we define device polarity to be the existence (or non-existence) of an inverter on the output of a PLD's logic array. A device output that has an inverter is *negative polarity*, while an output with no inverter is *positive polarity*. PLDs have outputs that can be either positive polarity, negative polarity, or programmable polarity. These three types of device outputs are shown in Figures 8.17(a), 8.17(b) and 8.17(c).

PLD manufacturers have often referred to their devices as being active-high or active-low (the 16L8 and 16H8 PALs, for example). The devices with positive polarity outputs are referred to as being active-high, while the devices with negative polarity outputs are often called active-low. This is very confusing terminology, since a blank PLD isn't active-anything—it's the circuit implemented within a PLD that is active-high or active-low.

Figure 8.18 demonstrates this idea graphically. Let's say we wish to implement the active-high function $F = A \# B \# C \# D$ in a PLD. Naturally, we require

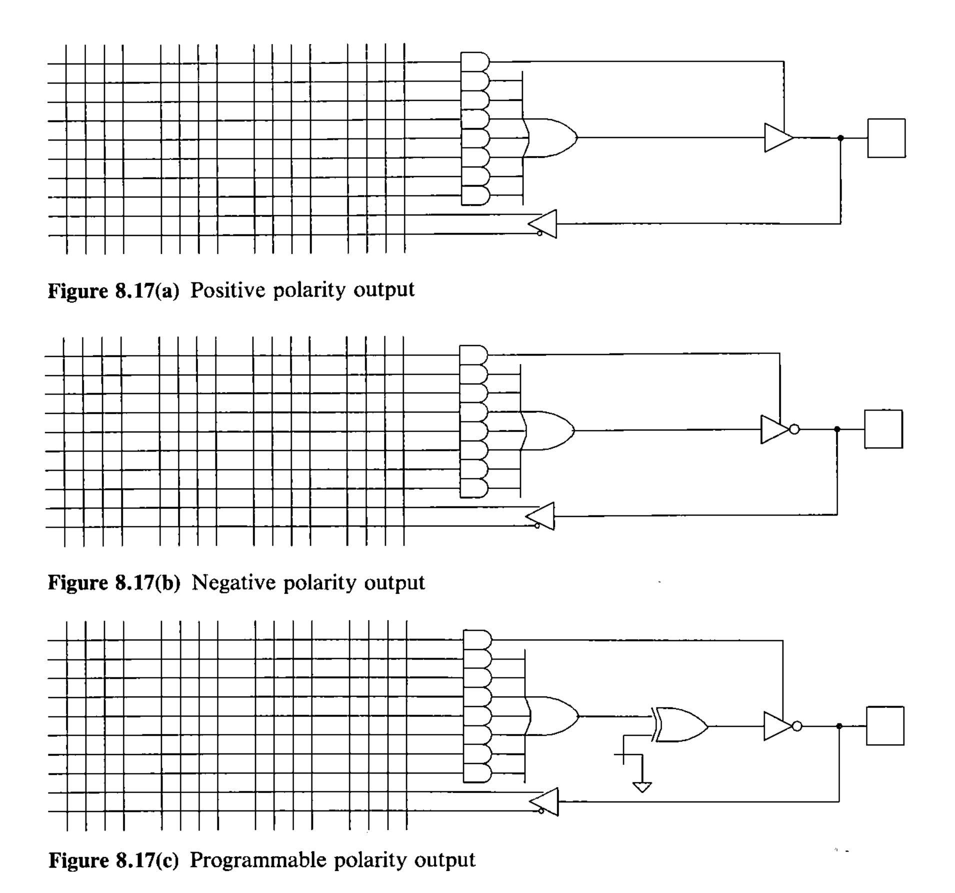

Figure 8.17(a) Positive polarity output

Figure 8.17(b) Negative polarity output

Figure 8.17(c) Programmable polarity output

four inputs and one output on our PLD, but do we require a particular output polarity for this function? As Figure 8.18 shows, we do not. As long as the target PLD is capable of providing sufficient product terms for the function, the actual implementation polarity is of no significance. The two circuits shown for each type of device are equivalent. As this demonstrates, there isn't necessarily a correlation between the device's polarity and the circuit's active level.

In ABEL-HDL (and many other languages), you specify the active level by declaring active-low signals with a ! (NOT) operator. We saw this in the decoder example presented previously. The equations for the decoder were written in terms of true logic although the actual circuit will produce low asserted signals under the conditions described by the equations.

Once again, it's important to realize that the active-low decoder design can be implemented with equal ease in a positive or negative polarity PLD. The design software (in this case ABEL) simply has to complement the design equation to fit the architecture of the target PLD. Of course, the complemented form of the equations will often result in a larger (or smaller) amount of logic (in terms of the

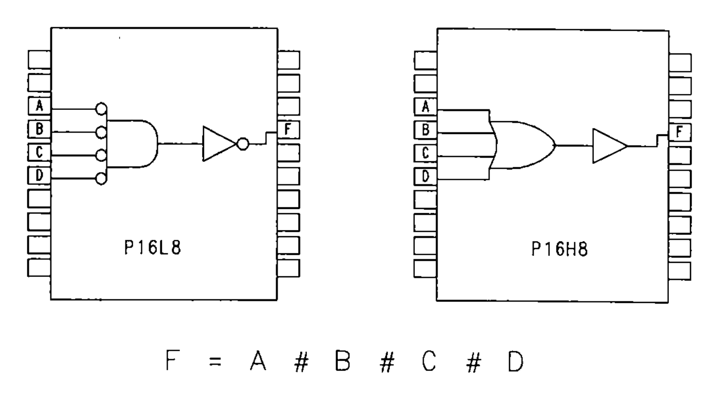

F = A # B # C # D

Figure 8.18 Implementing a function in positive and negative polarity PLDs

number of product terms) and this will often restrict the selection of devices in which the design can be implemented.

On-sets and Off-sets

How does this relate to the on-sets and off-sets discussed in Chapter 7? If the design software you are using doesn't support don't-care optimizations, you don't have to consider on-sets and off-sets at all. If don't-care conditions are significant, however, you must be careful not to confuse on-sets and off-sets with device polarities and circuit active levels. Consider the following equation:

$$!X = A \ \& \ B \ \# \ !C;$$

What does the inversion operator ! applied to *X* indicate? The use of a ! operator on the left side of the equation might be construed to mean that we are using active-low logic in our circuit, but does it really? An equation such as this can have subtly different meanings depending on the design tool that is being used, and the context in which the equation is used. It's very important to determine the semantics of such equations for the design language you will be using.

You might recall from our discussion in Chapter 7 that the use of a ! operator on the left side of an equation can be used to indicate an off-set equation. When you specify an off-set, you are specifying a condition under which the function is false. This can be quite different from specifying when a circuit is true and active-low, particularly if a dc-set is specified or implied by the use of multiple equations for a single output (multiple equations for a single output are supported in many PLD design languages in different ways; typically, multiple equations are simply appended together with an OR operator.)

In FutureDesigner's Gates language, for example, an off-set equation results if a ! operator is used on the left side of an equation. If on-set equations are also provided, the conditions not covered by on-sets or off-sets are assumed to be dc-sets. For example, consider the following pair of equations for the same output:

$$X = A \,\&\, B;$$

$$!X = \,!A \,\&\, !B;$$

A design tool that supports don't-care optimizations would assign a don't-care result to the conditions not specified in the above equations (*!A & B* or *A & !B*). If this isn't what you intended, then you may be surprised by the operation of the resulting circuit.

In most PLD design languages (including ABEL), an inversion operator used in this context is not significant; the language processor simply moves the operator to the right side of the equation before the equations for a common output are combined. The actual final form of the equation will depend on the polarity of the target device; the design tool will choose the form that is appropriate for that polarity. If the target PLD features programmable polarity, some of these design tools (including ABEL) will choose the polarity and corresponding equation format that results in the fewest product terms used. In some PLD design languages, you must place the inversion operator on the left side of the equation if the device has negative polarity outputs, and no automatic equation complement or polarity selection is performed.

8.8 IMPLEMENTING COUNTERS IN PLDS

We've seen how a variety of simple combinational circuits can be implemented in PLDs. Now we'll look at how some common synchronous sequential circuits can be designed. Many of the most common sequential applications are based on counters. Since the typical PLD isn't optimized for counter applications, it's important to understand how a counter can be most efficiently implemented in a PLD's constrained architecture.

The Basic T-type Flip-flop Counter

Let's begin by examining how a counter functions. Figure 8.19 shows the bit patterns that are observed on the outputs of a 4-bit up counter for the sixteen states of its operation. These binary values correspond to the decimal values zero through fifteen.

Using flip-flops, there are a variety of ways that such a counter can be implemented. Users of standard logic devices will usually build such a counter using T-type flip-flops in a circuit such as the one shown in Figure 8.20.

This is the simplest implementation of a counter and exploits the edge-trig-

Binary	Decimal
0000	0
0001	1
0010	2
0011	3
0100	4
0101	5
0110	6
0111	7
1000	8
1001	9
1010	10
1011	11
1100	12
1101	13
1110	14
1111	15

Figure 8.19 Bit values of 4-bit up counter

gered clock of each flip-flop. In a ripple counter such as this, each flip-flop is clocked at a rate of one-half the rate of the previous flip-flop. This is a reflection of the behavior seen in the bit values presented in Figure 8-19; each bit position toggles when its less-significant neighbor transitions from one to zero.

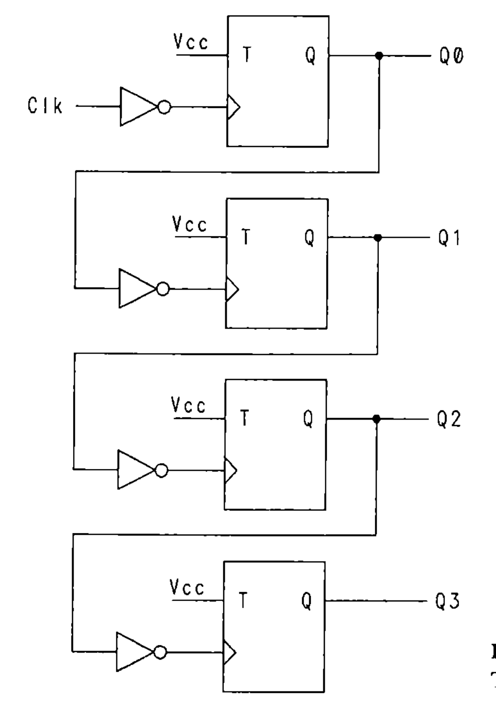

Figure 8.20 4-bit counter constructed out of T-type flip-flops

There are two problems with a counter such as this. The first problem is the fact that the current state of the counter can't be determined at any point in time; this is fine if the purpose of the counter is to divide a frequency, but is not acceptable if the counter is being used to control a state machine, cycle through memory addresses, or for some other application where the counter value is significant. The second problem is that the counter can't operate at high speeds due to the large number of propagation delays.

Furthermore, if we wish to use one or more PLDs to implement the counter circuit, we find that this circuit isn't appropriate for the average PLD, since most PLDs have flip-flops that are clocked from a common source (this may be a good circuit to consider, however, if the application is intended for an FPGA).

The T-type Flip-flop Look-ahead Counter

To implement this counter circuit in a PLD, we'll need to implement it using a different scheme, such as that shown in Figure 8.21. This 8-bit counter circuit doesn't require independently clocked flip-flops or multiple levels of logic to implement and can, therefore, be implemented in a PLD with T-type flip-flops. In this implementation of the counter, each succeeding bit of the counter is toggled whenever all of the preceding bits are true (as seen in Figure 8.19). This means that each of the counter's flip-flops requires just one product term with n inputs, where n is the number of less significant bits. The least significant bit of the counter is tied directly to V_{cc}.

The 8-bit counter can be described with the following Boolean equations for the flip-flops' T inputs:

$$Q7.T = Q6 \ \& \ Q5 \ \& \ Q4 \ \& \ Q3 \ \& \ Q2 \ \& \ Q1 \ \& \ Q0;$$

$$Q6.T = Q5 \ \& \ Q4 \ \& \ Q3 \ \& \ Q2 \ \& \ Q1 \ \& \ Q0;$$

$$Q5.T = Q4 \ \& \ Q3 \ \& \ Q2 \ \& \ Q1 \ \& \ Q0;$$

$$Q4.T = Q3 \ \& \ Q2 \ \& \ Q1 \ \& \ Q0;$$

$$Q3.T = Q2 \ \& \ Q1 \ \& \ Q0;$$

$$Q2.T = Q1 \ \& \ Q0;$$

$$Q1.T = Q0;$$

$$Q0.T = 1;$$

The T inputs to the counter's flip-flops are indicated through the use of *dot extension* suffixes appended to the name of the design's outputs. This method of specifying secondary signals related to a macrocell is common in PLD design languages. There are many different dot extensions available and these dot ex-

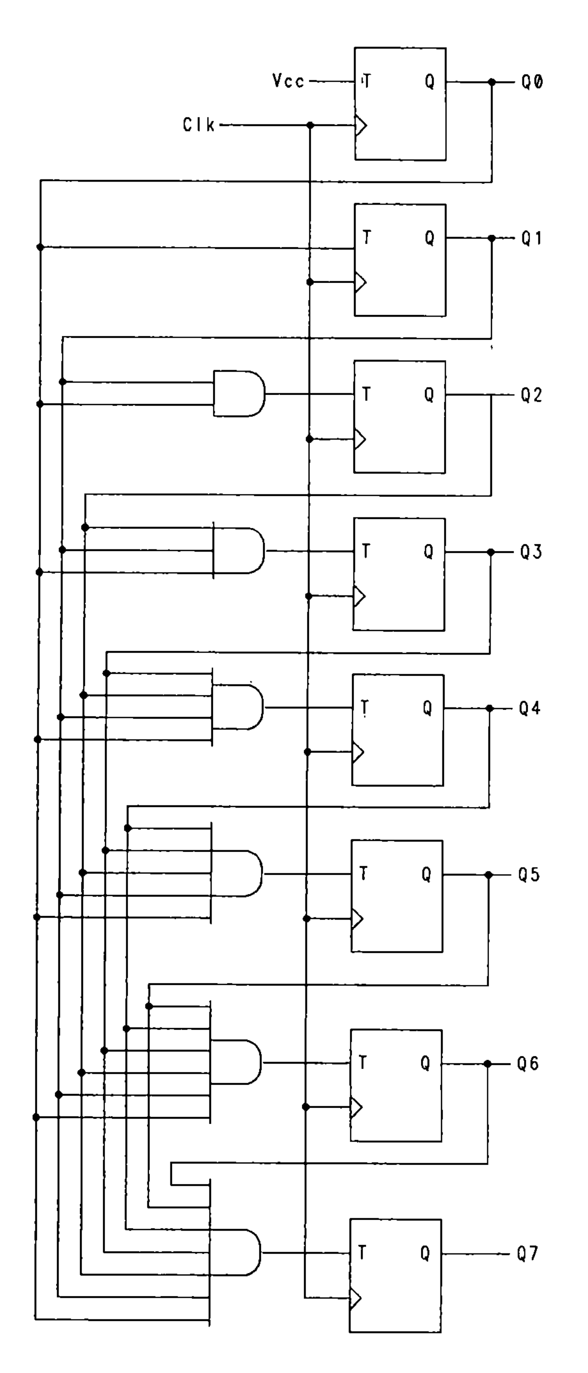

Figure 8.21 8-bit look-ahead carry counter

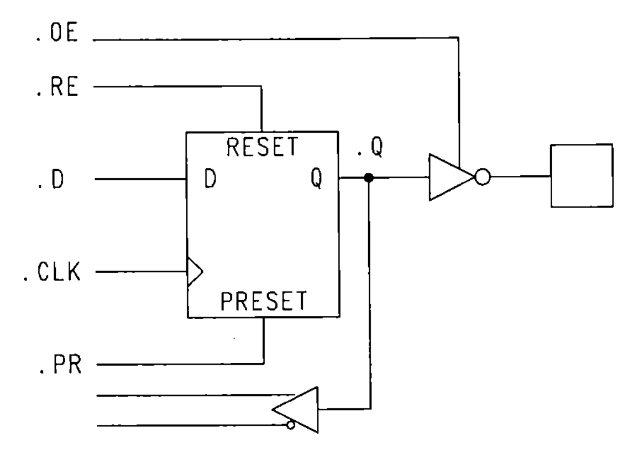

Figure 8.22 Typical output macrocell and corresponding ABEL dot extensions

tensions can be used to specify many of the signals related to a device's output macrocells or other internal signals.

A typical output macrocell is shown in Figure 8.22. This macrocell shows some of the dot extension signals associated with a D-type flip-flop. Other flip-flop types have different requirements and, correspondingly, different dot extension signals.

Using D-type Flip-flops for Counters

This counter circuit can be implemented in a PLD with T-type flip-flops, but these devices are typically more expensive than the simpler PALs with D-type flip-flops. How can a design that is described most naturally with T-type flip-flops be implemented in one of the lower cost PLDs that has only D-type flip-flops? The easiest way is to convert the circuit to a D-type flip-flop implementation by using the flip-flop emulation technique discussed in Chapter 7.

As we showed in that chapter, with the simple addition of an XOR gate, a D-type flip-flop can be made to function as a T-type flip-flop and a T-type flip-flop can be made to function as a D-type flip-flop. This is shown again in Figures 8.23 and 8.24.

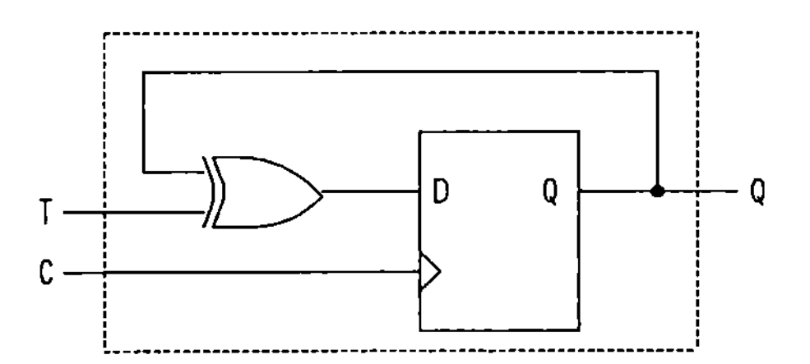

Figure 8.23 Emulating a T-type flip-flop with D-type

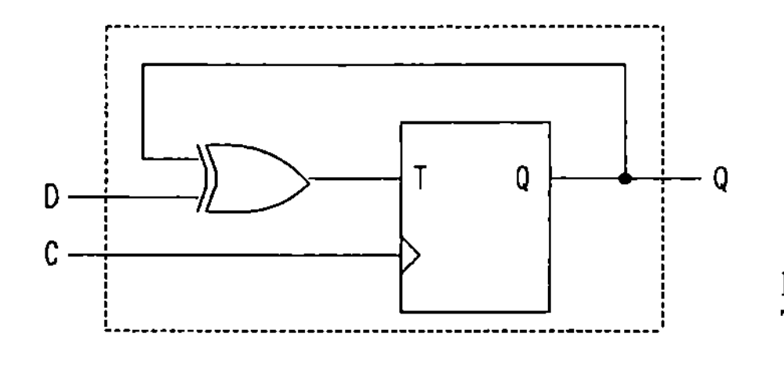

Figure 8.24 Emulating a D-type flip-flop with T-type

To implement the 8-bit counter in a device with D-Type flip-flops, we simply XOR the counter equations with the fedback flip-flop outputs as follows:

$$Q7 := Q7 \ \$ \ Q6 \ \& \ Q5 \ \& \ Q4 \ \& \ Q3 \ \& \ Q2 \ \& \ Q1 \ \& \ Q0;$$
$$Q6 := Q6 \ \$ \ Q5 \ \& \ Q4 \ \& \ Q3 \ \& \ Q2 \ \& \ Q1 \ \& \ Q0;$$
$$Q5 := Q5 \ \$ \ Q4 \ \& \ Q3 \ \& \ Q2 \ \& \ Q1 \ \& \ Q0;$$
$$Q4 := Q4 \ \$ \ Q3 \ \& \ Q2 \ \& \ Q1 \ \& \ Q0;$$
$$Q3 := Q3 \ \$ \ Q2 \ \& \ Q1 \ \& \ Q0;$$
$$Q2 := Q2 \ \$ \ Q1 \ \& \ Q0;$$
$$Q1 := Q1 \ \$ \ Q0;$$
$$Q0 := Q0 \ \$ \ 1;$$

Notice that we haven't used any dot extensions in these equations. Since the output of a D-type flip-flop simply follows the signal values applied to the flip-flop's input, there is no distinction required between the *D* flip-flop input and the corresponding output. ABEL-HDL's *clocked assignment operator* (:=) is used to indicate that the equations are being written for registered outputs of the PLD.

These equations can be implemented directly in a device with XOR gates (such as the 20X8 PAL) or implemented in a PAL device without XOR gates such as the 16R8. In the latter case, the XOR operators are converted into sum-of-products logic. The large number of product terms required to implement an 8-bit counter (the most significant bit of an *n*-bit counter will always require at least *n* product terms) may preclude the use of a simple 16R8 if additional logic is required for reset or data loading purposes.

Using High-level Equations for Counters

In ABEL and some other PLD design languages, counters for D-type or T-type flip-flops can be written using high-level equations. For example, the eight equations shown above for D-type flip-flops could be replaced by the following set declaration and high-level equation:

Declarations

Count = [*Q*7..*Q*0];

Equations

Count := *Count* + 1;

When writing high-level equations such as these, is it actually more convenient to think about the counter's behavior in terms of the state of its outputs. This means that a D-type flip-flop representation is more natural than the T-type representation.

The most popular registered PLDs feature D-type flip-flops on their outputs. To some extent this simplifies the job of writing equations, since the equations required to control a D-type flip-flop's *D* input are the same equations that you would expect to write if you wanted to describe the behavior of your design at the device outputs. For PLDs with D-type flip-flops, you can in most cases ignore the existence of the flip-flops when writing design equations. There are many exceptions to this, however; later we'll cover these situations in some detail.

If we wish to implement this simple counter in a device that has T-type flip-flops, we can convert the design back to a T-type representation with another XOR, resulting in the high-level equation:

Count.T = *Count* $ (*Count* + 1);

Figure 8.25 shows a complete ABEL design that utilizes high-level equations and explicit flip-flop conversion to implement a more complex up counter. This counter features synchronous hold, load, and clear inputs.

The ABEL design includes two equations: the counter equation already described, and an additional equation for the clock input to the design's flip-flops. The *.clk* dot extension is used to refer to the clock inputs to the design's flip-flops (*Q*7 through *Q*0).

Counter Reset Schemes

There are a variety of methods that can be used to reset a counter to an initial value, and the architecture of the target PLD is an important consideration when choosing the most efficient reset scheme. If you use a device such as a 22V10 that features a synchronous preset and asynchronous reset, you have many options from which to choose. If you are using a simpler device, you should consider the impact of different reset strategies.

```
module CNT600A
title' Octal counter with load and clear'

        cnt600a device 'E0600';

        D0,D1,D2,D3,D4,D5,D6,D7            pin   3, 4, 5, 6, 7, 8, 9,10;
        Q7,Q6,Q5,Q4,Q3,Q2,Q1,Q0            pin  16,15,17,18,19,20,21,22;
        CLK,I0,I1                          pin  14,2,11;

        Q7,Q6,Q5,Q4,Q3,Q2,Q1,Q0            istype 'reg_T';

        H,L,X,Z,C = 1, 0, .X., .Z., .C.;

        Data  = [D7,D6,D5,D4,D3,D2,D1,D0];
        Count = [Q7,Q6,Q5,Q4,Q3,Q2,Q1,Q0];

        Mode  = [I1,I0];
        Clear = [ 0, 0];
        Hold  = [ 0, 1];
        Load  = [ 1, 0];
        Inc   = [ 1, 1];

equations
        Count.t =((Count.q + 1) & (Mode == Inc)
                 # (Count.q    ) & (Mode == Hold)
                 # (Data       ) & (Mode == Load)
                 # (0          ) & (Mode == Clear))
                 $ Count.q;

        Count.C  = CLK;

test_vectors 'test load and increment'
             ([CLK,Mode ,Data] -> Count)
              [ C ,Load , 1  ] ->   1  ;
              [ C ,Inc  , X  ] ->   2  ;
              [ C ,Inc  , X  ] ->   3  ;
              [ C ,Inc  , X  ] ->   4  ;
              [ C ,Inc  , X  ] ->   5  ;
              [ C ,Load , 3  ] ->   3  ;
              [ C ,Inc  , X  ] ->   4  ;
              [ C ,Load , 7  ] ->   7  ;
              [ C ,Inc  , X  ] ->   8  ;
              [ C ,Hold , X  ] ->   8  ;
              [ C ,Inc  , X  ] ->   9  ;
              [ C ,Load ,^h3F] -> ^h3F ;
              [ C ,Inc  , X  ] -> ^h40 ;
              [ C ,Load ,^h7F] -> ^h7F ;
              [ C ,Inc  , X  ] -> ^h80 ;
              [ C ,Inc  , X  ] -> ^h81 ;
              [ C ,Clear, X  ] -> ^h00 ;
              [ C ,Load ,^hFE] -> ^hFE ;
              [ C ,Inc  , X  ] -> ^hFF ;
              [ C ,Inc  , X  ] -> ^h00 ;
End
```

Figure 8.25 Explicit D-type to T-type flip-flop conversion using ABEL

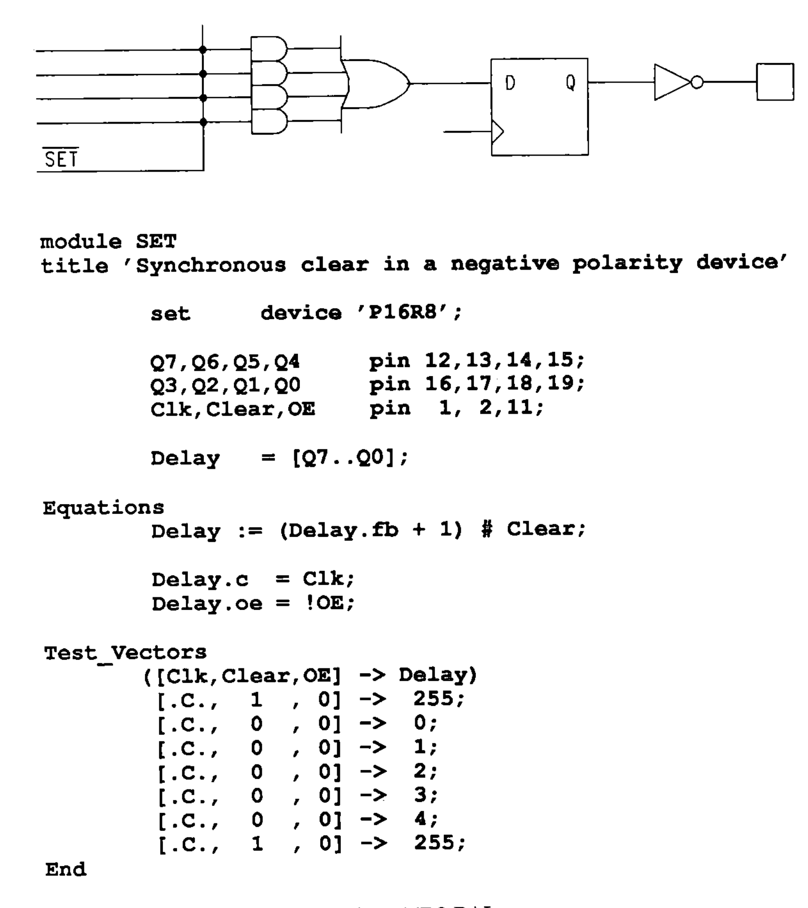


```
module SET
title 'Synchronous clear in a negative polarity device'

        set       device 'P16R8';

        Q7,Q6,Q5,Q4        pin 12,13,14,15;
        Q3,Q2,Q1,Q0        pin 16,17,18,19;
        Clk,Clear,OE       pin  1, 2,11;

        Delay    = [Q7..Q0];

Equations
        Delay := (Delay.fb + 1) # Clear;

        Delay.c  = Clk;
        Delay.oe = !OE;

Test_Vectors
       ([Clk,Clear,OE] -> Delay)
        [.C.,  1  , 0] ->  255;
        [.C.,  0  , 0] ->  0;
        [.C.,  0  , 0] ->  1;
        [.C.,  0  , 0] ->  2;
        [.C.,  0  , 0] ->  3;
        [.C.,  0  , 0] ->  4;
        [.C.,  1  , 0] ->  255;
End
```

Figure 8.26 Synchronous preset in a 16R8 PAL

The diagrams and corresponding ABEL design files shown in Figures 8.26 and 8.27, for example, show two different methods for counter initialization. These two methods result in different reset states and require different amounts of logic. The first method doesn't require any additional product terms to implement while the second method will consume one additional product term for each output of the design. This is because the first method exploits the inherent default state of the D-type flip-flop. While this default flip-flop state can be used to simplify the design of a circuit requiring a reset state, it must be realized that the specific configuration of the target PLD will affect the operation of the resulting circuit.

As a general rule, devices such as the 16R8 that have D-type flip-flops and fixed output inverters (that are located between the flip-flops and the device outputs) will require one additional product term for each output if a reset to an all-low state is desired, while no additional product terms will be required for reset to the state in which all outputs are high. You should, therefore, take into con-

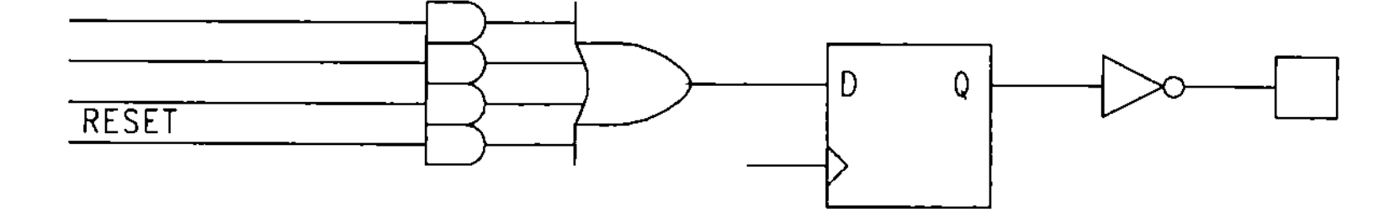


```
module RESET
title 'Synchronous clear in a negative polarity device'

        reset   device 'P16R8';

        Q7,Q6,Q5,Q4     pin 12,13,14,15;
        Q3,Q2,Q1,Q0     pin 16,17,18,19;
        Clk,Clear,OE    pin  1, 2,11;

        Delay   = [Q7..Q0];

Equations
        Delay := (Delay.fb + 1) & !Clear;

        Delay.c  = Clk;
        Delay.oe = !OE;

Test_Vectors
       ([Clk,Clear,OE] -> Delay)
        [.C.,  1  , 0] ->  0;
        [.C.,  0  , 0] ->  1;
        [.C.,  0  , 0] ->  2;
        [.C.,  0  , 0] ->  3;
        [.C.,  0  , 0] ->  4;
        [.C.,  1  , 0] ->  0;
End
```

Figure 8.27 Synchronous reset in a 16R8 PAL

sideration the architecture of the target PLD when designing your circuit's reset scheme.

Polarity Considerations for Counters

Because of the different configurations of registered PLDs' outputs, it's difficult to design efficient sequential circuits such as counters without being aware of the constraints of specific PLD architectures. Differences in architectures from one PLD to another can have profound effects on the behavior of seemingly simple circuits. This is particularly true of devices featuring programmable polarity.

First, consider the problem of register preset and reset. If your design requires the use of a preset or reset, you will need to determine whether the required features exist in the device chosen and how the configuration of the device outputs will affect the operation of those features. As an example, Figures 8.28 shows an output macrocell from a 22V10 PAL.

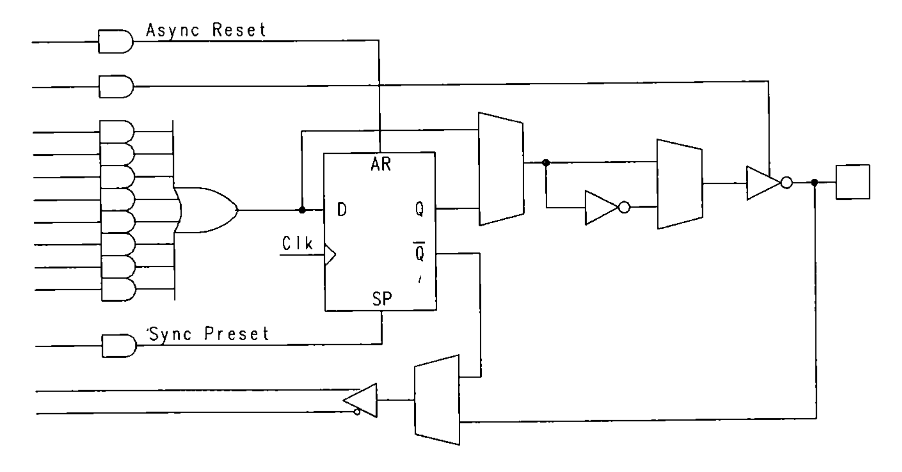

Figure 8.28 22V10 output macrocell

In a 22V10, the polarity for each registered output is controlled by selecting or bypassing an inverter located between the *Q* output of the D-type flip-flop and its associated output pin. This means that there may or may not be an inversion between the outputs of the flip-flops and the corresponding device output. The result is that the behavior of the asynchronous reset and synchronous preset features will be different depending on whether positive or negative polarity is selected for the programmable inversion. If the reset and preset were both synchronous or both asynchronous, this difference in behavior might be accommodated by simply swapping the reset and preset logic, but this isn't possible in the 22V10. Even if such swapping is possible in your design, the nature of the preset and reset places limitations on how much benefit can be gained from this, since all of the flip-flops are preset and reset from common product terms.

In many devices, the programmable inverter is located before the flip-flops, on the outputs of the OR gates. This simplifies the use of programmable polarity. In all fairness, however, the 22V10 style output macrocell can be used to advantage for resetting a state machine to a default state. To design a circuit that will reset to an arbitrarily encoded default state, simply choose a sequence of output polarity configurations that will result in the desired state encoding on the device outputs when the global preset or reset is activated.

Implementing Counters with SR-type Flip-flops

We have seen how counters can be implemented using D-type and T-type flip-flops. These two types of flip-flops are straightforward to use for this application and lend themselves well to high-level design description methods. Designing counters for other flip-flop types, most notably SR-type flip-flops, is somewhat more complex.

To design a counter that utilizes SR-type flip-flops, we must first analyze the design requirements by once again examining the bit patterns presented previously in Figure 8.19. When we examined these bit values to determine a T-type flip-flop implementation, we were attempting to determine the conditions that would indicate when each bit of the counter should toggle. This thinking reflected the behavior of the T-type flip-flop; the behavior of an SR-type flip-flop requires us to think about our counter in yet another way.

When we look at the counter's bit patterns again with SR-type flip-flops in mind, we are looking for those conditions under which each bit should be turned on (set) or turned off (reset). The required circuitry for each bit of the counter can be generalized: for any given clock cycle, each bit of the counter is set if it was off in the previous state and all lower-order (less significant) bits were on. Similarly, each bit is reset whenever the next higher-order (more significant) bit is turned on, or when the bit itself and all lower-order bits were previously on.

The circuit that implements this concept for a 4-bit counter is shown in Figure 8.29. This circuit requires a total of five distinct product terms when implemented in a product-term sharing PLD such as the PLS105.

The 4-bit counter circuit could be expressed directly as Boolean equations as shown below:

$$
\begin{aligned}
Q0.S &= !Q0; \\
Q0.R &= !Q3 \,\&\, Q2 \,\&\, Q1 \,\&\, Q0 \\
&\quad \#\; Q3 \,\&\, Q2 \,\&\, Q1 \,\&\, Q0 \\
&\quad \#\; !Q2 \,\&\, Q1 \,\&\, Q0 \\
&\quad \#\; !Q1 \,\&\, Q0; \\
Q1.S &= !Q1 \,\&\, Q0; \\
Q1.R &= !Q3 \,\&\, Q2 \,\&\, Q1 \,\&\, P0 \\
&\quad \#\; Q3 \,\&\, Q2 \,\&\, Q1 \,\&\, Q0 \\
&\quad \#\; !Q2 \,\&\, Q1 \,\&\, Q0; \\
Q2.S &= !Q2 \,\&\, Q1 \,\&\, Q0; \\
Q\,2.R &= !Q3 \,\&\, Q2 \,\&\, Q1 \,\&\, Q0 \\
&\quad \#\; Q1 \,\&\, Q2 \,\&\, Q1 \,\&\, Q0; \\
Q3.S &= !Q3 \,\&\, Q2 \,\&\, Q1 \,\&\, Q0; \\
Q3.R &= Q3 \,\&\, Q2 \,\&\, Q1 \,\&\, Q0;
\end{aligned}
$$

Alternatively, this same design can be expressed in ABEL-HDL as shown in Figure 8.30. This design description enumerates all of the possible counter

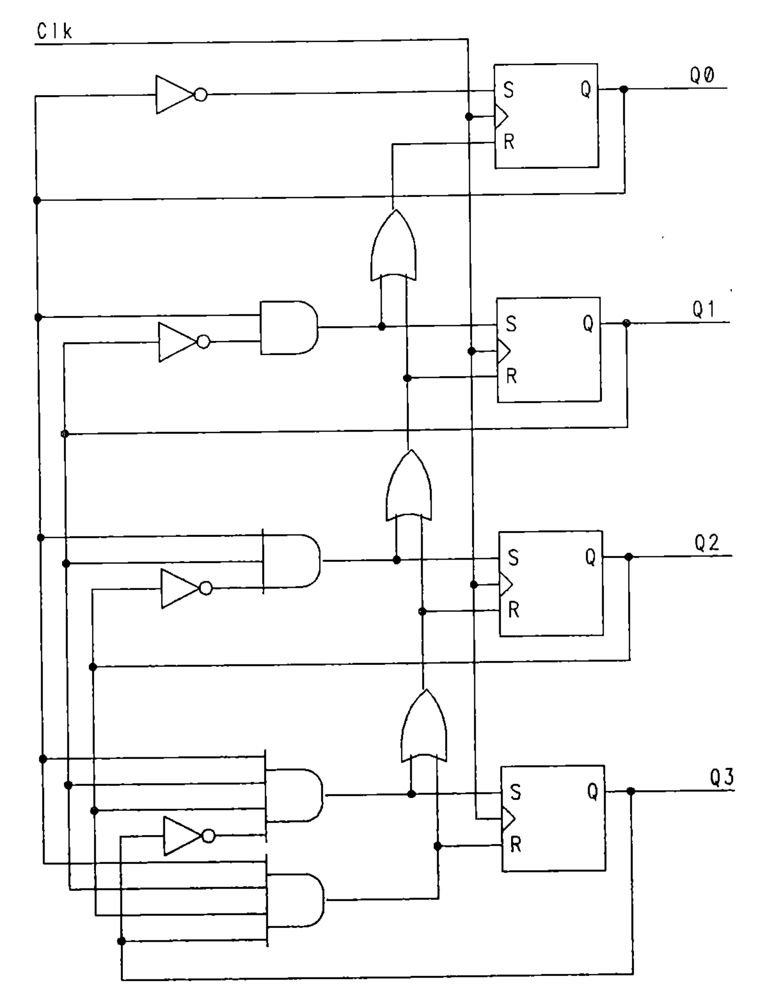

Figure 8.29 4-bit SR-type flip-flop counter circuit

values and, for each counter value, sets or resets the individual counter outputs based on the required next counter value.

Up Counters and Down Counters

The previous counter designs were all examples of up counters—counters that increment their values with each clock cycle. In PLDs with negative polarity outputs and D-type flip-flops, you will find that it's usually more efficient to implement a delay or event counter circuit by using a down counter (the 16R8, for example, is a natural device for down counting, but is rather inefficient for

```
module CNT105
title '4-bit Counter'

        cnt105          device 'F105';

        Clk,PR          pin  1,19;
        Q3,Q2,Q1,Q0     node istype 'reg_SR,buffer';

        Q               = [Q3,Q2,Q1,Q0];  "State Registers

equations
    Q.AP = PR; "Async Preset
    Q.C  = Clk;

    [                 Q0.S] = (Q ==  0);
    [            Q1.S,Q0.R] = (Q ==  1);
    [                 Q0.S] = (Q ==  2);
    [       Q2.S,Q1.R,Q0.R] = (Q ==  3);
    [                 Q0.S] = (Q ==  4);
    [            Q1.S,Q0.R] = (Q ==  5);
    [                 Q0.S] = (Q ==  6);
    [Q3.S,Q2.R,Q1.R,Q0.R] = (Q ==  7);
    [                 Q0.S] = (Q ==  8);
    [            Q1.S,Q0.R] = (Q ==  9);
    [                 Q0.S] = (Q == 10);
    [       Q2.S,Q1.R,Q0.R] = (Q == 11);
    [                 Q0.S] = (Q == 12);
    [            Q1.S,Q0.R] = (Q == 13);
    [                 Q0.S] = (Q == 14);
    [Q3.R,Q2.R,Q1.R,Q0.R] = (Q == 15);

test_vectors    ([Clk,PR] -> Q)
                 [ 0 , 0] -> .X.;
                 [ 1 , 1] -> 15; " Preset high
                 [ 1 , 0] -> 15; " Preset low
                 [.C., 0] ->  0;
                 [.C., 0] ->  1;
                 [.C., 0] ->  2;
                 [.C., 0] ->  3;
                 [.C., 0] ->  4;
                 [.C., 0] ->  5;
                 [.C., 0] ->  6;
                 [.C., 0] ->  7;
                 [.C., 0] ->  8;
                 [.C., 0] ->  9;
                 [.C., 0] -> 10;
                 [.C., 0] -> 11;
                 [.C., 0] -> 12;
                 [.C., 0] -> 13;
                 [.C., 0] -> 14;
                 [.C., 0] -> 15;
                 [.C., 0] ->  0;
                 [.C., 0] ->  1;
                 [.C., 0] ->  2;
end
```

Figure 8.30 4-bit SR-type counter described with ABEL

Up		Down	
Decimal	Binary	Binary	Decimal
0	0000	1111	15
1	0001	1110	14
2	0010	1101	13
3	0011	1100	12
4	0100	1011	11
5	0101	1010	10
6	0110	1001	9
7	0111	1000	8
8	1000	0111	7
9	1001	0110	6
10	1010	0101	5
11	1011	0100	4
12	1100	0011	3
13	1101	0010	2
14	1110	0001	1
15	1111	0000	0

Figure 8.31 Comparison of up counter and down counter values

up counting). As Figure 8.31 illustrates, converting an up counter to a down counter is a simple matter of inverting the counter circuit's outputs. The initial state then becomes the one's complement of the original initial state and the counter sequences down, rather than up.

For further reductions, you might find that inverting just some of the counter's outputs will result in the most efficient logic. This is most easily done in PLDs that feature programmable polarity for the device outputs. If your counter circuit will be feeding the inputs to other PLDs, you can often simply invert the active level (or *sense*) of the counter outputs and then re-invert them at the other PLD's inputs as illustrated in Figure 8.32.

Designing an Arbitrary Length Counter

Quite often, counters are required that are of an arbitrary length. Using high-level equations, such a counter can be easily developed. For example, if a 111-state

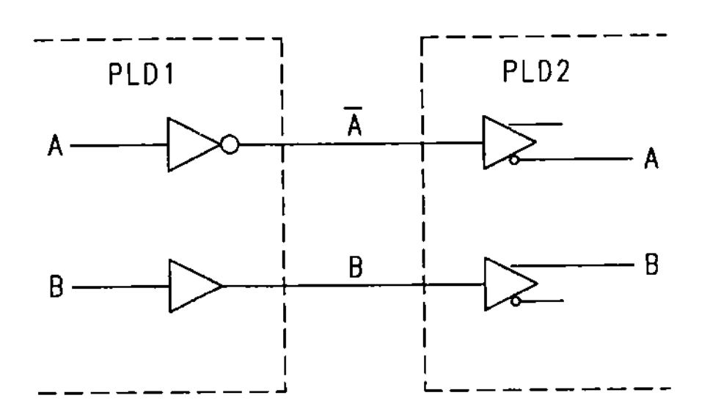

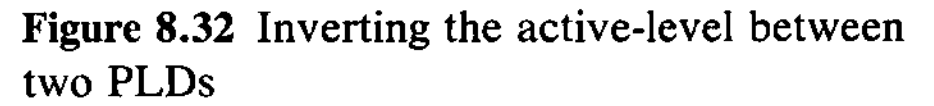
Figure 8.32 Inverting the active-level between two PLDs

counter is required that counts from zero to a value of 110, we could write a high-level equation of the form:

$$Count{:}= (Count + 1) \,\&\, (Count <= 110) \,\&\, !Clr;$$

This equation accurately describes a 111-state counter, but, as written, it produces sum-of-products equations that are too large to fit in a simple PAL device.

When D-type flip-flops are used, a counter with no terminating value (a 256-state 8-bit counter, for example) will always require n product terms for the most significant bit of the counter, where n is the number of counter bits. For an

Figure 8.33 111-state short counter implemented in a 22V10 PAL

```
Module COUNT111
Title 'Arbitrary length counter, 0 to 110'

        Count111 device 'P22V10';

        Clk,Clr                      pin 1,2;
        Q6,Q5,Q4,Q3,Q2,Q1,Q0         pin 16,17,18,19,20,21,22;
        Q6,Q5,Q4,Q3,Q2,Q1,Q0         istype 'invert';

        Count    = [Q6..Q0];

Equations

        Count := (Count.fb + 1) & !Clr;

        Count.sp = (Count.fb == 110); "synchronous preset term

        Count.clk = Clk;

test_vectors
       ([Clk,Clr] -> Count)
        [.C., 1 ] ->   0;
        [.C., 1 ] ->   0;
        [.C., 0 ] ->   1;
        [.C., 0 ] ->   2;
        [.C., 0 ] ->   3;
        [.C., 0 ] ->   4;
        [.C., 0 ] ->   5;
        [.C., 1 ] ->   0;

@const i=1; @repeat 107 {
        [.C., 0 ] -> @expr i;; @const i=i+1;}

        [.C., 0 ] -> 108;
        [.C., 0 ] -> 109;
        [.C., 0 ] -> 110;
        [.C., 0 ] ->   0;
        [.C., 0 ] ->   1;
        [.C., 0 ] ->   2;
        [.C., 0 ] ->   3;
        [.C., 1 ] ->   0;
        [.C., 1 ] ->   0;
end
```

arbitrary length counter, however, the size of the minimized equations is dependent on the specified terminating value. To get this counter to fit into a simple PAL device, some design changes are required.

One possible approach is to design the counter with a short count comparator by using a device with a synchronous reset term. The PAL 22V10 is one such device. This method is shown in the ABEL source file of Figure 8.33. An equation is written for the synchronous preset term of the 22V10 that compares the counter value with the desired terminating value. This design requires that the 22V10's configurable outputs all have a negative polarity, so that when the registers are preset, the desired counter reset value (all low) will appear on the outputs. This

Figure 8.34 111-state short counter implemented in a 20V8 GAL

```
Module COUNT111
Title 'Arbitrary length counter, 0 to 110'

        Count111 device 'P20V8R';

        Clk,Clr,Short              pin 1,2,15;
        Q6,Q5,Q4,Q3,Q2,Q1,Q0       pin 16,17,18,19,20,21,22;

        Count    = [Q6..Q0];

Equations

        Count := (Count.fb + 1) & !Short;

        Short  = (Count.fb == 110) # Clr;

        Count.clk = Clk;

test_vectors
        ([Clk,Clr] -> Count)
         [.C., 1 ] ->   0;
         [.C., 1 ] ->   0;
         [.C., 0 ] ->   1;
         [.C., 0 ] ->   2;
         [.C., 0 ] ->   3;
         [.C., 0 ] ->   4;
         [.C., 0 ] ->   5;
         [.C., 1 ] ->   0;

@const i=1; @repeat 107 {
         [.C., 0 ] -> @expr i;; @const i=i+1;}

         [.C., 0 ] -> 108;
         [.C., 0 ] -> 109;
         [.C., 0 ] -> 110;
         [.C., 0 ] ->   0;
         [.C., 0 ] ->   1;
         [.C., 0 ] ->   2;
         [.C., 0 ] ->   3;
         [.C., 1 ] ->   0;
         [.C., 1 ] ->   0;
end
```

is ensured by the use of an ISTYPE statement. The ISTYPE statement in ABEL is used to configure specific device features. In the absence of the *ISTYPE 'invert'* statement, ABEL would choose the output polarity that implemented the design equations most efficiently and this polarity may or may not be the polarity required for correct operation of the counter's reset function.

If you want to use a device without a preset feature, you can implement the short count circuit by writing a short count comparator equation for an unused combinational output of the device, and routing that output back into the equations. The ABEL design shown in Figure 8.34 uses this method to implement the

Figure 8.35 111-state short counter implemented in a 20R8 PAL

```
Module COUNT111
Title 'Arbitrary length counter, 0 to 110'

        Count111 device 'P20R8';

        Clk,Clr,Short              pin 1,2,15;
        Q6,Q5,Q4,Q3,Q2,Q1,Q0       pin 16,17,18,19,20,21,22;

        Count    = [Q6..Q0];

Equations

        Count := (Count.fb + 1) & !Short.fb;

        Short := (Count.fb == 109) # Clr;

        [Count,Short].clk = Clk;

test_vectors
       ([Clk,Clr] -> Count)
        [.C., 1 ] ->   0;
        [.C., 1 ] ->   0;
        [.C., 0 ] ->   0;
        [.C., 0 ] ->   1;
        [.C., 0 ] ->   2;
        [.C., 0 ] ->   3;
        [.C., 0 ] ->   4;
        [.C., 1 ] ->   5;
        [.C., 1 ] ->   0;
        [.C., 0 ] ->   0;

@const i=1; @repeat 107 {
        [.C., 0 ] -> @expr i;; @const i=i+1;}

        [.C., 0 ] -> 108;
        [.C., 0 ] -> 109;
        [.C., 0 ] -> 110;
        [.C., 0 ] ->   0;
        [.C., 0 ] ->   1;
        [.C., 0 ] ->   2;
        [.C., 0 ] ->   3;
        [.C., 1 ] ->   4;
        [.C., 1 ] ->   0;
end
```

design in a 20V8 GAL device. The architecture of the GAL allows seven of its eight outputs to be configured with D-type flip-flops while one of the outputs is combinational.

To use a simpler device, you can route the short count comparator to a registered output, and decode the counter's terminal count one state early. This is done in the design shown in Figure 8.35. This design implements the short counter in a 20R8, which has eight D-type flip-flops and no combinational outputs.

If a larger counter is required, one that counts to a value greater than can be accommodated in a single PLD, multiple counters of various terminating values can be cascaded by providing carry signals between succeeding counter blocks. This technique is used to advantage in Chapter 11.

8.9 WAVEFORM GENERATION USING COUNTERS

One common application of counters is in the construction of waveform generator circuits. A simple waveform generator consists of a counter and waveform decoding logic as shown in the block diagram of Figure 8.36.

When a repeating pattern of arbitrary waveforms (such as the one shown in Figure 8.37) is required, you must first determine the number of counter states that will be required to accurately produce all of the desired waveform events. The counter's terminal value depends on how many events there are in the waveform, as well as when the events are to occur in relation to clock edges that occur as a result of different possible clock speeds. In our sample waveform, the pattern repeats after just twelve clock cycles, so a counter that increments (or decrements) through twelve states is sufficient.

Analyzing the required events, we find that signal *A* should go high after clock cycle 1, and go low after clock cycle 3. Signal *B* goes high after cycle 2, low after cycle 5, and goes high again for one clock cycle during cycle 9.

Figure 8.36 Waveform generator block diagram

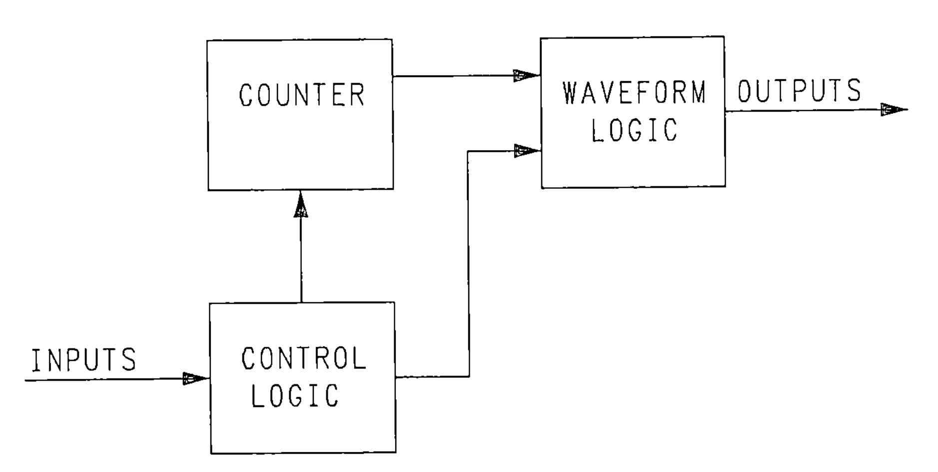

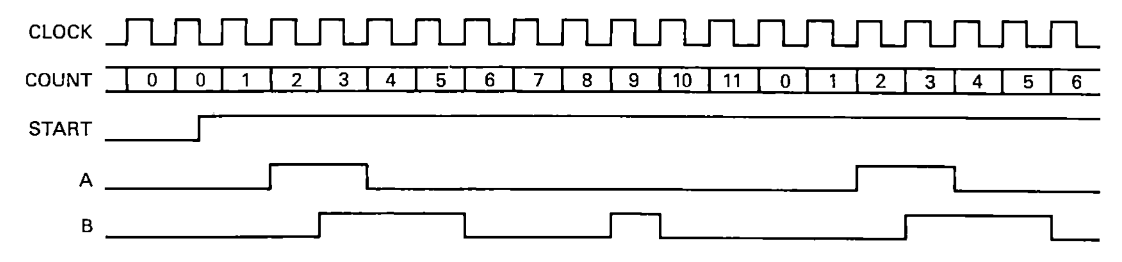

Figure 8.37 Sample waveform

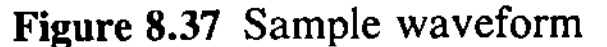

Figure 8.38 Waveform generator described with ABEL

```
module WAVE
title 'Waveform Generator'

        wave              device 'F105';

        Clk,Start,PR       pin   1, 8,19;
        A,B                pin  10,11          istype 'reg_SR';
        Q3,Q2,Q1,Q0        node 40,39,38,37    istype 'reg_SR';
        COMP               node 49;

        Q       = [Q3,Q2,Q1,Q0];  "Counter Registers

equations
    [Q,A,B].R  = !COMP; "Clear Illegal states
    [Q,A,B].C  = Clk;
    [Q,A,B].AP = PR;

   "Counter Equations
    [COMP,                   Q0.S]  = (Q ==  0) & Start; " 0 to  1
    [COMP,              Q1.S,Q0.R]  = (Q ==  1) & Start; " 1 to  2
    [COMP,                   Q0.S]  = (Q ==  2) & Start; " 2 to  3
    [COMP,         Q2.S,Q1.R,Q0.R]  = (Q ==  3) & Start; " 3 to  4
    [COMP,                   Q0.S]  = (Q ==  4) & Start; " 4 to  5
    [COMP,              Q1.S,Q0.R]  = (Q ==  5) & Start; " 5 to  6
    [COMP,                   Q0.S]  = (Q ==  6) & Start; " 6 to  7
    [COMP,Q3.S,Q2.R,Q1.R,Q0.R]      = (Q ==  7) & Start; " 7 to  8
    [COMP,                   Q0.S]  = (Q ==  8) & Start; " 8 to  9
    [COMP,              Q1.S,Q0.R]  = (Q ==  9) & Start; " 9 to 10
    [COMP,                   Q0.S]  = (Q == 10) & Start; "10 to 11
    [COMP,Q3.R,Q2.R,Q1.R,Q0.R]      = (Q == 11) & Start; "11 to  0 (redundant)

   "Output Waveform equations
   A.S  = (Q==1) & Start;
   A.R  = (Q==3) & Start;
   B.S  = (Q==2) & Start;
   B.R  = (Q==5) & Start;
   B.S  = (Q==8) & Start;
   B.R  = (Q==9) & Start;
```

```
test_vectors     ([Clk,PR,Start] -> [ Q,A,B])
                  [ 0 , 0,  0  ] -> [.X.,.X.,.X.];
                  [ 1 , 1,  0  ] -> [15,1,1]; " Preset high
                  [ 1 , 0,  0  ] -> [15,1,1]; " Preset low
                  [.C., 0,  0  ] -> [ 0,0,0];
                  [.C., 0,  0  ] -> [ 0,0,0];
                  [.C., 0,  0  ] -> [ 0,0,0];
                  [.C., 0,  1  ] -> [ 1,0,0];
                  [.C., 0,  1  ] -> [ 2,1,0];
                  [.C., 0,  1  ] -> [ 3,1,1];
                  [.C., 0,  1  ] -> [ 4,0,1];
                  [.C., 0,  1  ] -> [ 5,0,1];
                  [.C., 0,  1  ] -> [ 6,0,0];
                  [.C., 0,  1  ] -> [ 7,0,0];
                  [.C., 0,  1  ] -> [ 8,0,0];
                  [.C., 0,  1  ] -> [ 9,0,1];
                  [.C., 0,  1  ] -> [10,0,0];
                  [.C., 0,  1  ] -> [11,0,0];
                  [.C., 0,  1  ] -> [ 0,0,0];
                  [.C., 0,  1  ] -> [ 1,0,0];
                  [.C., 0,  1  ] -> [ 2,1,0];
                  [.C., 0,  1  ] -> [ 3,1,1];
                  [.C., 0,  0  ] -> [ 0,0,0];
                  [.C., 0,  0  ] -> [ 0,0,0];
end
```

Figure 8.38 *(continued)*

We have described the waveform in terms of transitions and this description maps quite naturally into an SR-type flip-flop implementation. If we use the PLS105 device, which features SR-type flip-flops, we can quite easily implement this design. Figure 8.38 lists the ABEL design description for the waveform generator.

This design uses the buried registers of the PLS105 for the counter function and registers associated with output pins for the waveform decoding function. The counter segment of the design is described in a manner similar to the earlier SR-type flip-flop counter example. This counter is enhanced with the addition of the PLS105's complement array feature for short count and power-up reset functions. The complement array (found on all Signetics FPLS devices) provides a method for detecting illegal states and resetting the SR-type flip-flops to a known state. In this design, the complement array is used to initialize the counter to state 0 on power-up and also guarantees that the counter can never advance beyond state eleven (as a result of this, the last of the counter equations is actually not necessary, and can be eliminated).

The waveform decoding is done separately and uses simple equality comparisons to trigger the required output events. Separating the counter logic from the waveform generation logic in this way makes it easier to modify the waveform generator if needed. The FPLA architecture allows the product terms generated for the counter and waveform decode logic to be shared, so the entire design requires just ten of the PLS105's 48 available product terms.

8.10 EXPLOITING XORS IN PLDS

PLDs with built-in XORs have been in existence for many years. These devices can be used to implement complex functions that won't fit into standard sum-of-products PLDs. We examined some of the uses for XOR gates in Chapter 7 and in the designs presented in this chapter.

Many PLD design languages include some way of specifying an XOR operator that is to be preserved in the minimized design equations and implemented in the device. Methods for this include the use of special XOR operators (sometimes called *hard XORs*), or the use of special XOR nodes that are accessed through dot extensions attached to output signal names. These methods allow the user to be very specific about how an XOR gate is utilized. In some cases, XORs are utilized automatically by the PLD design software, so there is little or no need to specify their use in the design language.

8.11 EXPLOITING UNUSED PLD OUTPUTS

As we have seen, high-level design techniques can be used to advantage for PLDs. At times, however, the inherent limitations of PLDs force you to do things not conceived of by the designers of high-level design tools. At these times, it may be necessary to roll up your sleeves, abandon high-level design techniques, and stuff your design into the desired device in whatever way you can. An example of this is when unused outputs are used to increase the effective size of other outputs.

Figure 8.39 Using an output enable term to wire-OR two output pins

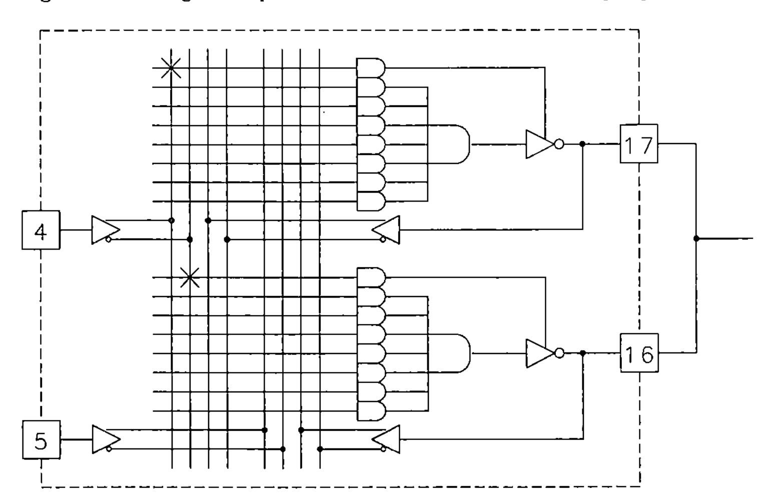

One method of utilizing an unused output pin is factoring which was discussed in Chapter 7. When we decomposed the earlier comparator functions into smaller pieces, we were performing *design-level factoring*. Factoring can be an effective method for splitting up a large equation, but has speed penalties when applied to a PAL-type architecture. Another method that is useful for combinational designs is to use the output enable term to wire-OR two PLD outputs as illustrated in Figure 8.39. An ABEL design file that demonstrates this technique is shown in Figure 8.40.

The technique shown splits a large Boolean equation into two smaller equations by factoring out one of the input variables and applying this variable to the output enable terms for the two resulting outputs. These two outputs can now be

Figure 8.40 Output enable wire-OR expressed in ABEL

```
Module WireOR
title 'Using output enable to wire-OR two output pins'

        wireor  device  'P16L8';

        A,B,C,D,E,F,G,H,I          pin 1,2,3,4,5,6,7,8,9;
        !Y1,!Y2                    pin 16,17;

"       Original Equation
"
"       Y  = !A &  E &  F &  H
"          # !A &  F & !G &  H
"          # !A & !F &  H & !I
"          # !A & !B & !C &  H
"          #  F &  G &  H &  I
"          # !B &  F &  H & !I
"          #  A &  B &  C &  D
"          #  A &  F &  G &  H;

equations

        Y1.OE = !A;

        Y1 = !A &  E &  F &  H
           # !A &  F & !G &  H
           # !A & !F &  H & !I
           # !A & !B & !C &  H
           #  F &  G &  H &  I
           # !B &  F &  H & !I ;

        Y2.OE = A;

        Y2 =  F &  G &  H &  I
           # !B &  F &  H & !I
           #  A &  B &  C &  D
           #  A &  F &  G &  H;

end
```

tied together externally, in essence creating a wired-OR. The totem-pole outputs of the PLD are protected by the output enables, which can never both be active at the same time. You should be aware, however, that there is a potential for glitches in this circuit, since the output enable timing (t_{ea} and t_{er}) may differ from the pin-to-pin (t_{pd}) timing. In addition to the glitch potential, there are potentially serious testability problems (discussed in Chapter 10) that can result from this type of circuit.

8.12 DESIGNING WITH TRUTH TABLES

Truth tables are useful for describing circuits such as decoders in which the relationship of inputs to outputs doesn't follow a regular pattern. They can also be used effectively for describing state machines. A truth table is composed of a header that specifies the ordering of the input and output entries in the table, along with truth table entries that specify input to output relationships. A sample ABEL truth table is shown in Figure 8.41. This truth table describes a logic function that converts binary coded decimal numbers into the appropriate set of outputs to drive a seven segment decoder.

As with Boolean equations, truth tables can be written more concisely if set notation is used. Figure 8.42 shows a complete ABEL source file with set notation simplifications, for the seven-segment display driver. This design is an example of a truth table that is incompletely specified; not all of the possible combinations of inputs are listed on the left side of the truth table. What sort of circuit is created when an incompletely specified truth table is processed? This depends on the level of logic minimization used.

Referring back to Chapter 7, you may recall that the unspecified entries in a truth table can be evaluated as don't-cares and used to help minimize the circuitry required. This level of logic minimization has historically been lacking in PLD design tools. In recent years, however, minimization algorithms such as Espresso have been applied in PLD design tools, allowing don't-cares to be ef-

Figure 8.41 ABEL truth-table syntax

```
truth_table
      ([D3,D2,D1,D0] -> [ a,  b,  c,  d,  e,  f,  g])
       [ 0, 0, 0, 0] -> [ 1,  1,  1,  1,  1,  1,  0];
       [ 0, 0, 0, 1] -> [ 0,  1,  1,  0,  0,  0,  0];
       [ 0, 0, 1, 0] -> [ 1,  1,  0,  1,  1,  0,  1];
       [ 0, 0, 1, 1] -> [ 1,  1,  1,  1,  0,  0,  1];
       [ 0, 1, 0, 0] -> [ 0,  1,  1,  0,  0,  1,  1];
       [ 0, 1, 0, 1] -> [ 1,  0,  1,  1,  0,  1,  1];
       [ 0, 1, 1, 0] -> [ 1,  0,  1,  1,  1,  1,  1];
       [ 0, 1, 1, 1] -> [ 1,  1,  1,  0,  0,  0,  0];
       [ 1, 0, 0, 0] -> [ 1,  1,  1,  1,  1,  1,  1];
       [ 1, 0, 0, 1] -> [ 1,  1,  1,  1,  0,  1,  1];
```

```
module BCD7
title 'BCD to seven segment display driver'

"Segments    -a-
"          f|   |b
"            -g-
"          e|   |c
"            -d-

        bcd7     device  'P16L8';

        D3,D2,D1,D0      pin 2,3,4,5;
        a,b,c,d,e,f,g    pin 13,14,15,16,17,18,19;
        OE               pin 11;

        bcd      = [D3,D2,D1,D0];
        led      = [a,b,c,d,e,f,g];

        ON,OFF   = 0,1;   " for common anode LEDs

equations
        led.oe = !OE;

@DCSET
truth_table     (bcd -> [  a,    b,    c,    d,    e,    f,    g])
                  0  -> [ ON,   ON,   ON,   ON,   ON,   ON,  OFF];
                  1  -> [OFF,   ON,   ON,  OFF,  OFF,  OFF,  OFF];
                  2  -> [ ON,   ON,  OFF,   ON,   ON,  OFF,   ON];
                  3  -> [ ON,   ON,   ON,   ON,  OFF,  OFF,   ON];
                  4  -> [OFF,   ON,   ON,  OFF,  OFF,   ON,   ON];
                  5  -> [ ON,  OFF,   ON,   ON,  OFF,   ON,   ON];
                  6  -> [ ON,  OFF,   ON,   ON,   ON,   ON,   ON];
                  7  -> [ ON,   ON,   ON,  OFF,  OFF,  OFF,  OFF];
                  8  -> [ ON,   ON,   ON,   ON,   ON,   ON,   ON];
                  9  -> [ ON,   ON,   ON,   ON,  OFF,   ON,   ON];

test_vectors     ([OE,bcd] -> [  a,    b,    c,    d,    e,    f,    g])
                  [ 0, 1 ] -> [OFF,   ON,   ON,  OFF,  OFF,  OFF,  OFF];
                  [ 0, 2 ] -> [ ON,   ON,  OFF,   ON,   ON,  OFF,   ON];
                  [ 0, 3 ] -> [ ON,   ON,   ON,   ON,  OFF,  OFF,   ON];
                  [ 0, 4 ] -> [OFF,   ON,   ON,  OFF,  OFF,   ON,   ON];
                  [ 0, 5 ] -> [ ON,  OFF,   ON,   ON,  OFF,   ON,   ON];
                  [ 0, 6 ] -> [ ON,  OFF,   ON,   ON,   ON,   ON,   ON];
                  [ 0, 7 ] -> [ ON,   ON,   ON,  OFF,  OFF,  OFF,  OFF];
                  [ 0, 8 ] -> [ ON,   ON,   ON,   ON,   ON,   ON,   ON];
                  [ 0, 9 ] -> [ ON,   ON,   ON,   ON,  OFF,   ON,   ON];
                  [ 0, 0 ] -> [ ON,   ON,   ON,   ON,   ON,   ON,  OFF];
                  [ 1, 5 ] -> [.Z., .Z., .Z., .Z., .Z., .Z., .Z.];
end
```

Figure 8.42 Seven-segment display driver truth table

fectively utilized. It should be pointed out, however, that not all PLD design tools using Espresso exploit that algorithm's don't-care features. The *@DCSET* directive used in this design specifies that don't-care optimization should be performed by ABEL.

8.13 STATE MACHINE DESIGN FOR PLDS

PLDs are excellent vehicles for the implementation of state machines. There are a wide variety of devices available with architectures designed specifically for state machine applications. In addition, high-level design entry and simulation tools for PLDs help simplify the specification and verification of complex state machines.

Using ABEL to Describe State Machines

As we briefly described in Chapter 2, truth tables were actually one of the first methods used for describing PLD logic. Truth tables are particularly well suited for describing state machines that contain a large number of similar transitions. Truth tables written for the purpose of state machine specification are often referred to as state tables.

To demonstrate how a state machine is expressed using truth tables, consider the state graph illustrated in Figure 8.43. The state machine illustrated by the state graph has four states (represented by the bubbles) and nine possible state transitions (represented by the arrows). This Moore-model state machine has no outputs other than the state register itself (which could consist of a pair of D-type flip-flops).

To describe this simple state machine in an ABEL-HDL truth table, all possible transition conditions are listed along with their resulting next states. The ABEL HDL design file is shown in Figure 8.44. In the design file, the four states

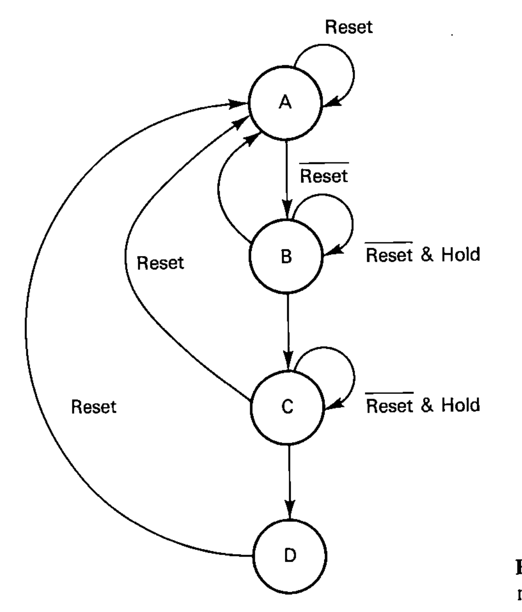

Figure 8.43 Four-state Moore-model state machine

```
module TT2
title 'Truth table flow control example'

        tt2      device   'P16R4';

        Clk,OE            pin 1,11;
        Hold,Reset        pin 2,3;
        Q1,Q0             pin 15,16;

        A = 0; B = 1; C = 2; D = 3;
        Bits = [Q1,Q0];

Equations
        [Q1,Q0].clk = Clk;

Truth_table
       ([Reset,Hold,[Q1,Q0]] :> [Q1,Q0])
        [  1  , .X.,   .X.  ] :>    A ;
        [  0  , .X.,    A   ] :>    B ;
        [  0  ,  1 ,    B   ] :>    B ;
        [  0  ,  0 ,    B   ] :>    C ;
        [  0  ,  1 ,    C   ] :>    C ;
        [  0  ,  0 ,    C   ] :>    D ;
        [ .X. , .X.,    D   ] :>    A ;

Test_Vectors
       ([Clk,OE,Reset,Hold] -> [Q1,Q0])
        [.C., 0,  1  ,  0 ] ->    A;
        [.C., 0,  1  ,  1 ] ->    A;
        [.C., 0,  0  ,  1 ] ->    B;
        [.C., 0,  0  ,  1 ] ->    B;
        [.C., 0,  0  ,  0 ] ->    C;
        [.C., 0,  0  ,  1 ] ->    C;
        [.C., 0,  0  ,  0 ] ->    D;
        [.C., 0,  0  ,  0 ] ->    A;
        [.C., 0,  0  ,  0 ] ->    B;
        [.C., 0,  1  ,  0 ] ->    A;
        [.C., 0,  0  ,  0 ] ->    B;
        [.C., 0,  0  ,  0 ] ->    C;
        [.C., 0,  0  ,  0 ] ->    D;
        [.C., 0,  0  ,  0 ] ->    A;

End
```

Figure 8.44 Four-state Moore-model state machine described with truth table

of the machine are given the values of zero through three and the symbolic names of A, B, C, and D, respectively. The state values decode to the following values for the two state bits:

State	$Q1$	$Q0$
A	0	0
B	0	1
C	1	0
D	1	1

Next, a truth table is used to specify the state machine's operation. The first line of the truth table specifies that the machine is to return to state *A* unconditionally when the *Reset* input is asserted. The second line specifies that, when *Reset* is false, the machine should advance to state *B* (without regard to the value of the *Hold* input). The third through sixth entries describe the behavior of the state machine in states *B* and *C*. In these states, an asserted *Hold* input results in the machine holding in its current state, while a false *Hold* results in the machine advancing to the next state. The final line of the truth table specifies that the machine should return unconditionally to state *A* after one clock cycle in state *D*.

The truth table is a convenient method for describing state machines that have relatively few transitions, or transition logic that is common to many states. More complex state machines, however, are often better described using alternative methods. One such method is the state diagram. A state diagram is a relatively simple method of describing the operation of complex Mealy-model and Moore-model state machines. State diagrams may take the form of textual statements or graphic flow diagram representations. Regardless of the actual input format, all state diagrams share similar basic elements: state register declarations, state values, state transitions, and state machine output logic.

Any state machine that can be described in a state diagram can also be described using a truth table (or equations, for that matter). The key difference between truth tables and state diagrams is that when you describe a state machine using a truth table, you describe the machine primarily in terms of its transitions. When you use a state diagram, you describe the machine more in terms of its possible states. Understanding this difference can be helpful when you are deciding which method to use for a given state machine.

ABEL State Diagrams

State diagram language syntax varies somewhat from one PLD language to another, but all of the commonly used state diagram languages share the same basic features. Figure 8.45 shows an ABEL state diagram. This state diagram describes the same state machine that we previously described using a truth table. Notice that the state diagram contains one state description for each of the four states in the machine. The state diagram, like the truth table, has a header which specifies which signals are to be used for the state register. Unlike a truth table, other state machine inputs and outputs are not included in the state diagram header. Another difference is that the state register declared in the state machine header is flip-flop type independent; the state machine is written the same regardless of the type of flip-flops used for the state register.

Following the header is a series of state descriptions. Each state description includes a value declaration, transition information, and optional equations for circuit I/O (not used in this example).

To describe the state machine transitions, each state description contains one or more transition statements. If you compare the state description for each

```
module SM2
title 'State machine flow control example'

        sm2     device  'P16R4';

        Clk,OE          pin 1,11;
        Hold,Reset      pin 2,3;
        Q1,Q0           pin 15,16;

        A = 0; B = 1; C = 2; D = 3;
        Bits = [Q1,Q0];

Equations
        [Q1,Q0].clk = Clk;
        [Q1,Q0].oe  = !OE;

State_Diagram [Q1,Q0]

        State A: If !Reset Then B Else A ;

        State B: Case    Reset          : A;
                        !Reset &  Hold  : B;
                        !Reset & !Hold  : C;
                 Endcase;

        State C: If Reset Then A
                 Else If Hold Then C
                 Else D;

        State D: Goto A;

Test_Vectors
       ([Clk,OE,Reset,Hold] -> [Q1,Q0])
        [.C., 0,  1  ,  0 ] ->    A;
        [.C., 0,  1  ,  1 ] ->    A;
        [.C., 0,  0  ,  1 ] ->    B;
        [.C., 0,  0  ,  1 ] ->    B;
        [.C., 0,  0  ,  0 ] ->    C;
        [.C., 0,  0  ,  1 ] ->    C;
        [.C., 0,  0  ,  0 ] ->    D;
        [.C., 0,  0  ,  0 ] ->    A;
        [.C., 0,  0  ,  0 ] ->    B;
        [.C., 0,  1  ,  0 ] ->    A;
        [.C., 0,  0  ,  0 ] ->    B;
        [.C., 0,  0  ,  0 ] ->    C;
        [.C., 0,  0  ,  0 ] ->    D;
        [.C., 0,  0  ,  0 ] ->    A;
```

Figure 8.45 ABEL state diagram syntax

state to its equivalent bubble in the state graph, you can see how the state transitions are expressed. State *A* has a simple two-way branch which is described using an *IF-THEN-ELSE* statement. States *B* and *C* have identical transition logic, but were described using different methods to show the use of ABEL's *CASE* transition statement. State *D* has a single unconditional transition, represented by the *GOTO* statement.

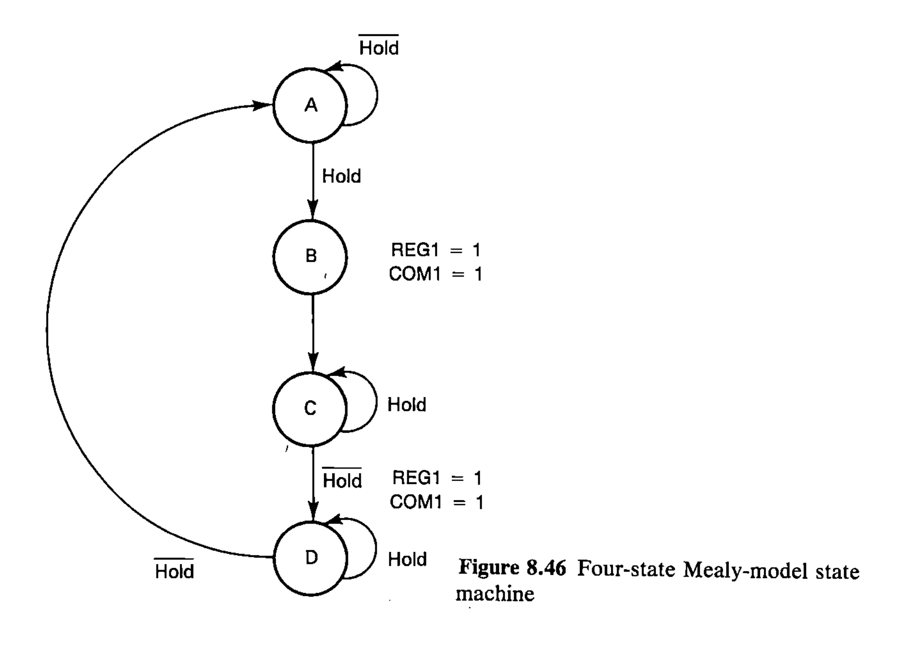

Figure 8.46 Four-state Mealy-model state machine

This simple state machine has demonstrated how state machine flow is described. There are other elements to state machines that must also be addressed, however, so let's examine a somewhat more complex state machine. Figure 8.46 shows a state graph of a simple Mealy-model state machine with the same four states as presented previously, but with different transition logic and with other state machine outputs.

The operation of this state machine begins in state *A* and remains there as long as the *Hold* input is false. When *Hold* is asserted, the machine advances to state *B*, and from there to state *C*, remaining in *C* until *Hold* is again false. In states *C* and *D*, an asserted *Hold* signal results in the machine staying in its present state, while a false *Hold* results in the machine advancing to the next state (*D* and *A*, respectively). The state graph also shows two outputs, and their corresponding values for each state in the machine. One of the outputs (*COM*1) is a combinational signal, while the other (*REG*1) is registered. If we map this flow diagram directly into an ABEL state machine description, we might write a design file such as the one shown in Figure 8.47.

As written, this state machine has a peculiar timing characteristic that needs attention: the registered output *REG*1 lags the state machine by a full clock cycle. This behavior is reflected in the test vectors written in the ABEL-HDL source file. Notice that, in the ninth test vector, *REG*1 is still false even though the machine has advanced to state *D*. Similarly, *REG*1 remains true for a full clock cycle in state *A* when the machine transitions from state *D*. This is because the input forming logic for *REG*1's D-type flip-flop relies on the current state as reflected in the *Q*1 and *Q*0 state registers. Since *REG*1, *Q*1 and *Q*0 are all clocked

```
module SM3
title 'State machine output control example'

        sm3     device  'P16R4';

        Clk,Hold,OE       pin 1,2,11;
        Q1,Q0             pin 15,16;
        REG1,COM1         pin 17,18;

        A = 0; B = 1; C = 2; D = 3;
        Bits = [Q1,Q0];

Equations
        [REG1,Bits].clk = Clk;
        [REG1,Bits].oe  = !OE;

State_Diagram Bits

        State A:
                REG1 := 0;
                COM1  = 0;
                If !Hold Then A Else B ;

        State B:
                REG1 := 1;
                COM1  = 1;
                Goto C;

        State C:
                REG1 := 0;
                COM1  = 0;
                If Hold Then C Else D;

        State D:
                REG1 := 1;
                COM1  = 1;
                If Hold Then D Else A;

Test_Vectors
       ([Clk,OE,Hold] -> [Bits,REG1,COM1])
        [.C., 0,  0 ] -> [.X. , .X., .X.];
        [.C., 0,  0 ] -> [.X. , .X., .X.];
        [.C., 0,  0 ] -> [.X. , .X., .X.];
        [.C., 0,  0 ] -> [ A  ,  0 ,  0 ];
        [.C., 0,  0 ] -> [ A  ,  0 ,  0 ];
        [.C., 0,  1 ] -> [ B  ,  0 ,  1 ];
        [.C., 0,  1 ] -> [ C  ,  1 ,  0 ];
        [.C., 0,  1 ] -> [ C  ,  0 ,  0 ];
        [.C., 0,  0 ] -> [ D  ,  0 ,  1 ];
        [.C., 0,  1 ] -> [ D  ,  1 ,  1 ];
        [.C., 0,  1 ] -> [ D  ,  1 ,  1 ];
        [.C., 0,  0 ] -> [ A  ,  1 ,  0 ];
        [.C., 0,  0 ] -> [ A  ,  0 ,  0 ];
End
```

Figure 8.47 Four-state Mealy-model state machine described in ABEL

```
module SM4
title 'State machine output control example'

        sm4     device  'P16R4';

        Clk,Hold,OE     pin 1,2,11;
        Q1,Q0           pin 15,16;
        REG1,COM1       pin 17,18;

        A = 0; B = 1; C = 2; D = 3;
        Bits = [Q1,Q0];

Equations
        [REG1,Bits].clk = Clk;
        [REG1,Bits].oe  = !OE;

State_Diagram Bits

        State A:
                REG1 := 0;
                COM1  = 0;
                If !Hold Then A Else B ;

        State B:
                REG1 := 1;
                COM1  = 1;
                Goto C;

        State C:
                COM1  = 0;
                If Hold Then C  With REG1 := 0 Endwith
                        Else D  With REG1 := 1 Endwith;

        State D:
                REG1 := 1;
                COM1  = 1;
                If Hold Then D Else A;

Test_Vectors
       ([Clk,OE,Hold] -> [Bits,REG1,COM1])
        [.C., 0,  0 ] -> [.X. , .X., .X.];
        [.C., 0,  0 ] -> [.X. , .X., .X.];
        [.C., 0,  0 ] -> [.X. , .X., .X.];
        [.C., 0,  0 ] -> [ A  ,  0 ,  0 ];
        [.C., 0,  0 ] -> [ A  ,  0 ,  0 ];
        [.C., 0,  1 ] -> [ B  ,  0 ,  1 ];
        [.C., 0,  1 ] -> [ C  ,  1 ,  0 ];
        [.C., 0,  1 ] -> [ C  ,  0 ,  0 ];
        [.C., 0,  0 ] -> [ D  ,  1 ,  1 ];
        [.C., 0,  1 ] -> [ D  ,  1 ,  1 ];
        [.C., 0,  1 ] -> [ D  ,  1 ,  1 ];
        [.C., 0,  0 ] -> [ A  ,  1 ,  0 ];
        [.C., 0,  0 ] -> [ A  ,  0 ,  0 ];
End
```

Figure 8.48 Modified ABEL-HDL state machine

from a common source, the current state information is not available to *REG*1 at the appropriate time.

This situation is corrected by decoding the logic for *REG*1 one state early. To do this, we write the equation for *REG*1 in terms of state transitions, rather than in terms of current states. This is done in ABEL-HDL by writing the equation in a *WITH-ENDWITH* section associated with a particular transition condition. The modified state machine description is shown in Figure 8.48.

8.14 STATE MACHINE POWER-UP AND ILLEGAL STATES

An ABEL-HDL state machine description is flip-flop type independent in regards to the state register. There are situations, however, when the type of flip-flop being used, and the architecture of the target PLD, can affect how the state machine is described. Consider, for example, a modified form of our first state machine. This state machine, shown in Figure 8.49, has only three significant states. The fourth state (state *D*) is shown but, as you can see, there are no transitions into this state. State *D* is, therefore, an illegal state.

Many (perhaps most) large state machines have a number of such illegal states. The task of the designer is to ensure that the machine is prevented from entering an illegal state, or is provided with appropriate logic for escaping from the illegal state. In a device with D-type flip-flops, the inherent default state (all registers low) can be used to advantage to protect against illegal states.

If we implement this state machine with the same state values as used previously (where state *D* had a value of 3, or binary 11) and the same 16R4 device, we will need to provide a transition out of state *D* since *D* represents the inherent default state (and possibly the power-up state as well) of the machine. The reset

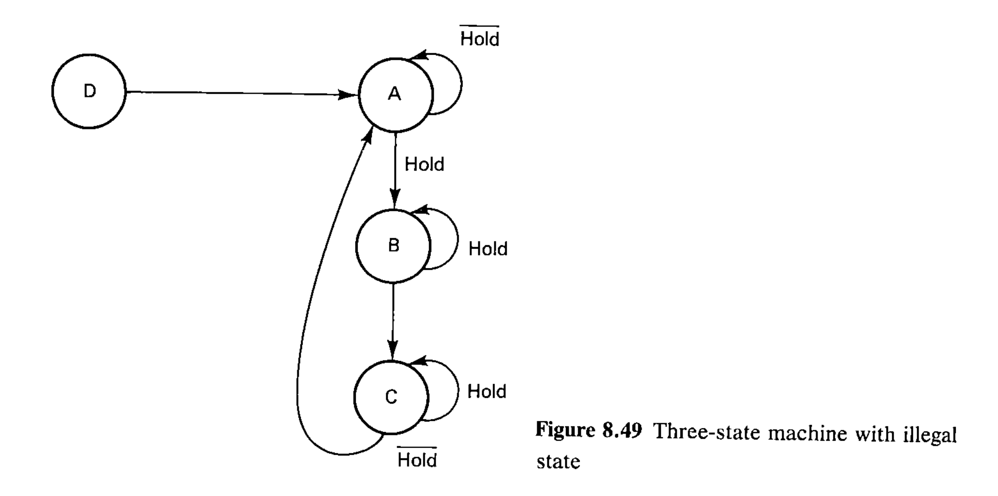

Figure 8.49 Three-state machine with illegal state

```
module SM5
title 'State machine illegal states example'

        sm5     device  'P16R4';

        Clk,OE          pin 1,11;
        Hold,Reset      pin 2,3;
        Q1,Q0           pin 15,16;

        A = 0; B = 1; C = 2; D = 3;
        Bits = [Q1,Q0];

Equations
        [Q1,Q0].clk = Clk;
        [Q1,Q0].oe  = !OE;

State_Diagram [Q1,Q0]

        State A: If !Hold Then A Else B ;

        State B: If  Hold Then B Else C;

        State C: If  Hold Then C Else A;

        State D: Goto A;

Test_Vectors
       ([Clk,OE,Hold] -> [Q1,Q0])
        [.C., 0,  0 ] ->   .X.;
        [.C., 0,  0 ] ->   .X.;
        [.C., 0,  0 ] ->    A ;
        [.C., 0,  0 ] ->    A ;
        [.C., 0,  1 ] ->    B ;
        [.C., 0,  1 ] ->    B ;
        [.C., 0,  0 ] ->    C ;
        [.C., 0,  1 ] ->    C ;
        [.C., 0,  0 ] ->    A ;
End
```

Figure 8.50 Three-state machine with illegal state described with ABEL

transition shown in the flow diagram (and in the ABEL design description shown in Figure 8.50) represents this method of circuit protection. For large state machines having many illegal states, however, the transition logic required to cover all illegal states may be impractical.

An alternative method, useful if the actual state values are not important, is to rearrange the values so that a valid state (either *A*, *B*, or *C*) has the default value. Doing so, however, may leave open the question of whether the power-up state of the machine is accounted for (PLDs from different manufacturers may power-up to a register-low state, a register-high state, or an unpredictable state).

If the state machine is being implemented in a device with another flip-flop type, particular care must be taken to ensure that all possible states are accounted for in some way, or that some form of global reset is provided. As we saw earlier, some PLA-type devices feature a special product term that can be used to detect and escape from illegal states. This *complement array,* as it is called, can also be emulated by using a device output in a manner similar to that used for counter termination earlier in this chapter.

8.15 ADDING STATE BITS TO SAVE LOGIC

We've seen many examples in which design modifications were made in order to fit into the constraints of specific PLD architectures. Another example where such modifications are frequently required is the situation when there are too many flip-flops or device outputs required by a design that uses classic state machine design philosophies. When a state machine is designed, it's often easiest to create the state transition circuitry first and then attach other circuitry (such as output decoding circuitry) to the state machine later. This typically results in the use of more I/O resources than are actually required to implement the circuit.

Let's look at a simple example of this situation. Consider the block diagram for a decade counter and display decoder shown in Figure 8.51. There are a number of possible methods of implementing this circuit. If the actual values of the state bits are important, the circuit would probably be implemented in two pieces with four flip-flops required for the counter state machine and seven outputs required for the display decoder circuit. This is the preferred method if you are limited in the number of flip-flops available. The ABEL source file of Figure 8.52 implements this counter/decoder in a large device using this method.

An alternative method, if the actual state of the counter isn't important, is to combine the function of the decoder with the state machine logic. In this implementation, the state machine uses seven state bits, the value of which correspond to the ten output conditions of the decoder circuit. Figure 8.53 shows this design, implemented in a single PAL device. Combining state bits with output logic can be an effective way to tailor a state machine application to the constraints of a particular device architecture.

Figure 8.51 Decade counter and display decoder

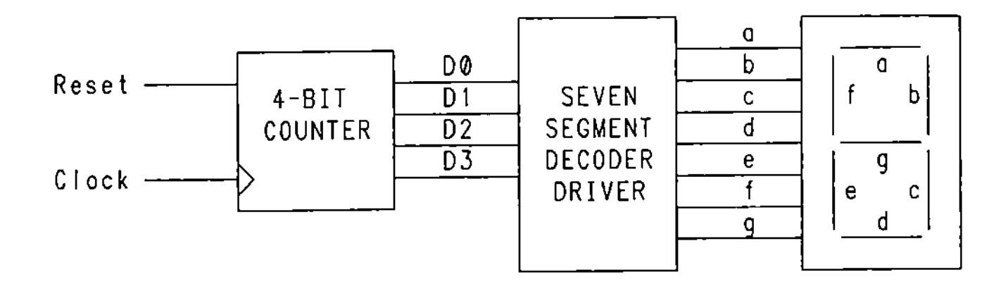

```
module CNTBCD
title 'Decade counter and seven segment decoder'

        cntbcd   device  'P26CV12';

        Clk,Reset,OE      pin;
        D3,D2,D1,D0       pin istype 'reg_D,invert';
        a,b,c,d,e,f,g     pin istype 'com,buffer';

        bcd      = [D3,D2,D1,D0];
        led      = [a,b,c,d,e,f,g];

        ON,OFF   = 0,1;   " for common anode LEDs

Equations
        bcd := (bcd.fb + 1) & (bcd.fb < 9) & !Reset;

        bcd.clk = Clk;

        led.oe  = !OE;

@DCSET
truth_table (bcd.fb -> [  a,   b,   c,   d,   e,   f,   g])
               0    -> [ ON,  ON,  ON,  ON,  ON,  ON, OFF];
               1    -> [OFF,  ON,  ON, OFF, OFF, OFF, OFF];
               2    -> [ ON,  ON, OFF,  ON,  ON, OFF,  ON];
               3    -> [ ON,  ON,  ON,  ON, OFF, OFF,  ON];
               4    -> [OFF,  ON,  ON, OFF, OFF,  ON,  ON];
               5    -> [ ON, OFF,  ON,  ON, OFF,  ON,  ON];
               6    -> [ ON, OFF,  ON,  ON,  ON,  ON,  ON];
               7    -> [ ON,  ON,  ON, OFF, OFF, OFF, OFF];
               8    -> [ ON,  ON,  ON,  ON,  ON,  ON,  ON];
               9    -> [ ON,  ON,  ON,  ON, OFF,  ON,  ON];

test_vectors ([Clk,OE,Reset] -> [bcd,  a,   b,   c,   d,   e,   f,   g])
              [.C., 0,  1  ] -> [ 0 , ON,  ON,  ON,  ON,  ON,  ON, OFF];
              [.C., 0,  0  ] -> [ 1 ,OFF,  ON,  ON, OFF, OFF, OFF, OFF];
              [.C., 0,  0  ] -> [ 2 , ON,  ON, OFF,  ON,  ON, OFF,  ON];
              [.C., 0,  0  ] -> [ 3 , ON,  ON,  ON,  ON, OFF, OFF,  ON];
              [.C., 0,  0  ] -> [ 4 ,OFF,  ON,  ON, OFF, OFF,  ON,  ON];
              [.C., 0,  0  ] -> [ 5 , ON, OFF,  ON,  ON, OFF,  ON,  ON];
              [ 0 , 1,  0  ] -> [.X.,.Z., .Z., .Z., .Z., .Z., .Z., .Z.];
              [.C., 0,  0  ] -> [ 6 , ON, OFF,  ON,  ON,  ON,  ON,  ON];
              [.C., 0,  0  ] -> [ 7 , ON,  ON,  ON, OFF, OFF, OFF, OFF];
              [.C., 0,  0  ] -> [ 8 , ON,  ON,  ON,  ON,  ON,  ON,  ON];
              [.C., 0,  0  ] -> [ 9 , ON,  ON,  ON,  ON, OFF,  ON,  ON];
              [.C., 0,  0  ] -> [ 0 , ON,  ON,  ON,  ON,  ON,  ON, OFF];
              [.C., 0,  0  ] -> [ 1 ,OFF,  ON,  ON, OFF, OFF, OFF, OFF];
end
```

Figure 8.52 Decade counter and decoder described with ABEL

```
module  CountLED
title 'Seven segment display decoder state machine'

        countled         device  'P16R8';

        Clk,Reset,OE     pin 1,2,11;
        a,b,c,d,e,f,g    pin 12,13,14,15,16,17,18;

        ON,OFF  = 0,1;          " for common anode LEDs

        led = [  a,   b,   c,   d,   e,   f,   g];
        S0  = [ ON,  ON,  ON,  ON,  ON,  ON, OFF];
        S1  = [OFF,  ON,  ON, OFF, OFF, OFF, OFF];
        S2  = [ ON,  ON, OFF,  ON,  ON, OFF,  ON];
        S3  = [ ON,  ON,  ON,  ON, OFF, OFF,  ON];
        S4  = [OFF,  ON,  ON, OFF, OFF,  ON,  ON];
        S5  = [ ON, OFF,  ON,  ON, OFF,  ON,  ON];
        S6  = [ ON, OFF,  ON,  ON,  ON,  ON,  ON];
        S7  = [ ON,  ON,  ON, OFF, OFF, OFF, OFF];
        S8  = [ ON,  ON,  ON,  ON,  ON,  ON,  ON];
        S9  = [ ON,  ON,  ON,  ON, OFF,  ON,  ON];
        S15 = [OFF, OFF, OFF, OFF, OFF, OFF, OFF];

Equations
        led.clk = Clk;
        led.oe  = !OE;

@DCSET
State_Diagram led
        State S0: IF Reset THEN S0 ELSE S1;
        State S1: IF Reset THEN S0 ELSE S2;
        State S2: IF Reset THEN S0 ELSE S3;
        State S3: IF Reset THEN S0 ELSE S4;
        State S4: IF Reset THEN S0 ELSE S5;
        State S5: IF Reset THEN S0 ELSE S6;
        State S6: IF Reset THEN S0 ELSE S7;
        State S7: IF Reset THEN S0 ELSE S8;
        State S8: IF Reset THEN S0 ELSE S9;
        State S9: IF Reset THEN S0 ELSE S0;
        State S15: GOTO S0;

Test_Vectors
       ([Clk,OE,Reset] -> led)
        [.C., 0,  1  ] -> S0;
        [.C., 0,  0  ] -> S1;
        [.C., 0,  0  ] -> S2;
        [.C., 0,  0  ] -> S3;
        [.C., 0,  0  ] -> S4;
        [.C., 0,  0  ] -> S5;
        [ 0 , 1,  0  ] -> .Z.;
        [.C., 0,  0  ] -> S6;
        [.C., 0,  0  ] -> S7;
        [.C., 0,  0  ] -> S8;
        [.C., 0,  0  ] -> S9;
        [.C., 0,  0  ] -> S0;
        [.C., 0,  0  ] -> S1;
        [.C., 0,  0  ] -> S2;
        [.C., 0,  0  ] -> S3;
        [.C., 0,  1  ] -> S0;
end
```

Figure 8.53 Modified decade counter and display decoder described with AEL

8.16 STATE MACHINE PARTITIONING

When we converted the state machine and decode logic of the previous example into an equivalent state machine, we increased the number of flip-flops used in the design while, at the same time, realizing a reduction in the total amount of logic required. It's also possible to simplify the overall logic required for a given state machine by partitioning that state machine into two or more smaller machines.

State machine partitioning is often required when the size of the machine prohibits its being implemented in PAL-type devices due to restricted numbers of output registers or product terms. Complex state machines with large numbers of transitions often result in extremely large amounts of logic for each state bit of the machine. Even if the number of required state register flip-flops is appropriate for a single PLD, the amount of input forming logic for those flip-flops may necessitate partitioning the state machine into multiple PLDs. At other times, the reduction in logic achieved through state machine partitioning is sufficient to allow the state machine to be implemented in a single PLD when it otherwise wouldn't have fit.

The result of partitioning a state machine is usually a larger number of required state register flip-flops and unique states, but this increase in resources is usually offset by reductions in the amount of required input-forming logic. Combining state machine partitioning with careful state assignments will result in the most efficient circuits.

To demonstrate how a state machine can be partitioned, we'll design a serial transmitter circuit. A serial transmitter is a circuit that accepts 8-bit wide parallel data and sends it in a serial stream at a rate corresponding to the selected baud (bits per second) rate (see Figure 8.54). To do this, the transmitter (which is actually one half of a UART circuit) must first store the eight bits of data, and then route each bit in turn to the output line. The eight bits of data are packaged with a start bit and two or more stop bits.

Figure 8.55 shows a state graph representing the operation of the serial transmitter circuit. A corresponding ABEL design description is shown in Figure 8.56. The ABEL design shown uses an EP900 PLD, which has sufficient I/O pins and flip-flop resources to implement the entire design as written.

The serial transmitter has eleven states corresponding to the start bit, the eight data bits, and the two stop bits of the transmitted data. The machine's default state is state *StopBit1*, and the design of the transition logic ensures that the

Figure 8.54 Serial transmission of 8-bit data

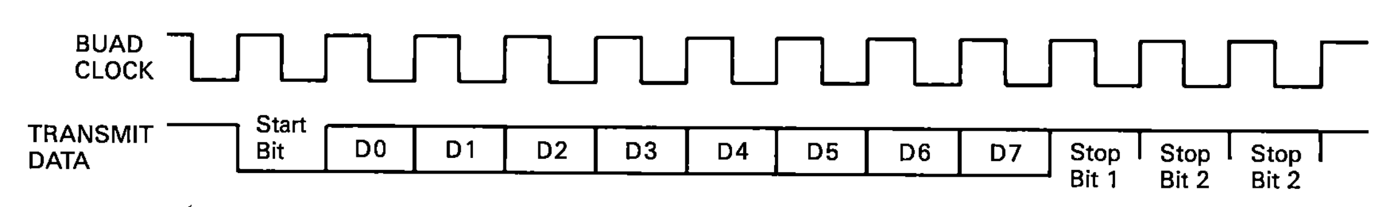

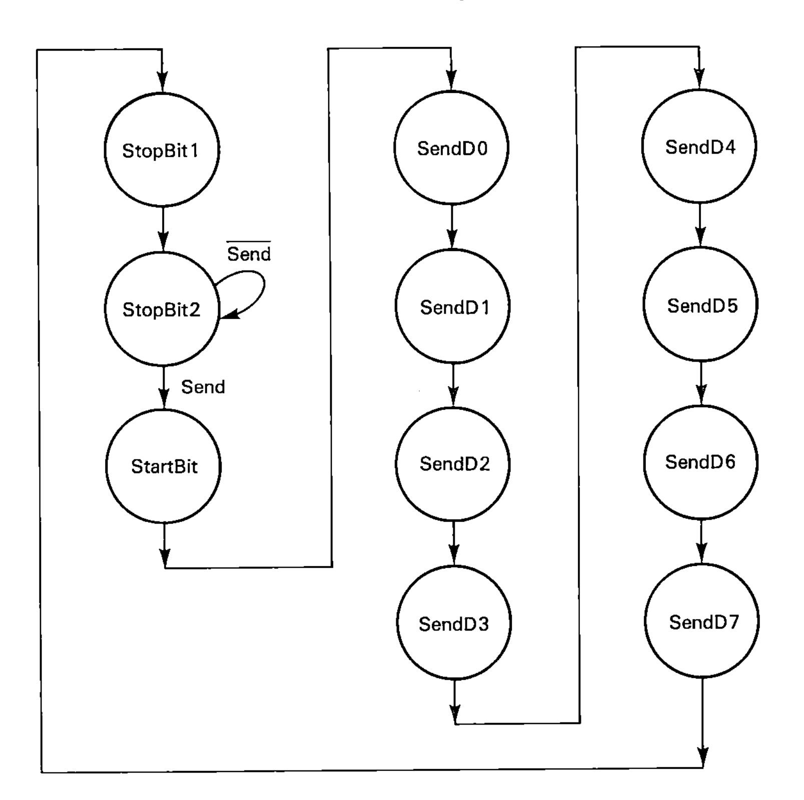

Figure 8.55 Serial transmitter state machine

machine will always transition to the initial state *StopBit2* within ten clock cycles of power-up regardless of the power-up state. In the absence of data to send, the machine remains in state *StopBit2* awaiting an assertion of the *Send* input. In the *StartBit* state, a start bit (logic level 0) is sent to the serial output (*TXD*) while the parallel data is loaded into the 8-bit wide data register. In each of the following eight states (*Send0* through *Send7*) the *D0* bit of the data is sent to *TXD* and the data register is shifted one bit. At the conclusion of the data output sequence, the *NextWord* output is strobed, indicating that the next byte of parallel data can be loaded.

Written in the form shown above, this state machine consumes an unnecessarily large amount of logic. While the EP900 device used is capable of implementing the entire machine, a simpler device would not be up to the task. Furthermore, if this design was being implemented in an FPGA, the wide input-forming logic that results from the above design description would be quite inefficient. Carefully chosen state values could result in a decrease in the amount of logic required for each state bit, but the most dramatic savings in logic is obtained by increasing the number of flip-flops used for the state register. This can be done either by adding flip-flops to the state register set (consisting, as

```
module XMIT1
title 'Serial transmiter'

        xmit1   device  'E0900';

"Inputs
        DI7,DI6,DI5,DI4,DI3,DI2,DI1,DI0 pin 2,3,4,17,18,19,22,23;
        Clk2,Clk1                       pin 21,1;
        Clr,Send                        pin 37,38;

"Outputs
        D7,D6,D5,D4,D3,D2,D1,D0         pin 25,26,27,28,29,30,31,32;
        SM3,SM2,SM1,SM0                 pin 5,6,7,8;
        NextWord,TXD                    pin 9,10;

"Sets
        Data            = [D7..D0];
        DataShift       = [D0,D7..D1];
        Sreg            = [SM3,SM2,SM1,SM0];
        DataIn          = [DI7..DI0];
        BaudClk         = [Clk1,Clk2];

"Register Types
        SM3,SM2,SM1,SM0         istype 'reg_d';
        D7,D6,D5,D4,D3,D2,D1,D0 istype 'reg_D';

Equations
        Sreg.C          = Clk1;
        Data.C          = Clk2;

        Sreg.RE         = Clr;
        Data.RE         = Clr;

Declarations
"States
        StopBit1        =  0;
        StopBit2        =  1;
        StartBit        =  2;
        SendD0          =  8;
        SendD1          =  9;
        SendD2          = 10;
        SendD3          = 11;
        SendD4          = 12;
        SendD5          = 13;
        SendD6          = 14;
        SendD7          = 15;

State_Diagram Sreg
        State StopBit1: NextWord  = 1;
                        TXD       = 1;
                        Data     := DataIn;
                        goto StopBit2;

        State StopBit2: NextWord  = 0;
                        TXD       = 1;
                        Data     := DataIn;
                        if(Send) then StartBit else StopBit2;

        State StartBit: NextWord  = 0;
                        TXD       = 0;
                        Data     := DataIn;
                        goto SendD0;
```

Figure 8.56 Serial transmitter state machine described with ABEL

```
State SendD0:   NextWord  = 0;
                TXD       = D0;
                Data     := DataShift;
                goto SendD1;

State SendD1:   NextWord  = 0;
                TXD       = D0;
                Data     := DataShift;
                goto SendD2;

State SendD2:   NextWord  = 0;
                TXD       = D0;
                Data     := DataShift;
                goto SendD3;

State SendD3:   NextWord  = 0;
                TXD       = D0;
                Data     := DataShift;
                goto SendD4;

State SendD4:   NextWord  = 0;
                TXD       = D0;
                Data     := DataShift;
                goto SendD5;

State SendD5:   NextWord  = 0;
                TXD       = D0;
                Data     := DataShift;
                goto SendD6;

State SendD6:   NextWord  = 0;
                TXD       = D0;
                Data     := DataShift;
                goto SendD7;

State SendD7:   NextWord  = 0;
                TXD       = D0;
                Data     := DataShift;
                goto StopBit1;

Declarations
        C = .C.;

test_vectors
       ([BaudClk,Clr,Send,DataIn] -> [Sreg     ,NextWord,TXD])
        [   C   , 1 ,  0 ,   0  ] -> [StopBit1,    1    , 1 ];
        [   C   , 1 ,  0 ,   0  ] -> [StopBit1,    1    , 1 ];
        [   C   , 0 ,  0 ,   0  ] -> [StopBit2,    0    , 1 ];
        [   C   , 0 ,  0 , ^h55 ] -> [StopBit2,    0    , 1 ];
        [   C   , 0 ,  0 , ^h55 ] -> [StopBit2,    0    , 1 ];
        [   C   , 0 ,  1 , ^h55 ] -> [StartBit,    0    , 0 ];
        [   C   , 0 ,  1 , ^h55 ] -> [SendD0  ,    0    , 1 ];
        [   C   , 0 ,  0 ,   0  ] -> [SendD1  ,    0    , 0 ];
        [   C   , 0 ,  0 ,   0  ] -> [SendD2  ,    0    , 1 ];
        [   C   , 0 ,  0 ,   0  ] -> [SendD3  ,    0    , 0 ];
        [   C   , 0 ,  0 ,   0  ] -> [SendD4  ,    0    , 1 ];
        [   C   , 0 ,  0 ,   0  ] -> [SendD5  ,    0    , 0 ];
        [   C   , 0 ,  0 ,   0  ] -> [SendD6  ,    0    , 1 ];
        [   C   , 0 ,  0 ,   0  ] -> [SendD7  ,    0    , 0 ];
        [   C   , 0 ,  0 ,   0  ] -> [StopBit1,    1    , 1 ];
        [   C   , 0 ,  0 , ^h0F ] -> [StopBit2,    0    , 1 ];
```

Figure 8.56 *(continued)*

```
      [   C   , 0 ,  0 , ^hOF ] -> [StopBit2,   0   , 1 ];
      [   C   , 0 ,  1 , ^hOF ] -> [StartBit,   0   , 0 ];
      [   C   , 0 ,  1 , ^hOF ] -> [SendD0  ,   0   , 1 ];
      [   C   , 0 ,  0 ,   0  ] -> [SendD1  ,   0   , 1 ];
      [   C   , 0 ,  0 ,   0  ] -> [SendD2  ,   0   , 1 ];
      [   C   , 0 ,  0 ,   0  ] -> [SendD3  ,   0   , 1 ];
      [   C   , 0 ,  0 ,   0  ] -> [SendD4  ,   0   , 0 ];
      [   C   , 0 ,  0 ,   0  ] -> [SendD5  ,   0   , 0 ];
      [   C   , 0 ,  0 ,   0  ] -> [SendD6  ,   0   , 0 ];
      [   C   , 0 ,  0 ,   0  ] -> [SendD7  ,   0   , 0 ];
      [   C   , 0 ,  0 ,   0  ] -> [StopBit1,   1   , 1 ];
end
```

Figure 8.56 *(continued)*

written, of *SM*0 through *SM*3) and reassigning the state values or by partitioning the state machine.

This state machine can be easily partitioned into smaller functional units resulting in an increase in the total number of states and decrease in the total amount of logic required. Figure 8.57 illustrates the three functional parts of this design that are candidates for partitioning.

For this redesign, we've chosen to isolate the eight states that correspond to the data shifting function, and implement these eight states as a semi-independent counter. Figure 8.58 shows state graphs for the resulting two segments of the partitioned state machine. The first state machine has four states: *StopBit*1, *StopBit*2, *StartBit,* and *Shift*. The *Shift* state replaces the eight shifter states of the previous state graph. During this state of the primary state machine, the counter state machine is active and cycles through its eight states. The data shifter is also split into a separate functional unit, further simplifying the design. The communication between the two machines is one-way; the primary state machine has complete control over the counter state machine, resetting the counter bits and monitoring their value during shift-and-transmit operations.

In the corresponding ABEL-HDL design description shown in Figure 8.59, the counter and shifter portions are described using high-level equations, while the remaining state machine logic is described using ABEL's state diagram language. The new design requires one additional registered output and one additional

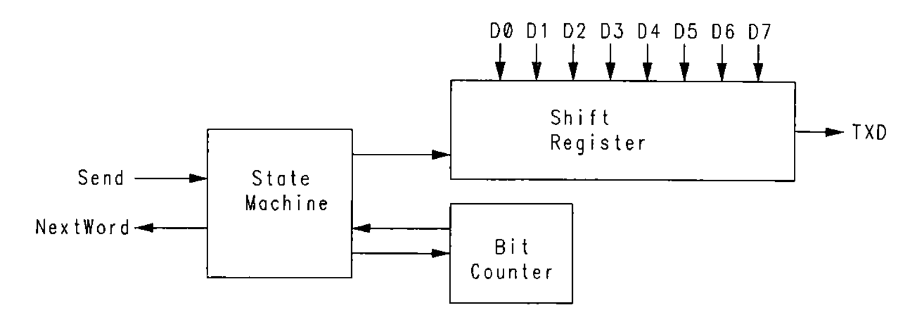

Figure 8.57 Serial transmitter functional blocks

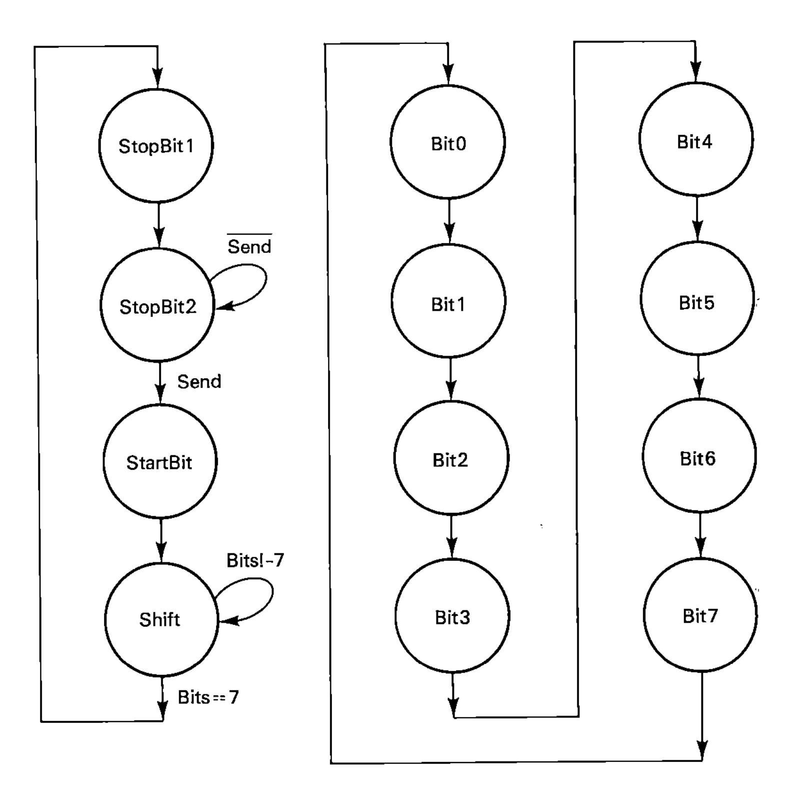

Figure 8.58 Partitioned serial transmitter state machine

fedback combinational output. The eight-state counter and the data shift register are both controlled from the primary state machine, which now has a total of four states. The counter value represented by *Bits* is reset in state *StartBit* (the expression *Bits.RE* = !0 results in ones being applied to the reset inputs of all registers in the *Bits* set) and increments during state *Shift* until the terminal count of seven is reached. The shifter logic is controlled by the additional fedback output signal *DataLoad,* which indicates when the input data should be loaded rather than shifted.

Where the original (non-partitioned) design would have required only four state register flip-flops to represent the eleven possible states, the partitioned design requires five: three for the eight-state counter, and two for the controlling state machine. The benefit to this partitioning strategy is that the input-forming logic is much less complex, and determining an optimal state encoding for the two state machines is much simpler. As an added benefit, the counter portion can be implemented using T-type flip-flops, while the primary state machine can be implemented using D-type flip-flops.

One side benefit to this redesign is that the primary state machine now

```
module XMIT2
title 'Serial transmiter with counter'

        xmit2   device  'E0900';

"Inputs
        DI7,DI6,DI5,DI4,DI3,DI2,DI1,DI0 pin 2,3,4,17,18,19,22,23;
        Clk2,Clk1                       pin 21,1;
        Clr,Send                        pin 37,38;

"Outputs
        D7,D6,D5,D4,D3,D2,D1,D0         pin 25,26,27,28,29,30,31,32;
        B2,B1,B0                        pin 33,34,35;
        SM1,SM0                         pin 5,6;
        NextWord,TXD,DataLoad           pin 9,10,11;

"Sets
        Bits            = [B2..B0];
        Data            = [D7..D0];
        DataShift       = [0,D7..D1];
        Sreg            = [SM1,SM0];
        DataIn          = [DI7..DI0];
        BaudClk         = [Clk1,Clk2];

"Register Types
        SM1,SM0                   istype 'reg_d';
        B2,B1,B0                  istype 'reg_T';
        D7,D6,D5,D4,D3,D2,D1,D0 istype 'reg_D';

Equations

"Data Shift Register
        Data    :=  DataLoad & DataIn           "Load
                 # !DataLoad & DataShift;       "Shift

"Bit Counter (T f/f)
        Bits.t  = Bits.fb $ (Bits.fb + 1);

        Sreg.C          = Clk1;
        Data.C          = Clk2;
        Bits.C          = Clk2;

        Sreg.RE         = Clr;
        Data.RE         = Clr;

Declarations
"States
        StopBit1        = 0;
        StopBit2        = 1;
        StartBit        = 2;
        Shift           = 3;


State_Diagram Sreg
        State StopBit1:
                        NextWord  = 1;
                        DataLoad  = 1;
                        TXD       = 1;
                        goto StopBit2;
```

Figure 8.59 Partitioned serial transmitter described with ABEL

```
        State StopBit2:
                        NextWord  = 0;
                        DataLoad  = 1;
                        TXD       = 1;
                        if(Send) then StartBit else StopBit2;

        State StartBit:
                        NextWord  = 0;
                        DataLoad  = 1;
                        TXD       = 0;
                        Bits.RE   = !0;
                        goto Shift;

        State Shift:
                        NextWord  = 0;
                        DataLoad  = 0;
                        TXD       = D0;
                        Bits.RE   = 0;
                        if (Bits.fb == 7) then StopBit1 else Shift;

test_vectors
([BaudClk,Clr,Send,DataIn] -> [Sreg     ,Bits,NextWord,DataLoad,TXD])
 [  .C.   , 1 ,  0 ,    0  ] -> [StopBit1, .X.,    1    ,    1    , 1 ];
 [  .C.   , 1 ,  0 ,    0  ] -> [StopBit1, .X.,    1    ,    1    , 1 ];
 [  .C.   , 0 ,  0 ,    0  ] -> [StopBit2, .X.,    0    ,    1    , 1 ];
 [  .C.   , 0 ,  0 , ^h55  ] -> [StopBit2, .X.,    0    ,    1    , 1 ];
 [  .C.   , 0 ,  0 , ^h55  ] -> [StopBit2, .X.,    0    ,    1    , 1 ];
 [  .C.   , 0 ,  1 , ^h55  ] -> [StartBit,  0 ,    0    ,    1    , 0 ];
 [  .C.   , 0 ,  1 , ^h55  ] -> [Shift   ,  0 ,    0    ,    0    , 1 ];
 [  .C.   , 0 ,  0 ,    0  ] -> [Shift   ,  1 ,    0    ,    0    , 0 ];
 [  .C.   , 0 ,  0 ,    0  ] -> [Shift   ,  2 ,    0    ,    0    , 1 ];
 [  .C.   , 0 ,  0 ,    0  ] -> [Shift   ,  3 ,    0    ,    0    , 0 ];
 [  .C.   , 0 ,  0 ,    0  ] -> [Shift   ,  4 ,    0    ,    0    , 1 ];
 [  .C.   , 0 ,  0 ,    0  ] -> [Shift   ,  5 ,    0    ,    0    , 0 ];
 [  .C.   , 0 ,  0 ,    0  ] -> [Shift   ,  6 ,    0    ,    0    , 1 ];
 [  .C.   , 0 ,  0 ,    0  ] -> [Shift   ,  7 ,    0    ,    0    , 0 ];
 [  .C.   , 0 ,  0 ,    0  ] -> [StopBit1, .X.,    1    ,    1    , 1 ];
 [  .C.   , 0 ,  0 , ^h0F  ] -> [StopBit2, .X.,    0    ,    1    , 1 ];
 [  .C.   , 0 ,  0 , ^h0F  ] -> [StopBit2, .X.,    0    ,    1    , 1 ];
 [  .C.   , 0 ,  1 , ^h0F  ] -> [StartBit,  0 ,    0    ,    1    , 0 ];
 [  .C.   , 0 ,  1 , ^h0F  ] -> [Shift   ,  0 ,    0    ,    0    , 1 ];
 [  .C.   , 0 ,  0 ,    0  ] -> [Shift   ,  1 ,    0    ,    0    , 1 ];
 [  .C.   , 0 ,  0 ,    0  ] -> [Shift   ,  2 ,    0    ,    0    , 1 ];
 [  .C.   , 0 ,  0 ,    0  ] -> [Shift   ,  3 ,    0    ,    0    , 1 ];
 [  .C.   , 0 ,  0 ,    0  ] -> [Shift   ,  4 ,    0    ,    0    , 0 ];
 [  .C.   , 0 ,  0 ,    0  ] -> [Shift   ,  5 ,    0    ,    0    , 0 ];
 [  .C.   , 0 ,  0 ,    0  ] -> [Shift   ,  6 ,    0    ,    0    , 0 ];
 [  .C.   , 0 ,  0 ,    0  ] -> [Shift   ,  7 ,    0    ,    0    , 0 ];
 [  .C.   , 0 ,  0 ,    0  ] -> [StopBit1, .X.,    1    ,    1    , 1 ];
End
```

Figure 8.59 *(continued)*

requires a maximum of only three clock cycles to stabilize after a random power-up rather than the ten cycles required previously. For clarity, the design is again shown implemented in an EP900, but the partitioning strategy used allows this circuit to be easily implemented in two simpler PALs or in an FPGA.

8.17 SUMMARY

In this chapter, we have shown how high-level design techniques can be used to implement a variety of circuits in PLDs. We have examined some of the design-level optimizations that are often required to fit applications into different PLD architectures. In the following chapter, we'll look at how designs are implemented in FPGAs. Specifically, we will show how high-level design techniques can be used when designing with the *logic cell array* (LCA) devices.

8.18 REFERENCES

Advanced Micro Devices, Incorporated, *PAL Device Handbook,* Advanced Micro Devices, Sunnyvale, CA, 1988.

Altera Corporation, *Applications Handbook,* Altera Corp., Santa Clara, CA, 1988.

Barna, Arpad and Porat, Dan I. *Integrated Circuits in Digital Electronics,* John Wiley and Sons, New York, NY, 1973.

Data I/O Corporation, *ABEL User's Guide,* Data I/O, Redmond, WA, 1988.

Data I/O Corporation, *Gates Logic Synthesizer User Manual,* Data I/O, Redmond, WA, 1989.

9

Designing With FPGAs

In this chapter, we'll examine how designs can be efficiently implemented in FPGA devices. Since each family of FPGAs has unique design requirements and design tools, we can't possibly describe the design process for all of them. Instead, we'll focus on a single type of FPGA—the logic cell array.

The LCA, which we described briefly in Chapter 4, was the first of the field programmable gate array devices. The LCA has gained widespread acceptance as a viable alternative to traditional PLDs and, in some cases, as a replacement for low-end gate arrays: While the LCAs represent just one family of FPGAs, the design requirements and tools are similar enough that many of the concepts discussed here are applicable to other FPGA devices as well.

9.1 LCA ARCHITECTURES

There are two families of LCAs in common use as of this writing—the 2000-series and the 3000-series of devices. A 4000-series was announced by Xilinx in late 1989. With the popularity of these devices, we can expect other LCA families to appear from Xilinx. In addition to the offerings of Xilinx, AT&T has announced that they too will be producing LCA devices.

The 2000-Series LCAs

Figure 9.1 shows a diagram of the Xilinx XC2064, the simplest device in the 2000-series LCA family. As we discussed in Chapter 4, the LCA contains a collection of general purpose *configurable logic blocks* (CLBs) surrounded by specialized *I/O blocks* (IOBs).

Configurable Logic Blocks

These CLBs are the heart of the LCA. An array of these simple elements allows the construction of complex logic circuits. The CLBs are arranged in a rectangular matrix in the interior of the device. The XC2064 device contains a total of 64 CLBs arranged in an eight by eight matrix. In the 2000-series LCA devices, each CLB contains a four-input combinational logic section that is capable of implementing any logic function of up to four variables. In addition, each CLB contains a configurable flip-flop and internal routing and control circuitry. A CLB is depicted in Figure 9.2.

Figure 9.1 XC2064 LCA block diagram

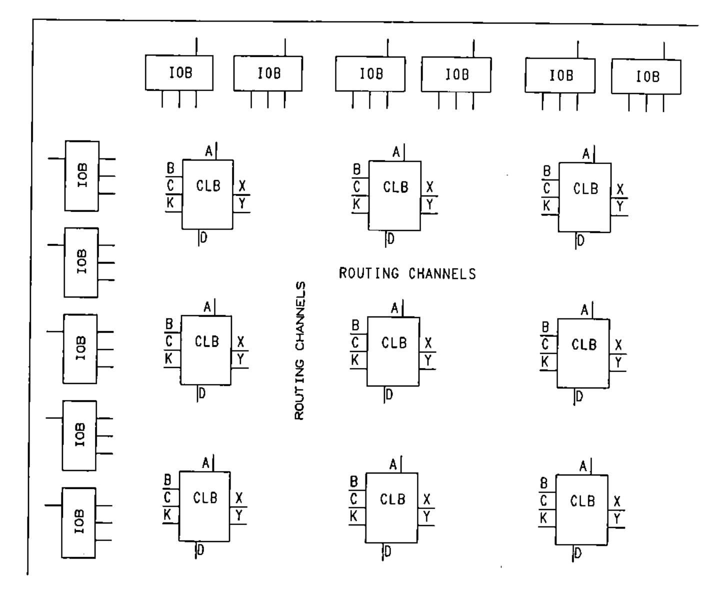

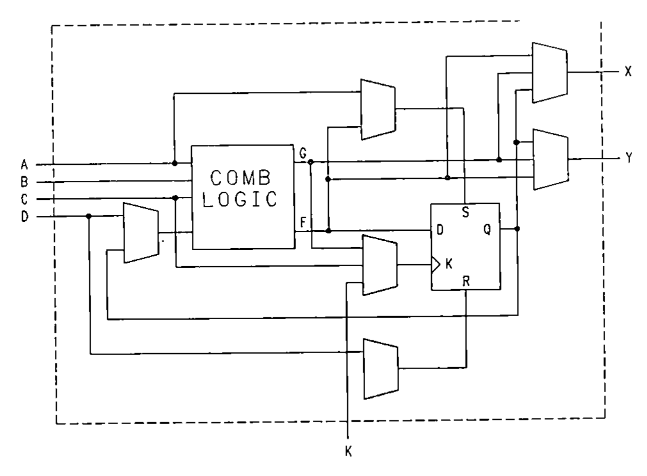

Figure 9.2 2000-series CLB

Each CLB has four general purpose inputs: *A, B, C,* and *D*. A special clock input (*K*) may be driven from the interconnect adjacent to the block. Two outputs are provided from each CLB and these outputs can be used to drive the interconnect to other CLBs and logic blocks, or may be used internally in the CLB.

The combinational logic section of the CLB is implemented as a lookup table, similar to a PROM. This means that any function with up to four inputs can be implemented. The delay through the logic cell is constant regardless of the logic function being implemented. In addition, the CLB can be configured for use as two three-input logic functions or, by selecting the flip-flop output feedback instead of input *D*, the CLB can be configured for multilevel applications including some five-input functions. When a flip-flop is fedback internally, the *D* input can be used as a reset line, but can't be used as a normal input.

If the four-variable form is selected, a single-output function is available from the CLB, as shown in Figure 9.3. Both outputs of the CLB implement the

Figure 9.3 Using CLB for single-output function

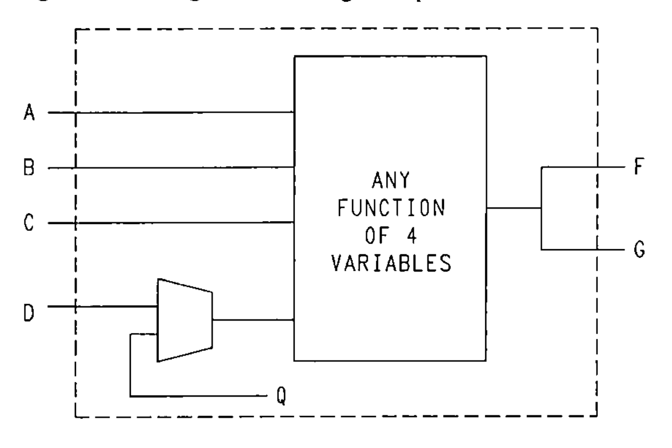

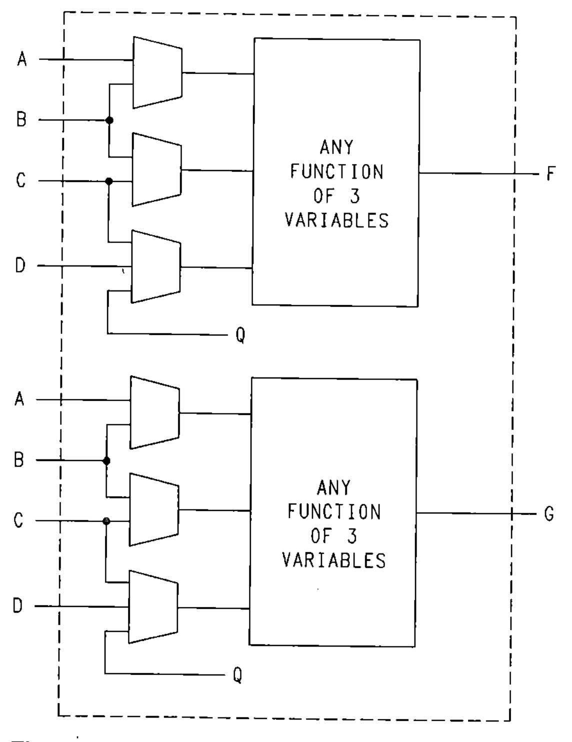

Figure 9.4 Using CLB for two-output function

same function. If the two-function form is selected, the two outputs are independent functions of three variables each (see Figure 9.4). The three input variables for each output may be selected from the four inputs to the cell, and from the fed back output of the CLB's flip-flop. The CLB is also capable of a third form, in which the output of the CLB dynamically switches between the two output functions. This multiplexer operation allows many five-input logic functions to be implemented in a single CLB, as shown in Figure 9.5.

The flip-flop of each CLB can be configured to be either an edge-triggered D-type flip-flop or a level sensitive latch. When not used, the storage element is disabled. Each storage element includes asynchronous set and reset functions. Clocking may be provided from the special clock input to the CLB from the general purpose input *C* or from the *G* output function. The clock may be inverted if needed.

The CLB flip-flop input is driven by the *F* combinational output signal. Asynchronous set and reset functions are provided for each flip-flop. The actual outputs of the CLB, *X*, and *Y* can be configured to be driven by either the combinational outputs or the output of the flip-flop.

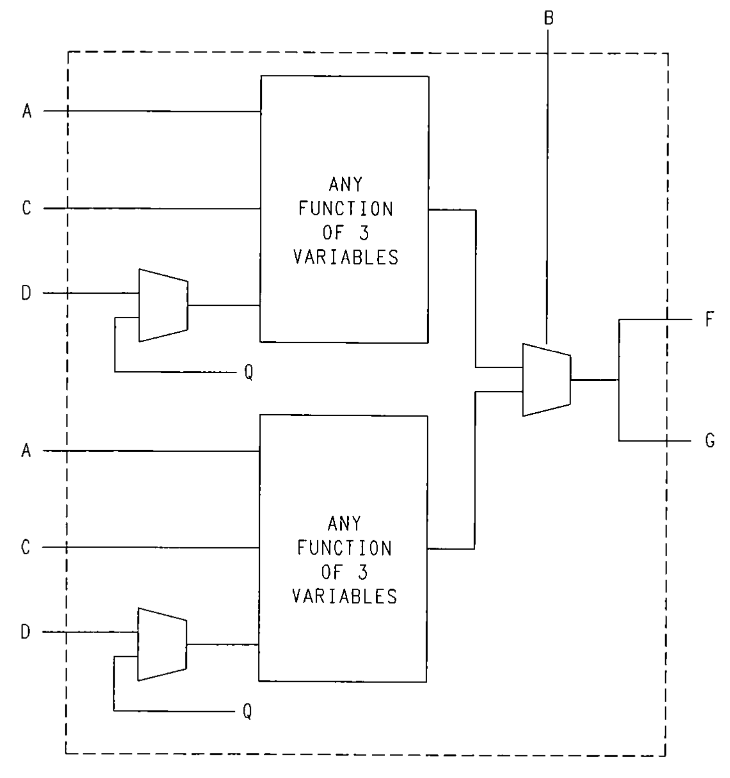

Figure 9.5 Using CLB for five-input function

I/O Blocks

The I/O blocks, 58 of them for the XC2064, are arranged around the periphery of the device. I/O blocks provide the interface between the external pins of the device and the internal logic. Each I/O block can be configured for operation as an input, an output, a three-state output, or a bidirectional buffer. Figure 9.6 shows the general architecture of the 2000-series I/O blocks. The I/O blocks are also configurable for operation with either TTL or CMOS level signals. A selectable edge-triggered flip-flop and a two-input multiplexer are provided in each I/O block to allow synchronization of input signals.

Each I/O block also allows for the inversion of the input or output signal and each has configurable slew rates and pull-up options. Each input provides threshold detection circuitry to translate external signals to internal logic levels. This threshold circuitry allows the device to be configured for operation with either TTL or CMOS levels.

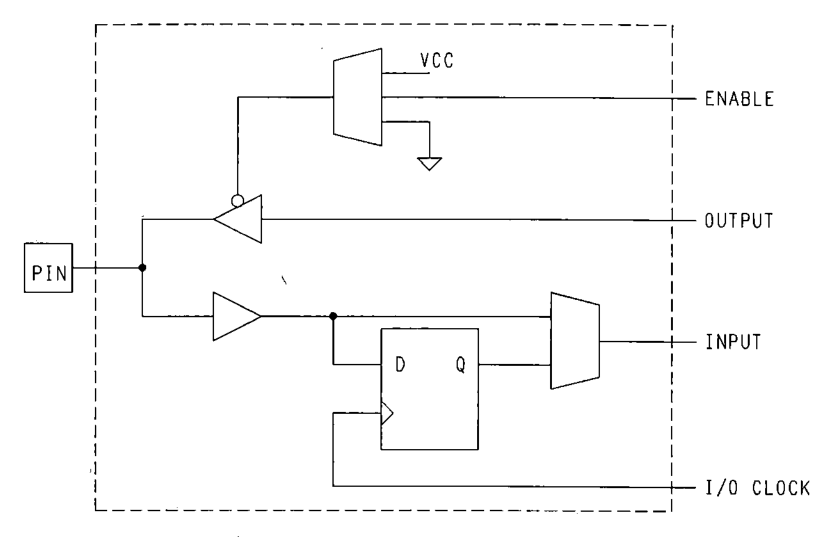

Figure 9.6 2000-series I/O block

Programmable Interconnections

The programmable interconnection channels in the LCA provide routing paths to connect inputs and outputs of the I/O blocks and logic cells in any configuration. The interconnection channels are programmed at crosspoints called *programmable interconnect points,* or PIPs. Interconnection channels are provided in three types: general purpose interconnects, long lines, and direct interconnects.

The *general purpose interconnections* run horizontally and vertically between the rows of columns and logic cells and I/O blocks. Where these segments would overlay each other at the intersection of a row and column, a switching matrix is provided to allow interconnection of adjacent row or column oriented segments. The segments themselves are only the height or width of a logic cell and terminate at the switching matrices.

Long lines, as shown in Figure 9.7, run vertically and horizontally completely across the device. They bypass the switch matrices and are used for signals that must travel a large distance across the device or require a low skew for multiple destinations. Long lines also serve as special interconnects for selected signals.

Direct interconnect lines, also shown in Figure 9.7, are the most efficient way to connect adjacent CLBs or I/O blocks. Signals routed in this way require the least amount of propagation time. For each CLB, direct interconnects may be used to connect the *X* output of the CLB to the *B* input of the CLB immediately to its right and to the *C* input of the CLB to its left. The *Y* output can be connected to the *D* input the CLB immediately above and to the *A* input of the CLB immediately below. When a CLB is adjacent to an I/O block, a direct interconnect is provided alternately to the I/O block inputs and outputs on all four edges of

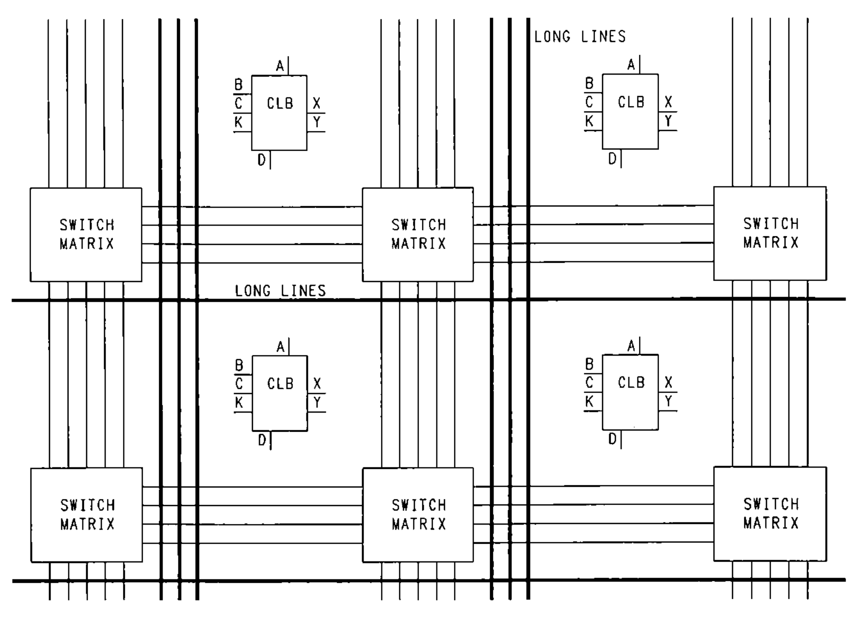

Figure 9.7 LCA interconnection channels

the device. The CLB outputs on the right edge of the device provide additional I/O block interconnects for adjacent I/O blocks.

3000-Series LCAs

The 3000-series LCAs differ from the 2000-series in the complexity of their CLBs and IOBs and, in the addition of an internal bus (specialized long line) that can be used to route multiplexed signals through the device.

The 3000-series CLB is illustrated in Figure 9.8. Each CLB contains two flip-flops, and has more input lines than found in the 2000-series CLBs. Five of these inputs may be used as combinational inputs, while the remainder are used for flip-flop reset, clocking, and data loading functions.

The 3000-series I/O blocks, depicted in Figure 9.9, include a D-type flip-flop for output register use as well as a configurable D-type flip-flop that can be alternatively used as an input latch.

4000-Series LCAs

The 4000 series of LCAs were announced in late 1989 and, like the 2000- and 3000-series LCAs, are static RAM-based devices containing CLBs and various types of interconnect resources. Improvements to the routing architecture and

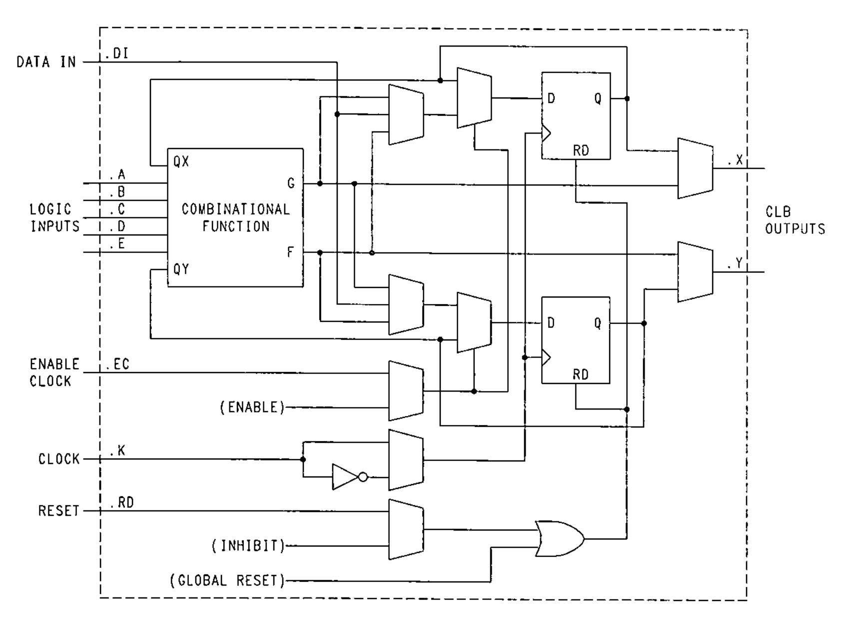

Figure 9.8 3000-series CLB

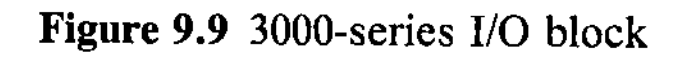
Figure 9.9 3000-series I/O block

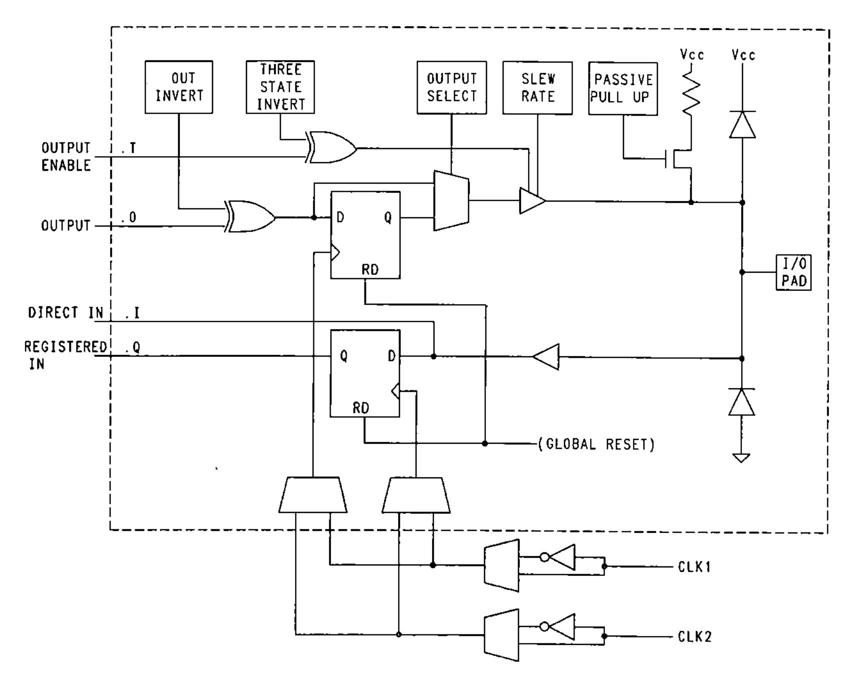

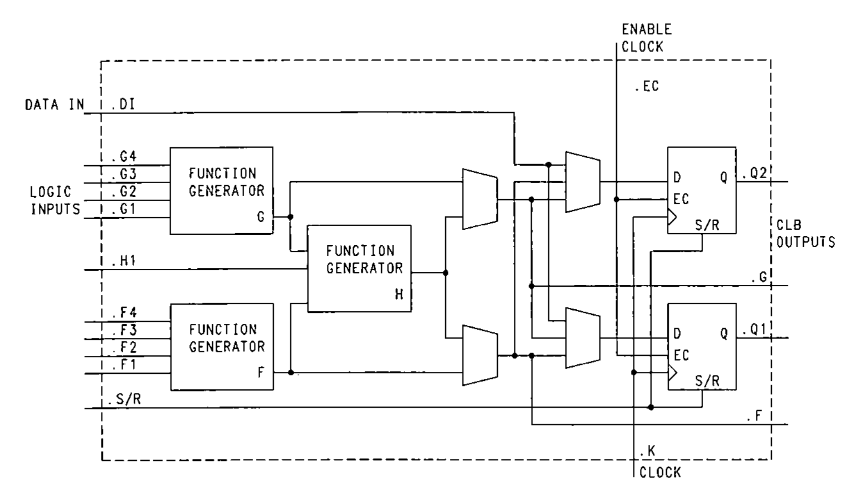

Figure 9.10 4000-series CLB

process improvements are said to increase the 4000-series LCAs' speed of operation to twice that of the 3000-series LCAs. Efficient routing of the 4000-series devices is improved by the addition of significantly more interconnect resources.

The CLBs in the 4000-series LCAs are similar to those found in the 3000 series. A 4000-series CLB is diagrammed in Figure 9.10. Unlike the 3000-series, the 4000-series CLBs have two independent combinational logic blocks with separate inputs. This means that a single CLB can implement logic functions with as many as nine inputs.

The 4000-series devices also feature an on-chip static RAM that can be configured to store up to 2,500 bytes of data. This memory area can be used for lookup tables, data buffers, or other data manipulation functions. The 4000-series devices are expected to be offered in sizes ranging from 2,000 to 20,000 equivalent gates.

9.2 THE LCA DESIGN PROCESS

Converting a design concept into a working LCA-based circuit is a process with many more steps and iterations than the equivalent process using PLDs. Designs intended for LCA implementation must be carefully analyzed to determine their suitability to the LCA architecture and design changes are frequently required to efficiently utilize the devices. The LCA design process consists of a number of steps. Using the Xilinx XACT design system as a model, we have diagrammed these steps in Figure 9.11.

The first phase of the design process is the design entry phase and involves the creation of one or more *external netlist format* (XNF) files. XNF is a common format developed by Xilinx that allows many different design entry tools to be

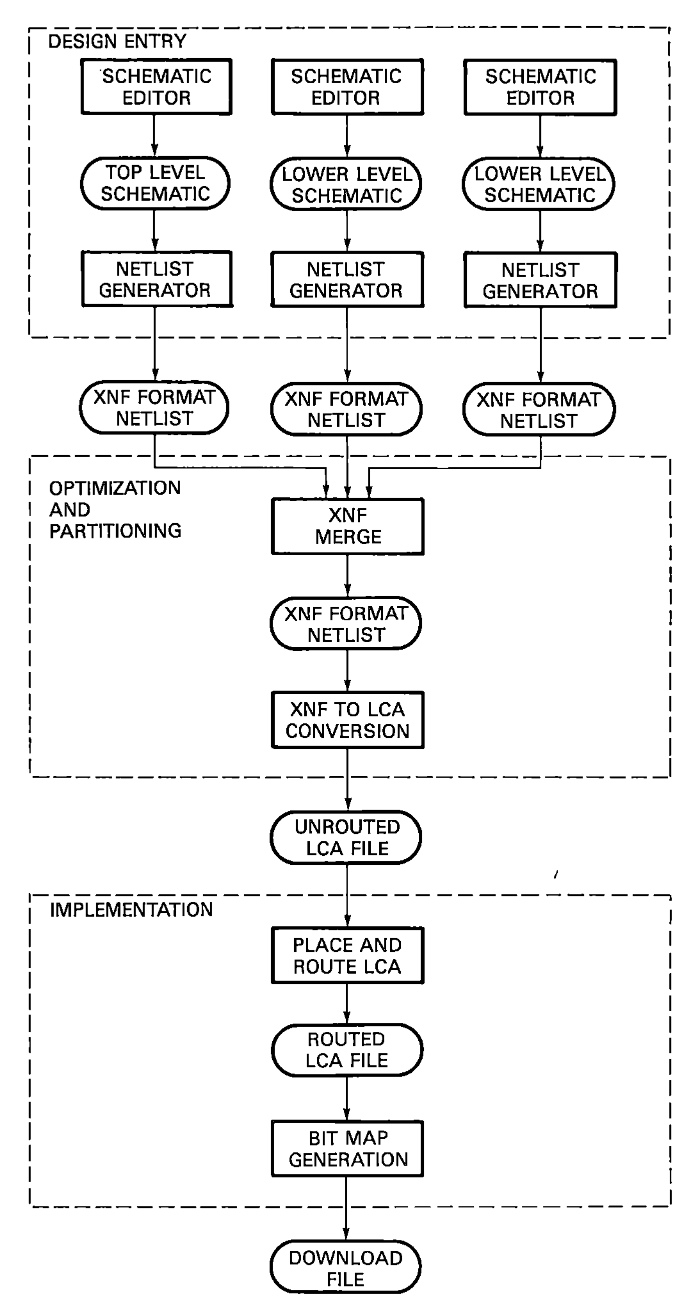

Figure 9.11 LCA design process using XACT tools

used with XACT. XNF design files have a one-to-one correspondence with designs entered in schematic form and are not LCA-specific. These files can be created from schematic drawings (as shown) or from other design entry tools (such as ABEL, CUPL, or PALASM) through the use of design translation utilities.

Once the design has been entered and converted to an XNF representation, it can be manipulated in a variety of ways to create an optimized circuit targeted specifically to a particular LCA. Multiple XNF files can be merged into a single circuit, and various circuit optimizations can be performed to better fit the circuit to the LCA's constrained architecture. At the conclusion of this design optimization phase, the circuit is in a form called an *LCA design file,* which is actually a script file that controls other XACT software features.

The final phase of the LCA design process is the implementation phase, in which the LCA design file is used to create an actual configuration bit pattern for the target device. The most significant part of the implementation phase is the placement and routing of circuit elements in the target LCA. For complex designs that utilize a large percentage (over 80 percent) of an LCA's resources, the placement and routing step can be long and difficult. At the successful conclusion of this phase, simulation tools can be used to verify the correct operation of the LCA circuit and actual devices can be tried out either in-circuit or through the use of an LCA verifier.

While design entry is normally accomplished using general purpose tools such as schematic entry systems, the optimization and implementation phases require specific tools provided by the LCA vendors. To demonstrate how this process works, we'll create a very simple circuit and use the XACT tools to implement the circuit in an LCA.

Design Entry

The first step is to describe the circuit. Figure 9.12 shows a schematic created using Data I/O's FutureNet schematic capture system. The circuit described by this schematic consists of a single D-type flip-flop fed by a single AND gate. The remaining symbols on the schematic represent the LCA's built-in clock and the input and output buffers of the device.

To transfer this schematic into the XACT design tool environment, we first create an XNF format file using the translation utilities provided by Data I/O and Xilinx. The resulting XNF format file is listed in Figure 9.13. In the XNF file, there is one entry for each symbol on the schematic, as well as symbol interconnect data.

Optimization

At this point it would be possible to invoke the *Xilinx circuit optimizer* (XNFOPT) if desired. XNFOPT is particularly useful if the design, as specified, has a structure

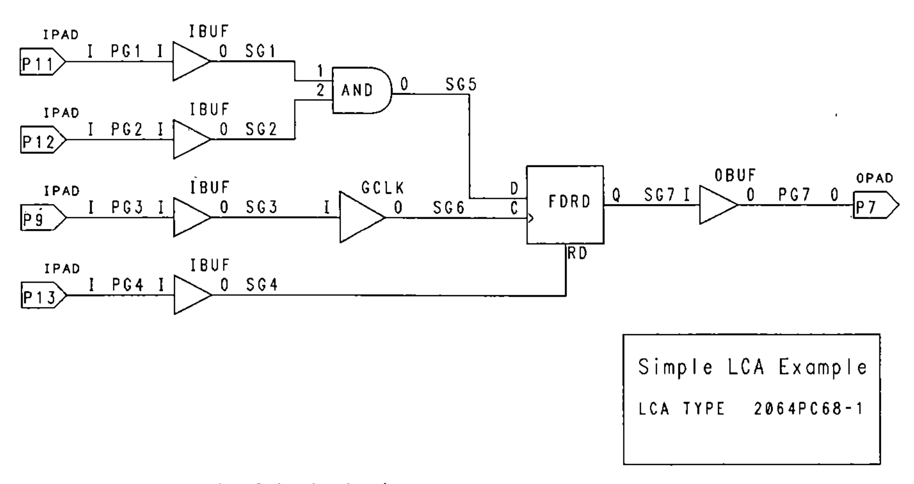

Figure 9.12 Schematic of simple circuit

that is not well suited to the LCA's architecture. For example, if a design is entered that uses wide (greater than five inputs) logic gates, then XNFOPT could be invoked to fragment the large gates into smaller elements more appropriate to the LCA's logic blocks.

This simple circuit can't benefit from any such optimization so we can proceed directly to the next step which is the creation of an LCA design file. The conversion of XNF files to LCA design files is actually a two-step process. XNF files are first converted to an intermediate form, called a *MAP file,* which represents a partitioned circuit. Each of the circuit partitions in the map file correspond to one CLB or IOB in the target device.

Depending on the complexity of the design, different design segments can be combined either before or after the MAP file phase. Combining before the MAP file phase can result in more efficient logic module utilization, since common sub-circuits can be identified and combined. The drawback to merging XNF files before the MAP file phase is the often greater need for routing resources.

After the MAP files have been generated, they can be converted to an LCA design file. The LCA design file is a command script that controls the XACT software during the placement and routing portion of the implementation phase. The entire design can be merged and converted into a single LCA design file or the various design segments can be maintained as separate LCA files and merged later. The decision as to which method to use is based on the size and type of circuit being converted. A single LCA design file that represents our simple unrouted circuit is shown in Figure 9.14.

Implementation

Now that we have an LCA design file, we can begin the placement and routing phase. The XNF design file representing our circuit is fed to the placement and

```
LCANET, 1
PROG, PIN2XNF, 2.01, Created from: small.pin
PART, 2064PC68-1
SYM, SG5, AND
PIN, O, O, SG5
PIN, 2, I, SG2
PIN, 1, I, SG1
END
SYM, SG1, IBUF
PIN, O, O, SG1
PIN, I, I, PG1
END
SYM, SG2, IBUF
PIN, O, O, SG2
PIN, I, I, PG2
END
SYM, SG4, IBUF
PIN, O, O, SG4
PIN, I, I, PG4
END
SYM, PG7, OBUF
PIN, O, O, PG7
PIN, I, I, SG7
END
SYM, SG3, IBUF
PIN, O, O, SG3
PIN, I, I, PG3
END
SYM, SG6, GCLK
PIN, O, O, SG6
PIN, I, I, SG3
END
SYM, SG7, DFF
PIN, RD, I, SG4
PIN, Q, O, SG7
PIN, D, I, SG5
PIN, C, I, SG6
END
EXT, PG1, I,, LOC=P11, BLKNM=PG1
EXT, PG2, I,, LOC=P12, BLKNM=PG2
EXT, PG3, I,, LOC=P9, BLKNM=PG3
EXT, PG4, I,, LOC=P13, BLKNM=PG4
EXT, PG7, O,, LOC=P7, BLKNM=PG7
EOF
```

Figure 9.13 XNF format file representing simple circuit

routing software (*automatic place and route,* or APR) and the result is a new LCA design file that includes the calculated logic block placements and signal routings.

If the automatically calculated logic block placements and signal routings are acceptable, the LCA design file can be converted into a bit pattern appropriate for downloading to the LCA device. Alternatively, the XACT design editor can be used to examine and modify the logic block placements and signal routings. The XACT design editor's floor plan display is shown in Figure 9.15. This screen display shows the signal routings from I/O blocks to the single logic module re-

```
;: small.lca, MAP2LCA 2.20
Design 2064PC68
Speed -1
Addnet SG7  AA.X P7.O
Addnet SG6  AA.K CLK.AA.O
Addnet SG4  AA.D P13.I
Addnet SG3  P9.I CLK.AA.I
Addnet SG2  AA.A P12.I
Addnet SG1  AA.B P11.I
Nameblk AA SG7
Editblk AA
BASE FG
CONFIG X:Q Y: Q:FF SET: RES:D CLK:K: F:A:B
EQUATE F=(A*B)
Endblk
Nameblk P9 PG3
Editblk P9
BASE IO
CONFIG I:PAD BUF:
Endblk
Nameblk P7 PG7
Editblk P7
BASE IO
CONFIG I: BUF:ON
Endblk
Nameblk P13 PG4
Editblk P13
BASE IO
CONFIG I:PAD BUF:
Endblk
Nameblk P12 PG2
Editblk P12
BASE IO
CONFIG I:PAD BUF:
Endblk
Nameblk P11 PG1
Editblk P11
BASE IO
CONFIG I:PAD BUF:
Endblk
```

Figure 9.14 LCA design file representing simple circuit

quired by this circuit. The screen display shown in Figure 9.16 illustrates how the logic function was mapped into one of the device's CLBs.

Configuring the LCA with the bit pattern created from the LCA design file is relatively straightforward. For prototyping, the bit pattern can be loaded directly into the target device through the use of a special download cable that connects the host computer to the actual device. Since the LCA device is volatile, production designs require that some other power-up configuration mechanism be provided. The typical method used is to store the LCA's bit pattern in a configuration PROM that is installed on the board along with the LCA device. A hex PROM format file representing the bit pattern for our simple circuit is shown in Figure 9.17.

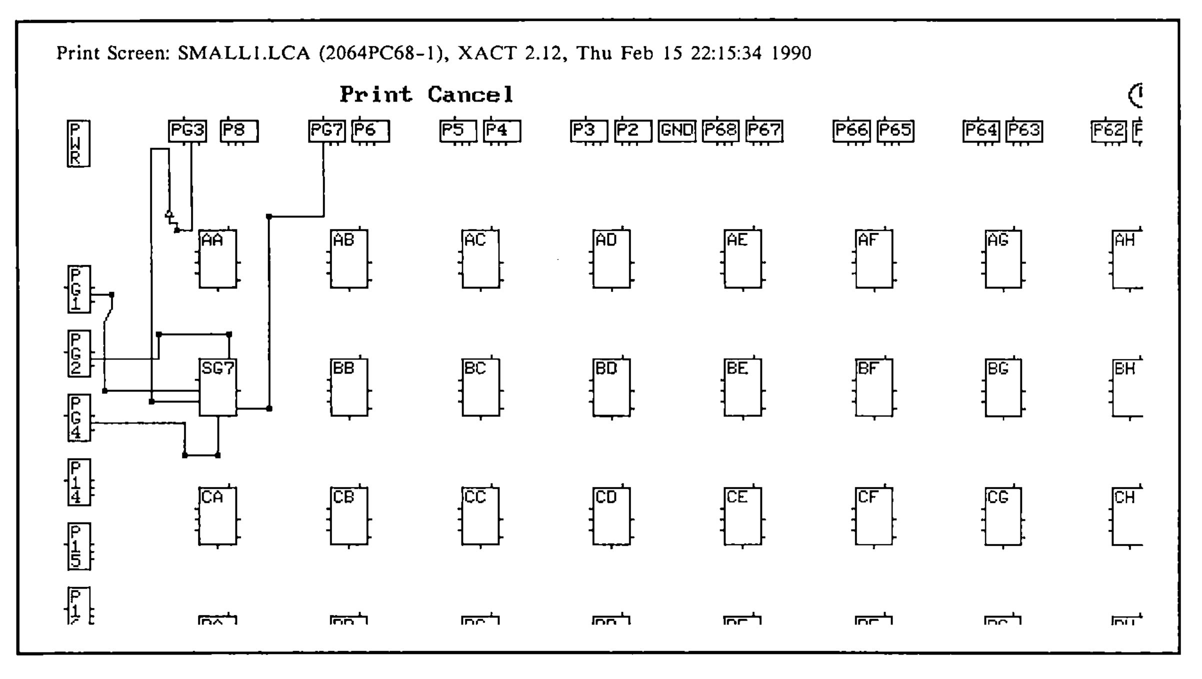

Figure 9.15 XACT design editor floor plan display

Figure 9.16 XACT design editor CLB display

Print Block BA: SMALLI.LCA (2064PC68-1), XACT 2.12, Thu Feb 15 22:16:10 1990

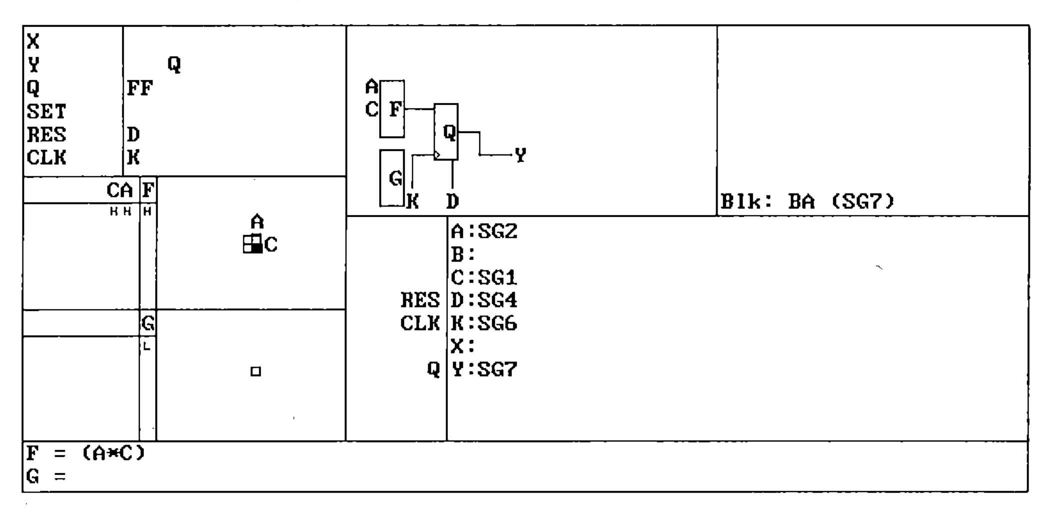

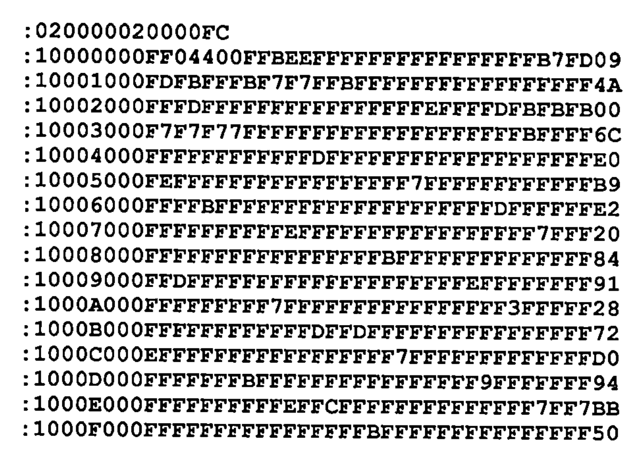

```
:020000020000FC
:10000000FF04400FFBEEFFFFFFFFFFFFFFFFB7FD09
:10001000FDFBFFFBF7F7FFBFFFFFFFFFFFFFFFFF4A
:10002000FFFDFFFFFFFFFFFFFFFFEFFFFDFBFBFB00
:10003000F7F7F77FFFFFFFFFFFFFFFFFFFFBFFFF6C
:10004000FFFFFFFFFFFFDFFFFFFFFFFFFFFFFFFFE0
:10005000FEFFFFFFFFFFFFFFFFFFF7FFFFFFFFFFB9
:10006000FFFFBFFFFFFFFFFFFFFFFFFFFDFFFFFFE2
:10007000FFFFFFFFFFEFFFFFFFFFFFFFFFFF7FFF20
:10008000FFFFFFFFFFFFFFFFFBFFFFFFFFFFFFFF84
:10009000FFDFFFFFFFFFFFFFFFFFFFFEFFFFFFFF91
:1000A000FFFFFFFFF7FFFFFFFFFFFFFFFF3FFFFF28
:1000B000FFFFFFFFFFFFDFFDFFFFFFFFFFFFFFFF72
:1000C000EFFFFFFFFFFFFFFFFFFF7FFFFFFFFFFFD0
:1000D000FFFFFFFBFFFFFFFFFFFFFFFF9FFFFFFF94
:1000E000FFFFFFFFFFEFFCFFFFFFFFFFFFFF7FF7BB
:1000F000FFFFFFFFFFFFFFFFBFFFFFFFFFFFFFFF50
```

Figure 9.17 PROM format file representing LCA bit pattern

9.3 USING HIGH-LEVEL DESIGN TOOLS FOR LCAs

To demonstrate how higher-level design tools can be used for LCA design, we'll implement another simple design. This design consists of a counter and a seven-segment display decoder. It can be implemented using the demo board supplied with the Xilinx PGA Development System. We'll use Data I/O's FutureDesigner to create the design and will use the Xilinx XACT software to implement the resulting circuit in an XC2064 LCA device. In order to demonstrate how designs expressed in different forms are processed, we'll use a mixture of schematics and behavioral description methods for design entry.

Figure 9.18 shows a flow chart of the LCA design process using Future-Designer and XACT. The diagram illustrates the many steps that are gone through to convert design concepts into working LCA devices. Each box in the diagram represents one step in the design process.

Creating a Top-level Circuit Diagram

The first step in the process is to create a top-level drawing of the overall circuit. This is done using the FutureNet schematic editor. The resulting drawing is shown in Figure 9.19. In the top-level drawing, certain parts of the design are expressed structurally. These portions include the oscillator and switch debounce circuitry. The remaining circuitry is expressed in the form of functional blocks. These circuit segments will be described behaviorally using FutureDesigner's Gates design editor.

Our top-level drawing for this circuit has preassigned pins that correspond to the predetermined pin layout provided on the Xilinx demo board. It's also possible to leave the LCA pin numbers unassigned and let the APR algorithm determine the numbering later.

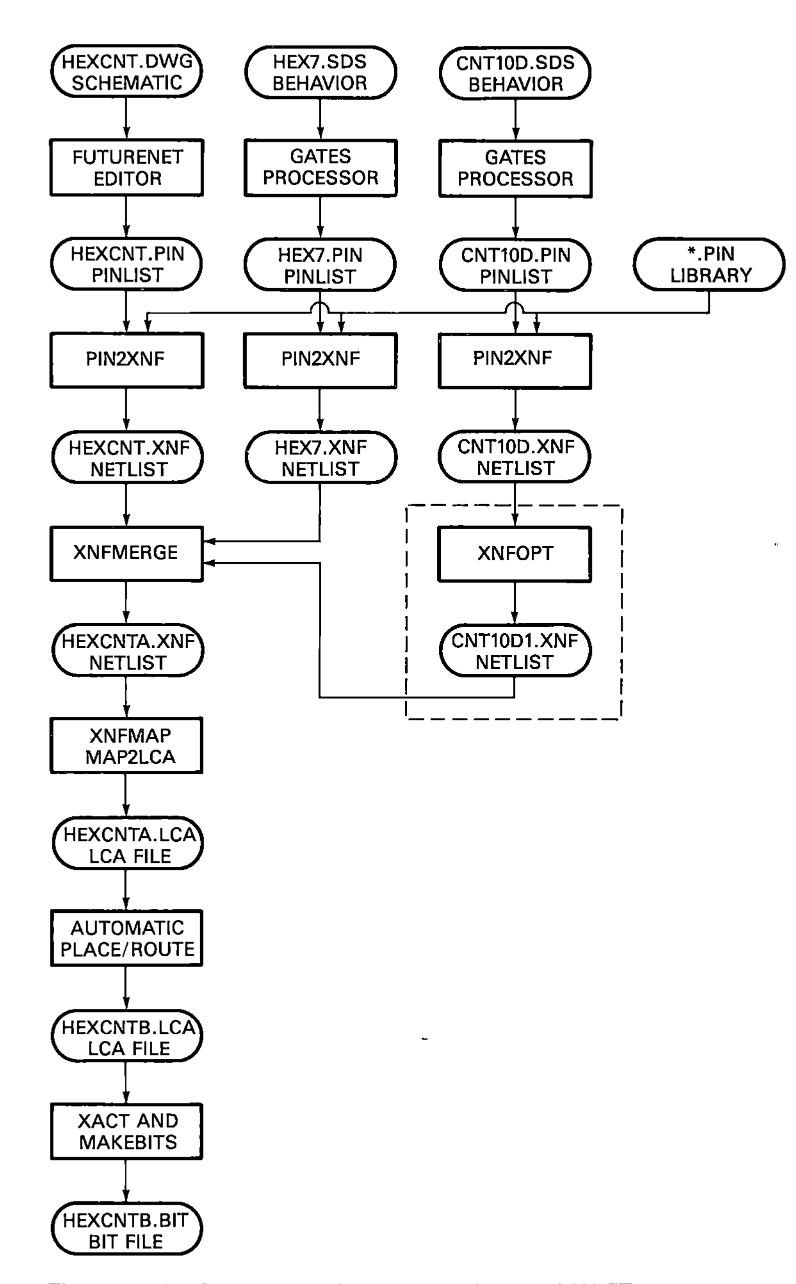

Figure 9.18 Design process using FutureDesigner and XACT

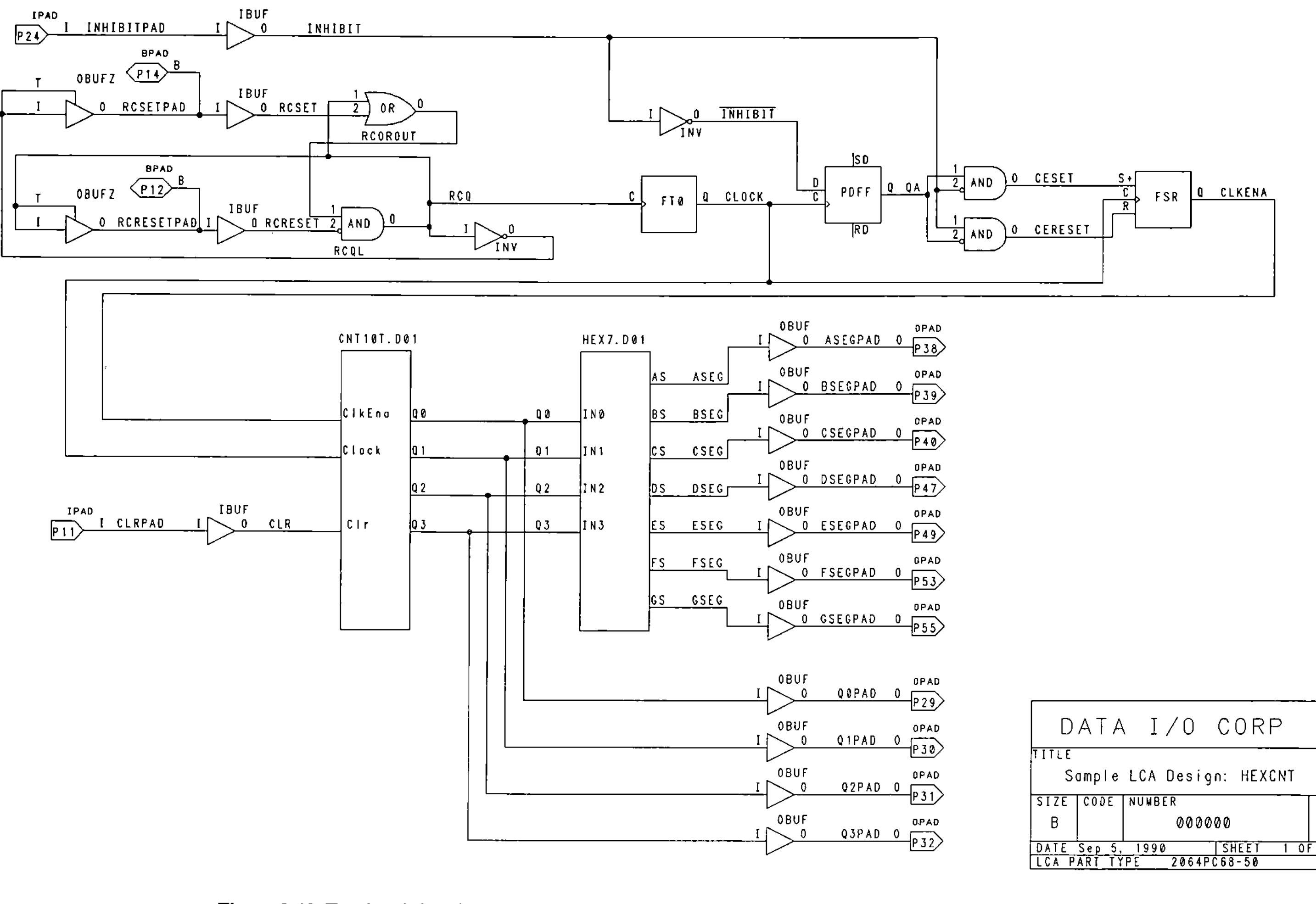

Figure 9.19 Top-level drawing of counter/decoder circuit

```
      Form type:  declaration      Form name:  cnt10d

Name                               Type        Definition/Comment
ClkEna,Clock,Clr                   input
Q3..Q0                             bidir       d
Count                              set         [Q3..Q0]


      Form type:  equation         Form name:  cnt10d

Count.d = (Count.q + 1) &  ClkEna
        # (Count.q    ) & !ClkEna

Count.clk = Clock

Count.ar = !Clr # (Count.q == 10)
```

Figure 9.20 Gates declaration and equation forms for counter

Describing the Counter and Decoder Modules

The next step is to invoke Gates (from within the schematic editor) to behaviorally describe the counter and display decoder portions of the design. We'll first describe the counter circuit. Figure 9.20 shows the Gates declaration and equation forms that fully describe a ten-state short counter. This counter is described using high-level equations as we described in the previous chapter and utilizes D-type flip-flops. This counter is simply a free-running counter with a hold function and asynchronous reset. Next, the display decoder circuit is developed. This display decoder is described using a Gates truth table, as shown in Figure 9.21.

Verifying the Design

After the counter and decoder circuits have been entered in the Gates design editor, they can immediately be functionally verified using Gates functional simulator. Early functional verification of a design is important, not only to save time that might otherwise be spent implementing incorrect designs, but also to simplify later debugging. Particularly in larger designs, apparent design problems can turn out to be caused by timing problems, rather than incorrect logic. Such timing-related problems are common in LCA applications. Verifying the logic of the design prior to implementation in the device will help to narrow down the search for causes of later circuit problems.

The simulation form for the counter circuit is shown in Figure 9.22. To simulate this portion of the design, we have entered a number of test vectors that verify the count, hold and clear functions of the counter. A similar simulation form could be written to test the operation of the counter and display decoder

Form type: declaration Form name: hex7

Name	Type	Definition/Comment
IN3,IN2,IN1,IN0	input	
AS,BS,CS,DS,ES,FS,GS	output	comb
BCD	set	[IN3,IN2,IN1,IN0]
ON	macro	1
OFF	macro	0

Form type: truth-table Form name: hex7

BCD	AS	BS	CS	DS	ES	FS	GS
0	ON	ON	ON	ON	ON	ON	OFF
1	OFF	ON	ON	OFF	OFF	OFF	OFF
2	ON	ON	OFF	ON	ON	OFF	ON
3	ON	ON	ON	ON	OFF	OFF	ON
4	OFF	ON	ON	OFF	OFF	ON	ON
5	ON	OFF	ON	ON	OFF	ON	ON
6	ON	OFF	ON	ON	ON	ON	ON
7	ON	ON	ON	OFF	OFF	OFF	OFF
8	ON	ON	ON	ON	ON	ON	ON
9	ON	ON	ON	ON	OFF	ON	ON
^hA	ON	ON	ON	OFF	ON	ON	ON
^hb	OFF	OFF	ON	ON	ON	ON	ON
^hC	ON	OFF	OFF	ON	ON	ON	OFF
^hd	OFF	ON	ON	ON	ON	OFF	ON
^hE	ON	OFF	OFF	ON	ON	ON	ON
^hF	ON	OFF	OFF	OFF	ON	ON	ON

Figure 9.21 Display decoder truth table form

together. For this design, however, we have separated the two design portions into different Gates design files to show different processing strategies.

Design Optimization

The Gates design editor includes logic synthesis features that can be used to improve the efficiency of LCA-based circuits. The first step in the logic synthesis process is converting the high-level design descriptions into sum-of-products Boolean equations. These equations are then processed by Gates, using the Espresso logic minimization algorithm, to produce a minimized set of Boolean equations.

For many applications, the constrained architecture of the LCA's logic blocks make it necessary to perform multilevel optimizations in addition to logic minimization. There are multilevel optimization algorithms found both in Gates and in the XNFOPT module that is supplied by Xilinx. Both of these algorithms perform factoring operations that decompose large sum-of-products circuits into a series of smaller logic circuits.

```
      Form type:  simulation     Form name:  cnt10d

      Clock ClkEna Clr | Count
     -------------------------
1     ~c     0      0     0
2     ~c     0      1     0
3     ~c     1      1     1
4     ~c     1      1     2
5     ~c     1      1     3
6     ~c     1      1     4
7     ~c     1      1     5
8     ~c     0      1     5
9     ~c     1      1     6
10    ~c     1      1     7
11    ~c     1      1     8
12    ~c     1      1     9
13    ~c     1      1     0
14    ~c     1      1     1
15    ~c     1      1     2
16    ~c     1      1     3
17    ~c     1      1     4
18    ~c     1      1     5
19    ~c     1      1     6
20    ~c     1      0     0

    Input signals                  width   orientation
   Clock                             5          h
   ClkEna                            6          h
   Clr                               3          h

    Output signals                 width   orientation
   Count                             5          h

    Run Mode: until-error         Equations: reduced
```

Figure 9.22 Gates simulation form

In most cases, the factoring algorithm available in Gates is sufficient to implement a circuit in the LCA's architecture. In some cases, however, XNFOPT's LCA-specific optimization features can result in a more efficient circuit. By experimenting with these two optimizers it's often possible to create more efficient circuits than either is capable of producing on their own.

Generating Schematics from Behavioral Descriptions

Before the counter and display decoder modules can be processed by the XACT tools or the XNFOPT module they must be converted to an appropriate format—the XNF format described earlier. There are a variety of methods for converting behavioral design descriptions into XNF format. Some design tools produce XNF files directly, while others use PALASM 2's PDS (*PAL design specification*) as an intermediate format.

The method used in FutureDesigner is to generate (automatically) a set of schematics that can be processed to produce the XNF files. This method allows designs to be expressed using a mixture of behavioral and schematic forms. If desired, changes to the generated schematics can be made before they are converted to XNF format, but this isn't recommended since there is no way to back-annotate the changes to the original design descriptions.

The resulting automatically generated schematic for the counter portion of the design is shown in Figure 9.23. As shown, this schematic could be translated

Figure 9.23 Automatically generated schematic for counter

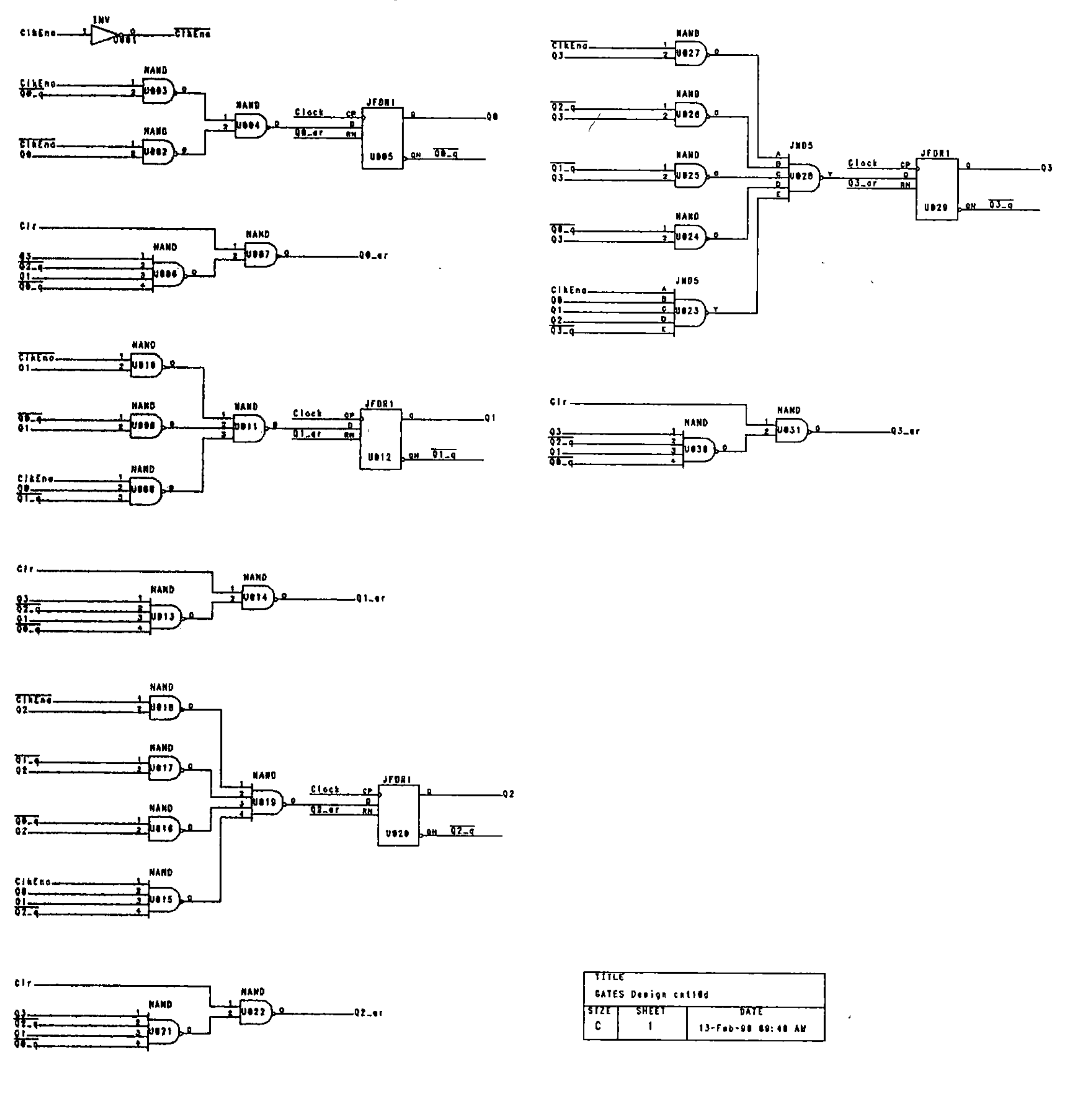

```
      Form type:  declaration      Form name:  cnt10t

Name                                Type        Definition/Comment
ClkEna,Clock,Clr                    input
Q3..Q0                              bidir       t
Count                               set         [Q3..Q0]
Short                               bidir       comb


      Form type:  equation         Form name:  cnt10t

Count.t = (   (Count.q + 1) &  ClkEna
          # (Count.q     ) & !ClkEna)
        $ Count.q

Count.clk = Clock

Count.ar = Short

Short = !Clr # (Count.q == 10)
```

Figure 9.24 Gates declaration and equation forms for improved counter

and implemented in an LCA. There are improvements that we can make to this circuit, however.

First, notice that the short count reset logic is duplicated for each of the four counter outputs. By assigning the reset logic to a common intermediate signal, we can reduce the amount of circuitry required for this function. In addition, we have specified this design using D-type flip-flops. As we examined in the previous chapter, a counter such as this can be more efficiently designed using T-type flip-flops. The XC2064's CLBs can be used for T-type flip-flop applications since they include XOR gates and an internal feedback path. If we make these design changes, we can dramatically decrease the amount of circuitry required. Figure 9.24 shows the updated Gates design forms describing the counter and Figure 9.25 shows the corresponding automatically generated schematic.

The display decoder portion of the design is, as written and processed, appropriate for implementation in the LCA. Figure 9.26 shows the automatically generated schematics for the decoder. Notice that this portion of the design, which was processed with the Gates factoring algorithm, is composed entirely of logic gates with fewer than four inputs each and includes multilevel logic (expressed with intermediate factor variables).

Converting Schematics into XNF Format Files

The conversion of the counter and decoder schematics into XNF produces three outputs: a constraints file listing the pin assignments that were made on the top-

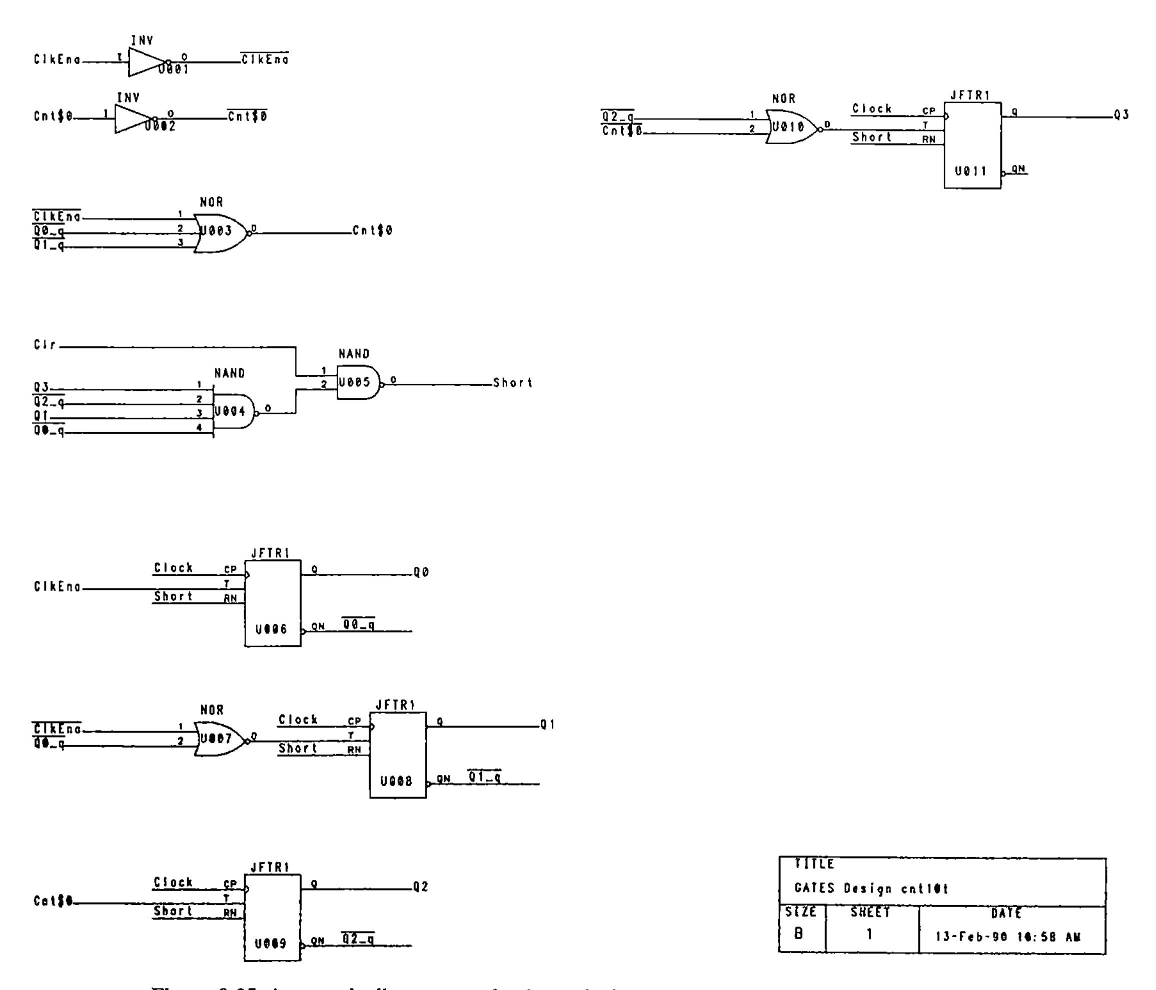

Figure 9.25 Automatically generated schematic for improved counter

level drawing and two XNF netlists representing the counter and decoder portions. The constraints file is illustrated in Figure 9.27. This file will be used during the automatic placement and routing step to keep the manual pin assignments from being overridden by the placement and routing software. Constraints files also allow previously placed and routed circuit elements to be frozen into specific sections of the device. This can be done to preserve critical timing paths or simply to speed later placement and routing iterations.

The two XNF files can be processed further with the XNFOPT optimizer or can be implemented in the LCA in their current form. For this small design, there is no benefit to using the optimizer, so we'll skip that step. From this point, the design process is much the same as we described earlier: we merge the two

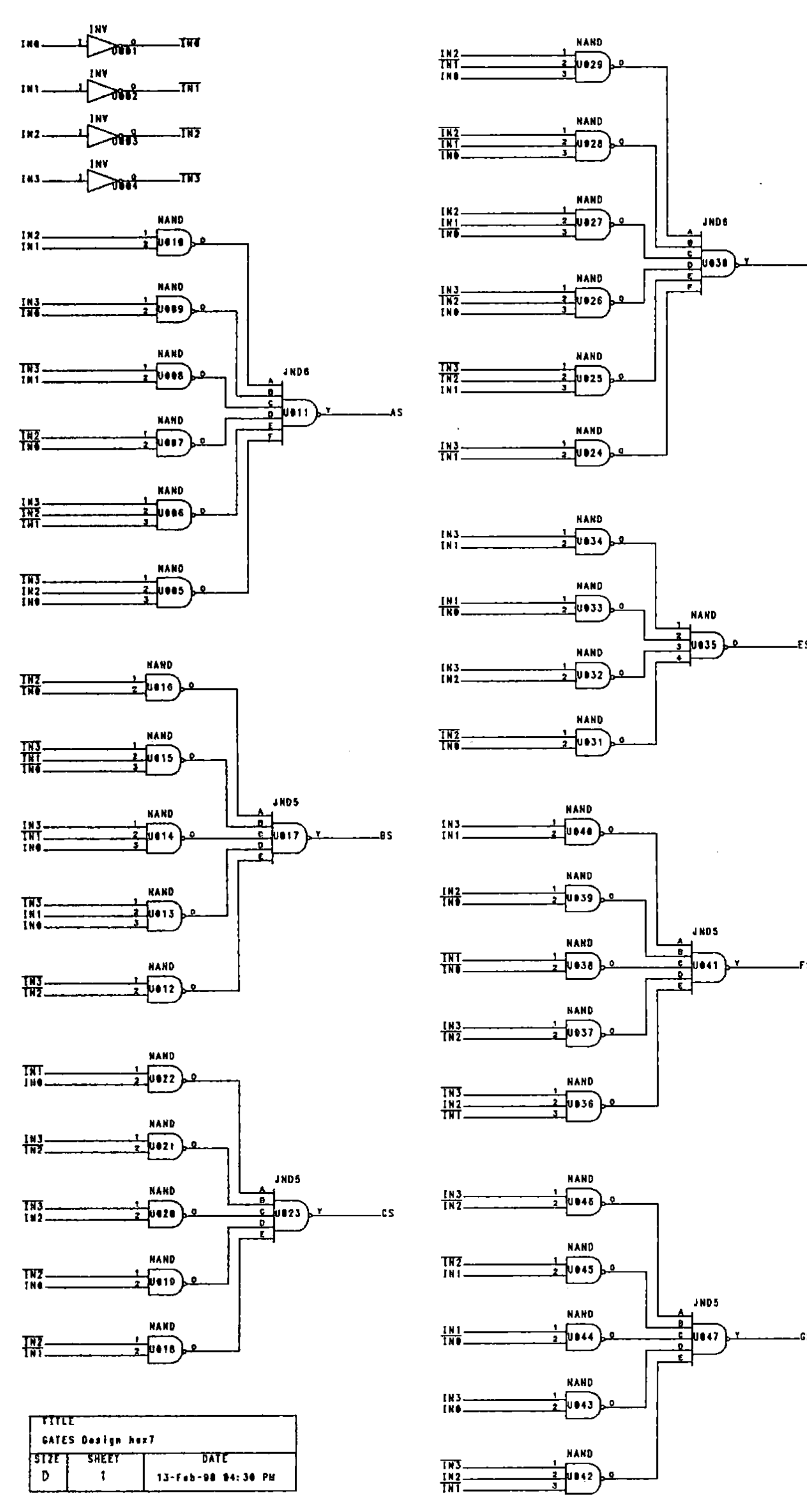

Figure 9.26 Automatically generated schematic of display decoder

```
; Design: hexcnt.lca, Created by MAP2LCA Ver 3.00 at Tue Sep 04 20:53:47 1990
;(NOTE: Don't edit this file. It is rewritten each time MAP2LCA is run)
place block GSEGPAD P55 ;
place block FSEGPAD P53 ;
place block ESEGPAD P49 ;
place block DSEGPAD P47 ;
place block CSEGPAD P40 ;
place block BSEGPAD P39 ;
place block ASEGPAD P38 ;
place block Q3PAD P32 ;
place block Q2PAD P31 ;
place block Q1PAD P30 ;
place block Q0PAD P29 ;
place block INHIBITPAD P24 ;
place block RCSETPAD P14 ;
place block RCRESETPAD P12 ;
place block CLRPAD P11 ;
```

Figure 9.27 Constraints file containing pin assignments

XNF files into a single file representing the complete design and convert this design into an LCA design file and corresponding download format file. The screen display illustrated in Figure 9.28 shows the final placement and routing of the circuit into the target LCA.

Although we have presented these design steps as a sequential process, many designs evolve over time requiring the repetition of groups of steps. For example, later changes may add new logic elements requiring additional iterations of automatic and manual placement and routing to determine the best layout.

Configuring the LCA

Once the design and layout are final, the LCA can be configured. This step has two parts. First, the Xilinx MAKEBITS utility is invoked. This utility converts the chip floor plan into a bit pattern. Then, with the demo-board cable connected to the host computer's parallel port, the DOWNLOAD utility is invoked causing the bit pattern to be loaded into the LCA's RAM.

The finished design can now be operated in the Xilinx demo board to verify its correct operation. Since LCAs have erasable logic, it's possible to go back and experiment with different design solutions. For example, when we first programmed this simple design into the LCA and turned it on, we discovered an unanticipated (although in hindsight obvious) problem—the numbers on the decoder changed so quickly that it was impossible to read them. We solved this problem by putting another counter in the LCA to divide the clock rate by sixteen.

Once a design has been determined to be satisfactory, the final LCA bit pattern can be programmed into a configuration PROM. The configuration PROM loads the LCA upon power-up, providing a permanent implementation of the design.

Print World: HEXCNT1.LCA (2064PC68-50), XACT 3.00, Tue Sep 04 21:41:14 1990

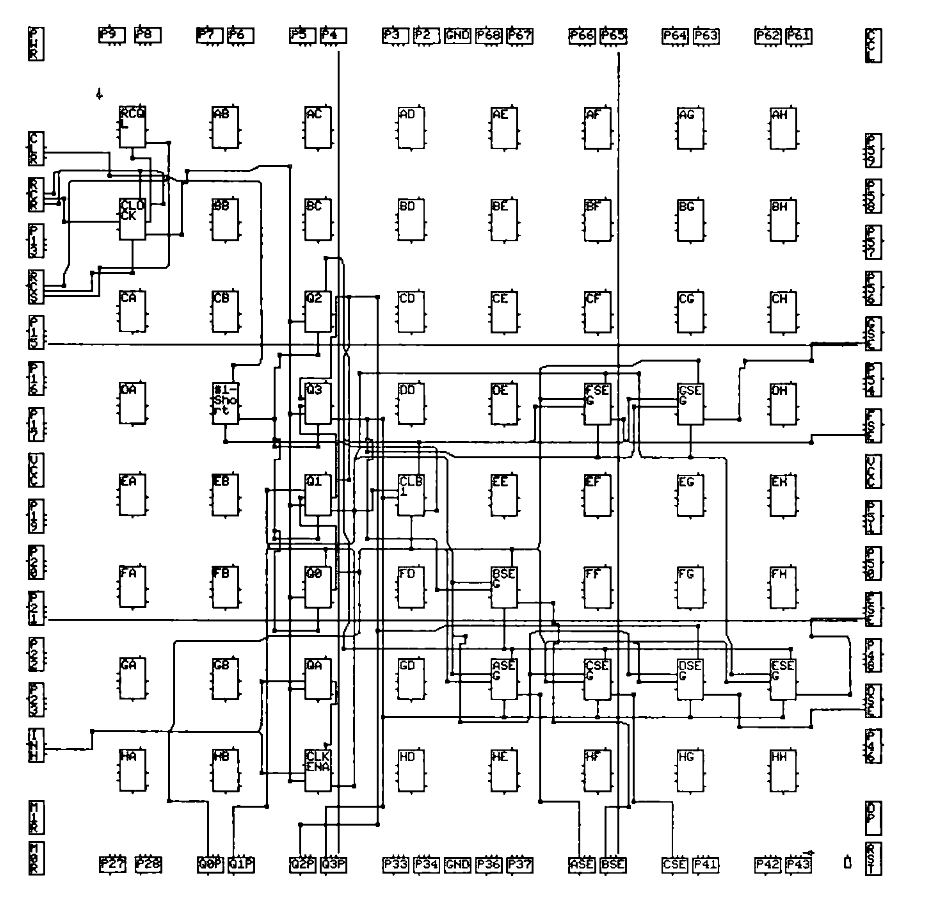

Figure 9.28 Final LCA floor plan for the counter/decoder

9.4 DESIGN CONSIDERATIONS FOR LCAs

As you might imagine from the above description, designing for LCAs is a much more involved process that designing for PLDs. With PLDs, once a design has been converted into Boolean equation form and minimized, it's a relatively straightforward task to implement that those equations into a PLD. If the design equations fit into the device, then the process is a success; if they don't fit, a redesign is called for.

The constrained architecture of the LCA's logic blocks create the need for careful consideration of various implementation options. Major design changes, or design-level optimizations, may be required in order to efficiently utilize an

LCA. LCA design tools can help to identify and perform simple circuit optimizations, but only the circuit designer can recognize and implement design-level optimizations.

Even after the circuit design is frozen, the device fitting process may take a significant amount of time. This is because there are many possible ways to place and route a given circuit in an LCA device and different implementation methods can result in quite different timing characteristics. For designs that utilize a large percentage of the available LCA resources, determining how the circuit should be placed and routed can take more time than the development of the circuit logic. Here again, automated tools can help, but may not be capable of completely fitting a large design into a device without some manual intervention.

9.5 DESIGN-LEVEL OPTIMIZATIONS FOR LCAs

There are many ways to get additional functionality out of the LCA. By working within the constraints imposed by the architecture, it's possible to dramatically improve the efficiency and performance of these devices.

Minimize Random Combinational Logic

Large combinational logic functions, particularly those with more than four or five inputs, are difficult to efficiently implement in an LCA. For this reason you should attempt to use flip-flops as much as possible to solve design problems, rather than relying on strictly combinational logic functions. For example, if you are designing a state machine that relies on complex combinational logic to determine transitions, you may be able to save circuitry and improve the skew characteristics of the state bits by partially decoding the transitions one state early and synchronizing this transition logic with an additional flip-flop.

Combinational functions with wide inputs will have to be partitioned (factored) into smaller pieces before being implemented in the LCA's CLBs. This can lead to unacceptable timing behavior and wasted CLB resources. Functions that require large numbers of inputs (address decoders, for example) are poorly suited to the LCA architecture, unless changes are made to break up the input signals into smaller groups.

Minimize the Number of Required CLB Inputs

Both the 2000- and 3000-series devices allow two independent logic functions to be implemented in a single CLB. Since the number of available inputs to each CLB is quite limited (typically four inputs to a 2000 series and five to a 3000-series CLB), you will achieve the best utilization by using functions that share a maximum number of inputs.

Minimize Interconnect Requirements Through Internal Feedback

Since interconnection resources are a critical constraint in LCA devices, it pays to minimize their use whenever possible. One way to do this is to maximize the use of the feedback provided in each CLB. By providing pairs of functions that share feedback requirements (for example, when the output of one logic function feeds the input of another), you will further reduce the amount of interconnect required. This is particularly useful in the 3000-series LCAs where the feedback is completely internal to the CLB.

Use T-type Flip-flops Whenever Possible

In earlier chapters we described how D-type flip-flops can be converted to T-type flip-flops with the addition of a simple XOR function. While the CLBs in an LCA don't feature fixed XOR gates, the architecture of the CLBs makes it possible to emulate T-type flip-flops with little or no penalty in logic.

The use of T-type flip-flops is one area where designs intended for simple PLDs differ from those intended for LCAs. Some of the most efficient circuits implemented in PAL-type devices are actually the least efficient circuits you can implement in an LCA. A counter that is constructed from D-type flip-flops is a prime example of a circuit that is easy to implement in a typical PLD, but quite inefficient when implemented in an LCA.

When emulating a T-type flip-flop in a 2000-series CLB, the only loss is the *D* CLB input (which can in any case be used as a flip-flop reset line). In the 3000 series, the register feedback is internal to the CLB, so no logic is consumed at all in converting D-type flip-flops to T-type.

If you require a counter circuit, you should use T-type flip-flops, rather than D-type. If the counter is being used for a frequency division function, you can use the CLB output of one counter bit as the clock input to the next counter bit, thereby freeing up large amounts of combinational CLB circuitry for other purposes.

Isolate Common Clock and Reset Logic

Many designs require that complex clocking or reset logic be provided to large numbers of registers. To save logic, you should attempt to share the logic for these functions by combining the clocking or reset logic into shared functions. For example, try to use a clocking scheme that can be shared by all registers, rather than duplicating similar combinational clocking logic for different flip-flops. If you are using high-level design tools, you may have to take precautions to ensure that common clocking or reset logic isn't duplicated needlessly during design processing.

9.6 IMPLEMENTATION-LEVEL OPTIMIZATIONS

After the logic for an LCA-based design has been determined, it's necessary to fit the logic into the architecture of the device. This is done through the process of placement and routing. For the most part, *automated place and route* (APR) software can be used for this part of the process. Quite often, however, the results are unacceptable timing delays or inefficient or incomplete interconnection strategies. The entire place and route process can be improved by following some simple rules.

Split Up Large Combinational Logic Functions

Since large functions don't map well into CLBs, it's important to split up any large pieces of circuitry into smaller functions. Since two logic functions can be implemented in a single CLB, it's best to split large functions into subfunctions that require fewer inputs than are actually available in each CLB (two or three inputs per subfunction for a 2000-series LCA, for example). Using very small subfunctions will make it more likely that subfunctions with similar input requirements can be identified and paired in the CLBs.

Functionally Partition the Circuit

Before the logic of your design is processed by APR software, it may help to partition the design by function (with the goal of minimizing interconnection requirements) and perform an initial manual placement of the circuitry into the device's CLBs. The APR software will perform better if it's given a suggested placement before beginning.

9.7 SUMMARY

Designing for FPGAs is a more involved process that designing for traditional PLDs. The architectures of these devices are powerful; however, the designer must be aware of how the specific architectural features affect the applications that are to be implemented in them. The requirement for placement and routing of logic functions into the devices' constrained architectures can mean that significantly more effort is required by the designer. Fortunately, design tools exist that can automate the more tedious parts of this process.

As the design tools mature, we can expect this process to improve, but the need for careful analysis of the design itself will still exist. Design-level optimizations (that can only be performed by the circuit designer) will remain the key to efficient FPGA utilization.

9.8 REFERENCES

Data I/O Corporation. "Developing LCA Designs with FutureDesigner." *Data I/O Application Note 303,* Data I/O Corp., Redmond, WA, 1988.

Data I/O Corporation, *Gates Logic Synthesizer User Manual,* Data I/O Corp., Redmond, WA, 1989.

Xilinx, Incorporated, *The Programmable Gate Array Data Book,* Xilinx, Inc., San Jose, CA, 1989.

10

Hazards, Metastability, and Testability

Up to this point, we've looked at techniques that simplify the design process for PLDs, and help us to efficiently implement circuits in these devices. We've seen how the use of automated tools can speed the design process by reducing, or eliminating entirely, many of the more tedious aspects of the design process. Unfortunately, the use of automated tools can mask the existence of design problems, leading to poor reliability or testability of the resulting PLDs. Before a design incorporating PLDs can be produced in quantity, some important steps must be taken to ensure that the design is robust and testable. In this chapter, we'll examine some of the common PLD problems, and concentrate on techniques for improving the reliability and testability of PLD-based designs.

10.1 HAZARDS

Hazards are unpredictable conditions that can exist within logic circuits leading to erroneous circuit operations. Hazards are not, by any means, limited to PLD-based circuits but the nature of PLD construction, and the design methods used for them, can lead to a larger than usual number of hazard conditions. A major problem with hazards, in terms of their existence on PLDs, is that a design that works reliably using one manufacturer's device may not work at all in a seemingly identical device obtained from another manufacturer. This is due to the different timing characteristics of various type of PLDs.

Different PLDs, and different signal paths within a PLD, often have differing propagation delays. If the circuitry implemented in a PLD (or any technology, for that matter) relies on equal propagations of signals, hazards can exist and *glitches* are likely to occur. A glitch is a spurious output signal often seen as a spike when viewed with an oscilloscope.

Figure 10.1 illustrates a circuit that contains such a hazard. This output of this circuit will glitch under the situation shown in the timing diagram included in the figure. This is due to the unequal propagation delay from the input signal *B* to the two AND gates. This type of hazard is called a *static hazard.* A static hazard is indicated when the initial and resulting values of the circuit are the same. In other words, the glitch occurs on an output signal that wouldn't otherwise change state. When a glitch occurs on an output signal that is changing states (from high to low or from low to high) it's referred to as a *dynamic hazard.*

This circuit is highly simplified, but if you simply substitute the inverter delay path with a delay that might be seen in a multilevel logic implementation

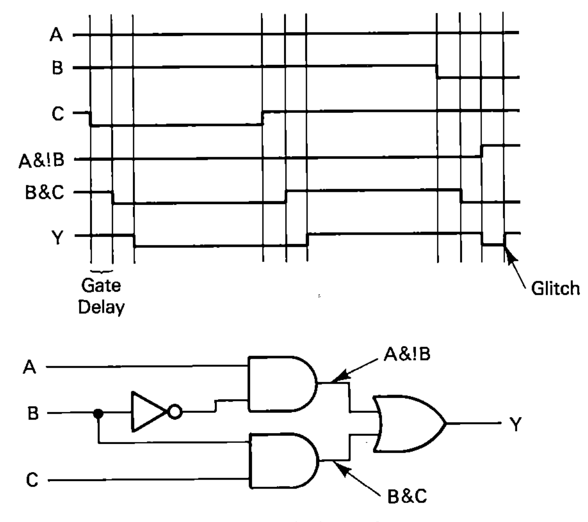

Figure 10.1 Circuit with static hazard

or from a signal routed outside a PLD, you can see how such hazards can be created in PLD-based designs.

Prevention of Hazards

Hazards come in two distinct flavors—*function hazards* and *logic hazards*. A function hazard exists whenever the stability of a circuit output depends on the simultaneous change in two or more circuit inputs. This can be illustrated in a truth table such as the one shown in Figure 10.2. The function described by the truth table is a simple XOR.

A	B	Y
0	0	0
0	1	1
1	0	1
1	1	0

Figure 10.2 Truth table containing function hazard

In the sequence of circuit stimuli shown in the timing diagram of Figure 10.3, there is a period of time during which the two circuit inputs are both changing state. During this time the output of the function can glitch.

As you might imagine from this simple example, hazards such as these are found everywhere in digital systems. The challenge to the design engineer is to identify those hazards that will adversely affect the reliability of the circuit as a whole and make whatever circuit modifications are necessary to protect the system. How do you protect your system from such hazards? In most cases you can't remove the hazards, so you must instead either avoid them or mask them.

An example of hazard avoidance is found in state value assignment. Consider a state machine with the four states shown in Figure 10.4. The current state of the machine is indicated by the two state bits whose values for each state are as indicated in the diagram. The transition from state *B* to state *C*, for example, creates the same glitch potential that we saw in the timing diagram of Figure 10.3. If the change in state of the two state bits doesn't occur at precisely the same time, the machine will briefly appear to be in state *A* or *D*. To avoid this hazard,

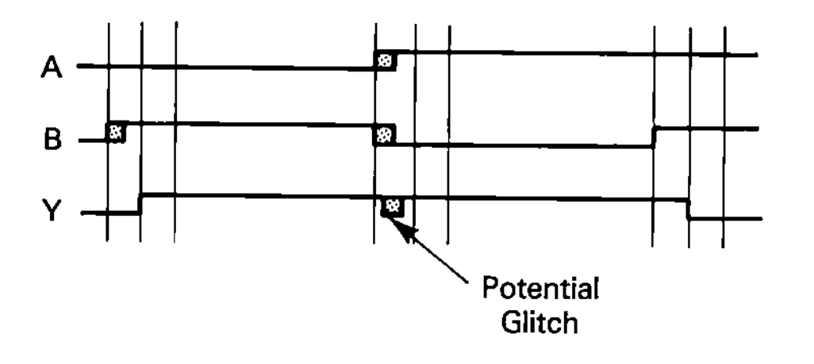

Figure 10.3 Condition under which glitch can occur

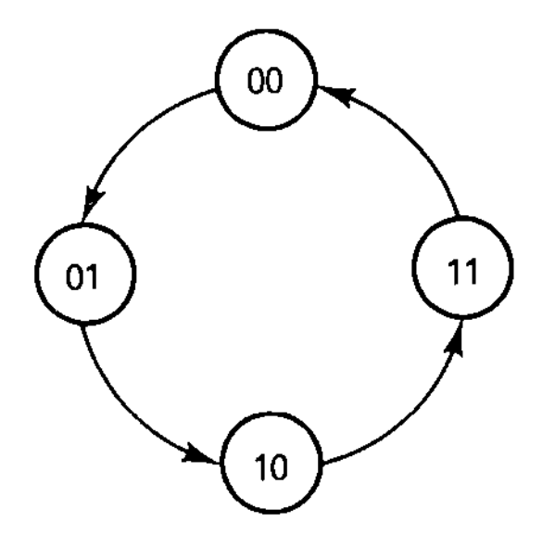

Figure 10.4 State machine with hazard

we can simply swap the state values for states C and D so that, for any state machine transition, there is only one bit change required.

There other classes of hazards that don't require that two or more inputs change at the same time. These hazards are called *logic hazards* because they are related to the physical characteristics of the actual logic circuitry. Logic hazards can be identified in simple circuits by examining a K-map representation of the logic.

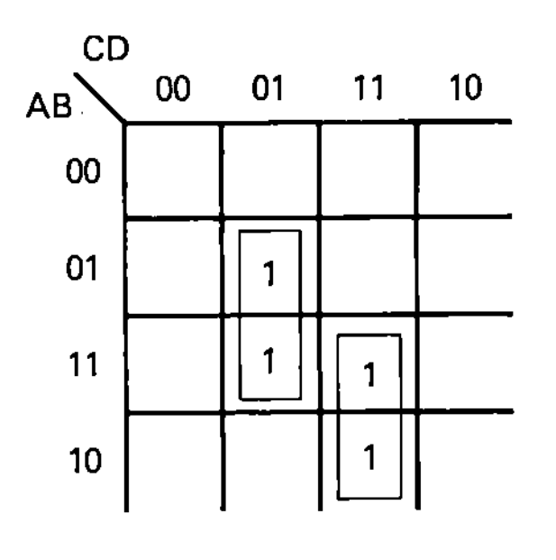

Figure 10.5 K-map showing hazard potential

Figure 10.5 illustrates a design in K-map form that may contain a logic hazard. The logic hazard is related to the C input, which is the single input variable that is specialized in the two product terms of the function (refer back to Chapter 7 for a detailed description of these terms). The two product terms that are indicated by the K-map groupings shown are as follows:

$$Y = B \;\&\; !C \;\&\; D \;\#\; A \;\&\; C \;\&\; D$$

Since the output of both of these product terms relies on C, the period during which C changes state can create a glitch under the circumstances illustrated in Figure 10.6. To mask this glitch, it's necessary to provide additional circuitry that will hold the output high during the change in value of C. The appropriate circuitry can be quickly identified from the K-map as shown by the new groupings in Figure 10.7. The additional product term (A & B & D) identified in the K-map serves to protect the output from the glitch.

As a general rule, if you wish to remove all logic hazards from a circuit,

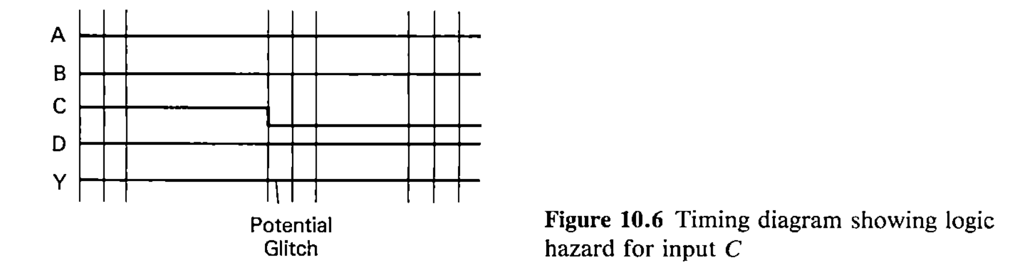

Figure 10.6 Timing diagram showing logic hazard for input C

you must ensure that all horizontally and vertically adjacent 1-cells of a K-map are grouped into a common product term. This, of course, is counter to the goals of logic minimization, and also has testability implications, as we'll examine later in this chapter.

The hazards covered here all result in spurious glitches that occur within a logic circuit. These hazards are most damaging in high-speed or asynchronous designs. For synchronous circuits (those with globally clocked output registers, for example), glitches in input forming logic are of little concern since the glitch is short and the signal will stabilize to the correct value long before the next clock transition occurs. In asynchronous systems, however, these hazards can become extremely troublesome and can defeat all efforts to effectively utilize automated design tools.

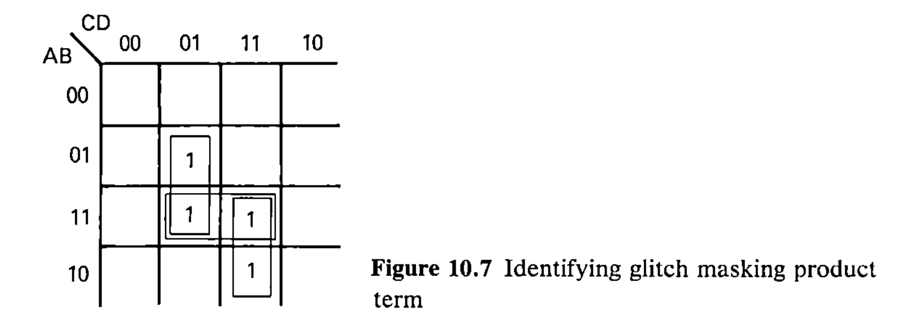

Figure 10.7 Identifying glitch masking product term

10.2 REDUCING NOISE IN PLD-BASED CIRCUITS

PLDs, particularly large devices with widely varying power consumption (such as CMOS devices) or high-speed operation, are prone to large transient currents and attendant noise during switching. The amount of noise generated in your circuit as a whole can be reduced by avoiding some common design errors. The specific requirements vary from one device type to another, but all PLDs share some common design requirements.

The most common source of noise is large transient currents caused by gate switching operations within one or more PLDs. There are two areas in which the problem of noise can be attacked—the design of the PLD circuitry and the physical characteristics of the board on which the PLDs are installed.

As far as the design goes, you can reduce transient current surges by designing your PLD circuits (particularly circuits utilizing registers) so that a minimum number of gates change value simultaneously. This is the same technique used in the hazard protection measures discussed earlier and is good design practice in general. As an example, consider the impact on the power consumption if a state machine circuit implemented in a large CMOS PLD changes state frequently from a state with all registers 1 to all registers 0 and back again. The current on the power pins can fluctuate between perhaps a few dozen microamps, and up to ten milliamps. Obviously, this can generate a significant amount of noise that must be dealt with in the design of the board.

Board-level physical characteristics often have more of an impact on noise. One common mistake is omitting the de-coupling capacitor from PLD power pins. Each power pin on a PLD should be connected directly to the ground pin with a capacitor. Failure to provide such a capacitor can result in erratic behavior in high-speed circuits. PLD manufacturers will generally recommend a particular size of capacitor to use for their devices.

Since the amount of power consumed by a CMOS device is dependent on the switching characteristics of its logic gates, you should ensure that all unused input pins on CMOS PLDs are tied to either ground or V_{cc}. If the inputs aren't tied off in this way, the inputs can float unpredictably between states causing erratic power consumption and noise.

The quality of the power supply is important. Many PLDs require that the power supply reach operating voltage within a specified time. This is because the devices must perform power-up reset operations. The power-up is detected by voltage comparators built into the PLDs. These comparator circuits require a minimum rise time for proper power-up detection.

Latch-up is a noise-related problem peculiar to CMOS devices and is caused by poor quality signals applied to device inputs. If an input above V_{cc} or below ground is applied to a CMOS PLD, a reverse bias can be created within the device leading to a direct short between power and ground. The device inputs will then become stuck in a high state, until the device is powered down or, in extreme cases, burns up. The possibility of latch-up can be reduced by providing PLD inputs with series resistors to reduce the AC currents. Good board design practice will also help to reduce the probability of latch-up.

Other problems can be created when PLDs are operated at high speeds. Large PLDs that have long signal paths (such as those packaged in 40-pin DIPs) can have problems with ground bounce. Ground bounce occurs when a large current surge is required and the distance between the ground pin and the V_{cc} pin is long. In some cases, ground bounce conditions can exist that momentarily produce as much as two volts on the ground plane of the PLD's die. In CMOS devices, this leads to the possibility of latch-up (since the input pins will appear to have negative voltages). Any PLD is prone to false clocking of registers under this condition. PLCC packages are less prone to ground bounce problems, since they have shorter pin to die distances and multiple ground and power pins.

If you are using a device that has multiple ground and power pins, be sure that they are all connected in the circuit. The device may appear to function just fine with single power and ground pins, but may not operate reliably due to inductance on the unconnected pins.

10.3 METASTABILITY

Metastability is a condition that plagues registered PLDs (and other synchronous circuit elements) when they are used in circuits that have asynchronous signals. In normal operation, digital storage elements such as flip-flops are said to be *bistable;* this means that they are stable in only two possible states (high or low). A *metastable state* is an unpredictable, symmetrically balanced state of operation that a clocked flip-flop can enter if the input to that flip-flop changes during the setup and hold period of the clock transition. Once in this state, the flip-flop will require an unknown amount of time to *relax* to a stable (but unpredictable) state.

With the increasing number of systems being developed that rely on asynchronous circuit interfaces, the problem of metastability is becoming more widespread. The metastability problem frequently exhibits itself when asynchronous signals are used as inputs to synchronously registered PLDs. Typical applications in which metastability appears include arbiter circuits, bus interfaces, synchronizers, and state machines utilizing asynchronous inputs or clocks.

PLD manufacturers have responded to the metastability problem by designing the flip-flops in their devices to be *metastable hardened,* meaning they will relax to a steady state quickly after a metastable event. Such claims should be examined carefully, however, particularly if high clock speeds are required and the application requires a high level of reliability.

If a high level of metastable protection is required, you can utilize additional flip-flops as shown in Figure 10.8. This circuit, a bus synchronizer, has two additional D-type flip-flops that are used to insulate the primary storage element from metastable triggering input conditions. These D-type flip-flops can be pro-

Figure 10.8 Metastable-protected bus synchronizer

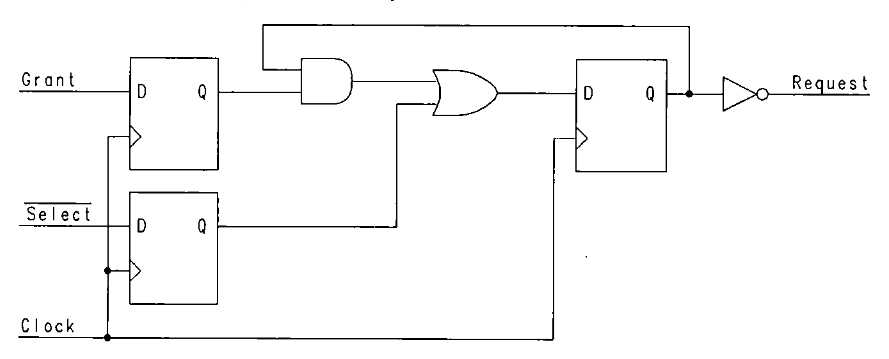

vided externally to the PLD in which the circuit is implemented, or borrowed from unused PLD outputs.

Many PLD manufacturers publish statistics on the metastable behavior of their devices. Statistics can be used to quantify the metastable behavior of particular PLD types and these statistics, if available, should be utilized to help determine the suitability of a particular PLD for asynchronous interface or arbitration applications. If a very high level of reliability is required, these statistics must be carefully applied.

10.4 TESTABILITY

Because of the nature of their architectures, PLDs present unique testing problems. These problems must be identified and resolved before a PLD based design can be reliably manufactured.

The Case for Testability

There is a well-accepted rule of thumb that may be used to estimate the costs of inadequate testing at various stages in the production process. This rule states that a defective PLD that costs $.50 to remove from inventory (at, say, an automated device programming and test station) will cost $5 if detected during board test, $50 if detected during final system test, and as much as $500 if detected in the field. Clearly, it's critical to detect defective or badly programmed devices as early as possible in the manufacturing process.

Incoming inspection and programming yield statistics can be misleading when dealing with large numbers of devices on a single board. Even if the acceptance rate is relatively high for a device, the acceptance rate for a large system utilizing those devices can be quite poor. For example, consider a case where the manufacturer's yield rate is ninety-nine percent. This is a reasonably good yield, but if the defective one percent of the parts aren't identified before the programmed devices are installed on boards each containing twenty such devices, the dropout rate for the boards will average 18.2 percent.

Perhaps more importantly, the insidious nature of PLD defects, coupled with inadequate board level test methods, often results in boards that pass board level test, but fail later during system test or in the field.

Fault Modeling of PLDs

Any potential failure in a PLD is referred to as a *fault*. Faults in PLDs can be classed as one of three types—*logic design faults, parametric faults,* and *device faults*. Logic design faults are related to the actual function that a properly programmed PLD is intended to perform. Logic faults can only be tested by applying test vectors that are supplied by the design engineer. These test vectors are often

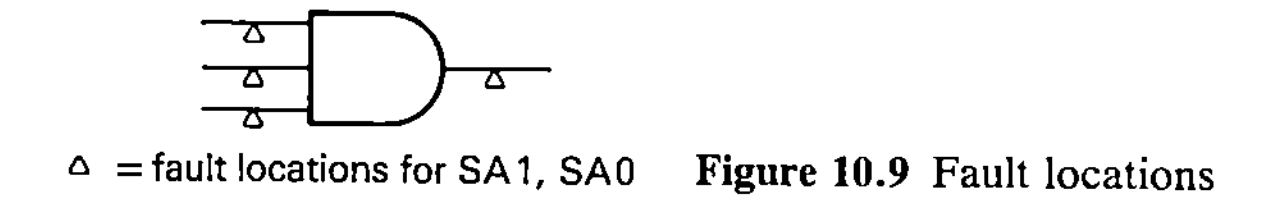

Figure 10.9 Fault locations

referred to as *design vectors* or *functional test vectors* and are used to verify that the circuit does what it's expected to do.

Parametric faults are those related to a device's physical characteristics or environment. This includes such problems as incorrect circuit speed, incorrect voltage or current parameters, or noise. Parametric problems such as these may be related to device manufacturing deficiencies, poor board design or poor quality board manufacturing. Hopefully, most device related parametric problems are taken care of in incoming inspection, while board related problems are dealt with in board test.

The third type of fault, the device fault, is the type of fault that causes the most trouble in PLDs. This is because PLDs have so many logic elements whose correct operation is difficult to observe directly. The device fault, therefore, is the area of most interest when designing PLDs for testability.

The Stuck-at-0/Stuck-at-1 Fault Model

The most common method used for analyzing the testability of a PLD in terms of device faults is the *stuck-at-0/stuck-at-1 model.* This model defines a fault as being any device node that can be *stuck-at* a logic level 1 or logic level 0. To demonstrate the large number of faults existing in a PLD, consider the simple three input AND gate shown in Figure 10.9.

This simple gate has a total of eight faults; four are *stuck-at-0* faults (henceforth referred to as *SA0* faults) and four are *stuck-at-1* faults (referred to as *SA1* faults). We can see from this example that the number of faults related to a single logic gate is based on the number of inputs to that gate. Since a typical PLD has a large number of AND gates, each one of which has many inputs, PLDs have a large number of faults that must be observable and testable during device test.

To adequately test this circuit, a circuit stimulus in the form of one or more test vectors must be developed that will allow us to observe on the circuit output whether any of the nodes are in a stuck-at condition. It isn't necessary to know exactly which node is stuck, since a failure of any node indicates a failure of the device.

To test this circuit for SA0 conditions, one test vector is all that is required, as shown in Figure 10.10. This vector applies a logic level 1 to all three inputs,

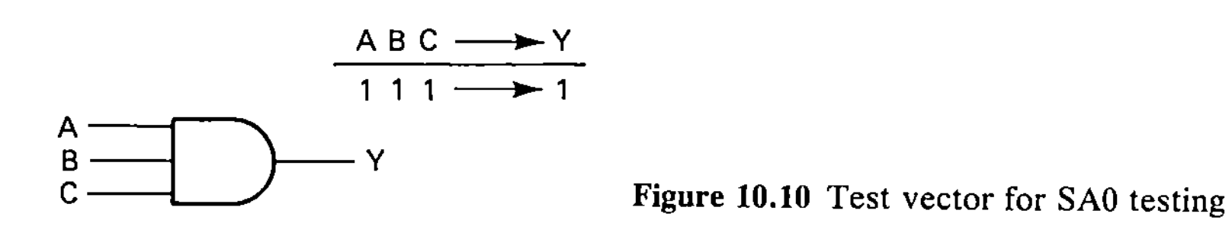

Figure 10.10 Test vector for SA0 testing

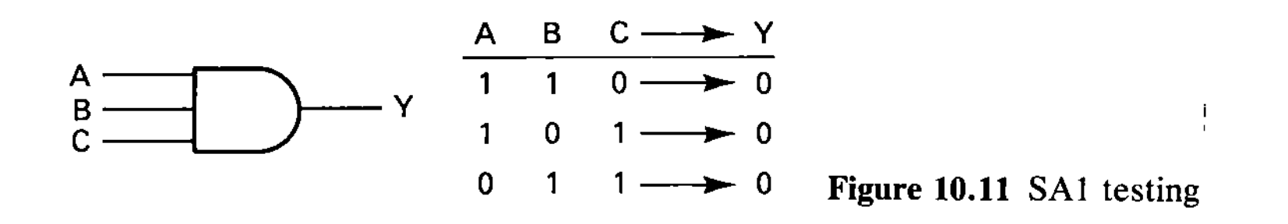

A	B	C → Y
1	1	0 → 0
1	0	1 → 0
0	1	1 → 0

Figure 10.11 SA1 testing

and if the result on the output is also logic level 1, there are clearly no circuit nodes stuck at a logic level 0. If any of the device nodes were stuck at logic level 0, the output would also be at a logic level 0 and the test vector would fail.

Figure 10.11 shows the same circuit being tested for SA1 conditions. SA1 testing for this circuit requires three test vectors; one for each AND gate input. Each test vector applies a logic level 0 to one of the inputs, while applying a logic level 1 to the remaining two. This stimulus should result in a logic level 0 appearing on the circuit output for each of the test vectors. The appearance of a logic level 1 on the output for any of these vectors is evidence of a SA1 condition at one of the circuit nodes.

This example demonstrates the concept of *fault equivalence*. In order to test and detect all eight faults in the device, only four test vectors were required. This is because all of the SA0 faults in the circuit are equivalent and the SA1 fault on the circuit output can be considered to be equivalent to any of the SA1 faults on the inputs. These equivalences hold true for an AND gate of any size; the number of faults requiring unique test stimulus is always $n + 1$, where n is the number of SA1 faults (and the number of inputs to the AND gate) and 1 represents the equivalent SA0 faults.

A similar concept holds true for the OR gates of a PLD. In an OR gate, all the SA1 faults are equivalent, and the SA0 fault on the output is equivalent to any of the SA0 faults on the OR inputs. PLDs, being sum-of-products (AND feeding OR) in design, can be further simplified for fault modeling purposes by identifying equivalences between the AND and OR faults. As Figure 10.12 shows, all of the SAO faults for one product term in a sum-of-products circuit are equivalent. Only one test vector is required to completely test each product term of a sum-of-products combinational circuit for SA0 faults.

Figure 10.12 also shows that the SA1 fault associated with the output of

Figure 10.12 Fault equivalence for SA0 faults in sum-of-products circuit

A	B	C	D	E	F → Y	
1	1	1	0	0	0 → 1	SA0 (A,B,C)
0	0	0	1	1	1 → 1	SA0 (D,E,F)
0	1	1	0	0	0 → 0	SA1 (A)
1	0	1	0	0	0 → 0	SA1 (B)
1	1	0	0	0	0 → 0	SA1 (C)
0	0	0	0	1	1 → 0	SA1 (D)
0	0	0	1	0	1 → 0	SA1 (E)
0	0	0	1	1	0 → 0	SA1 (F)

each AND gate is equivalent to the SA1 faults associated with the corresponding OR input and the OR output. This means that the total number of test vectors required for a one output combinational circuit is equal to the total number of inputs to the AND gates.

In PLDs, there are many additional nodes in the device that have SA0 and SA1 faults associated with them. These include device registers and other output macrocell circuitry. The percentage of these SA0 and SA1 faults that are covered by the test vector suite used in production is a good measure of how effectively devices are being tested.

Testability Grading

There are many design practices that can result in undetectable faults in a PLD circuit. Later in this chapter we'll examine automated tools that can be used to grade a PLD design and report on the number of undetectable faults in that design. These tools can also be used to generate test vectors to cover the detectable faults. It's also possible to generate this data by careful hand analysis of the circuit, but this is a tedious and time-consuming task.

10.5 THE PLD TESTING PROCESS

There are three distinct phases in the process of testing PLDs for production. The first is the *design for testability* phase. PLDs that haven't been designed with testability in mind will cause grief in production no matter how thorough the production testing methods are.

The second testing phase is *test suite generation and evaluation*. PLDs, unlike PROMs, require a disciplined approach to test suite generation. A PROM can be adequately tested simply by applying all of the PROM addresses and observing the data that results on the device outputs.

PLDs, on the other hand, have many complex internal elements that must be tested for proper operation. The nature of their design often makes it difficult to verify (by observation of the device operation when under test) that these internal nodes are functioning properly.

The need for the third phase of the testing process is determined by the results obtained in the second phase. In this phase, design changes are made to correct testability deficiencies identified during the test suit generation and evaluation phase.

10.6 DESIGN FOR TESTABILITY

Testability has historically been a problem with PLD based designs. It's important to realize that testability can only be properly addressed at the design stage; no amount of expertise on the part of a test engineer can overcome basic testability

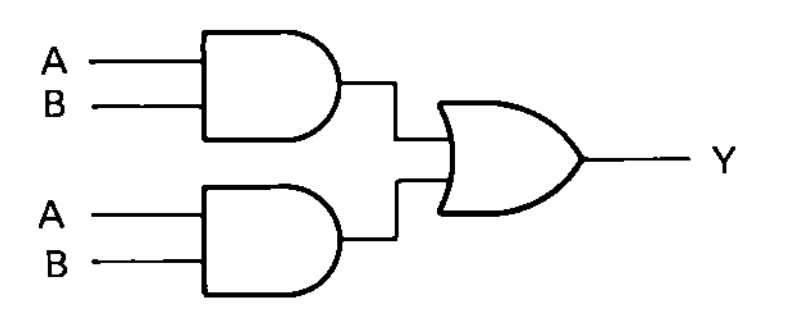

Figure 10.13 Redundant circuit

limitations in a PLD design. Therefore, it's important that designs are analyzed for testability before they are released into production. Often, major changes are required in the design to provide an acceptable level of testability. Being aware of the common causes of testability problems is the first step in eliminating those problems.

Redundant Logic

Redundant logic is a frequent barrier to testability. Figure 10.13 illustrates an obvious case of redundancy in a simple circuit. Notice that the two product terms represented in the circuit are identical. In reality, circuit redundancies tend to be more insidious. For example, the circuit shown in Figure 10.14 also has redundancy. This can be determined by examining a K-map for the function. As the K-map shows, the term *!B & C* is redundant.

Defects in PLD circuits can be masked (and therefore remain undetected) if redundant circuitry exists. In these cases, there are no test vectors that can be applied to the PLD that will expose a defect in one of the redundant circuits. Unintentional circuit redundancies are often found in PLDs that have been designed without the use of automated logic minimization tools. Minimizing logic by hand (using Karnaugh maps, for example) is a tedious process, so if a set of Boolean equations can be mapped into a PLD as originally written, a designer may simply neglect to perform logic minimization.

In this case, is it important that the redundant logic be removed? The case might be made that since the redundant logic gives that part of circuit a built-in fault tolerance, the existence of a defect in one of the redundant circuits is of no importance. This is true to some extent, but in practice the existence of a defect anywhere in a PLD is strong indicator that the remaining circuitry is of questionable integrity. Furthermore, such assumptions on the part of the designer place undue burdens on the test engineer, who must determine if redundant logic in a circuit has been designed in for a purpose.

Figure 10.14 Less obvious redundant circuit

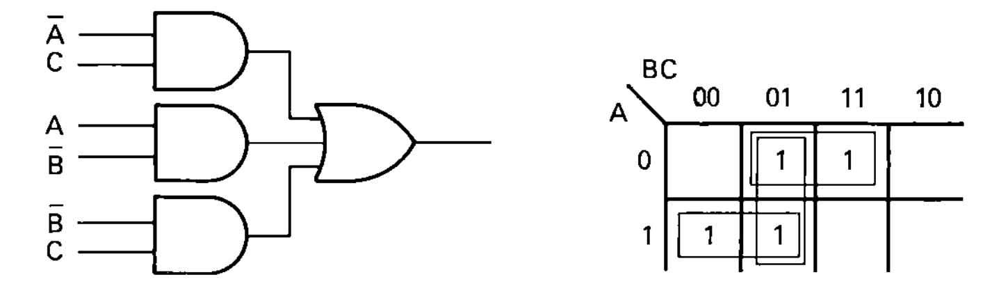

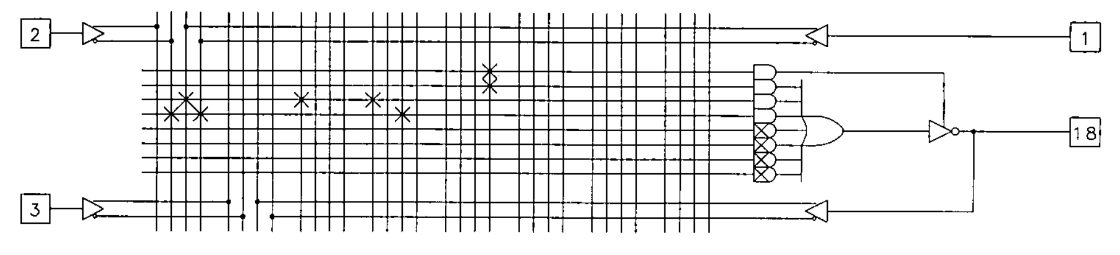

Figure 10.15 Circuit with fault-masking output enable

Unintentional redundancy is easy to remove from a design, if logic minimization routines are available and are used. Logic minimization routines remove redundant terms from logic equations. Even PLDs that have been designed without the use of logic minimization tools can be redesigned with a minimum of effort. Converting a JEDEC format fuse file into Boolean equations is a straightforward process (de-compiler programs exist for this purpose), and these equations can be quite easily reprocessed using the logic reduction tools supplied with PLD design software to generate a new set of minimized equations.

Another type of unintentional redundancy that can mask faults is common to designs that use a term-controlled output enable feature. Figure 10.15 shows one output of a circuit that uses a simple function (of one input) to control the output enable. The circuit contains masked SA0 and SA1 faults because the input that controls the output enable is also used in the logic for the output itself.

This type of redundancy is more difficult to detect than the conditions described earlier, and can't be eliminated by normal logic minimization algorithms. To expose the masked faults, the designer must make changes to the design. Since the device shown allows only one product term for the output enable control, it isn't possible to add a test input that will override the enable control during testing. Instead, the equations for the output itself must be changed to eliminate all product terms that are common to both the output enable and the output equation.

Intentional Redundancy

Redundancy in a circuit is often put in intentionally to cover possible glitch conditions, as we presented earlier in this chapter. If circuit redundancy is required for the reasons described, it may be possible to isolate the redundant logic through the use of an additional device input, as shown in Figure 10.16. This input line

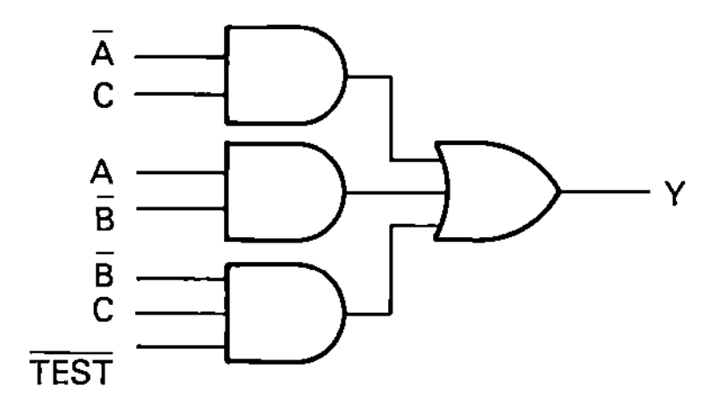

Figure 10.16 Providing test input to isolate redundant logic

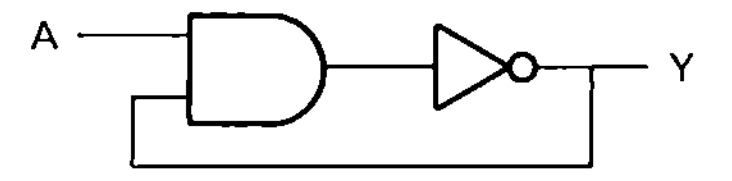

Figure 10.17 A simple oscillator circuit

(which, in a PLD, requires a previously unused input pin) isn't used when the PLD is operating normally; it's only used when test vectors are applied to the device.

Determining Whether Redundancy Exists

There are algorithmic methods for determining whether a circuit contains redundant logic, and these methods can be used by test engineers directly or implemented in testability analysis software. These techniques are rather tedious to apply manually, but lend themselves well to automated tools. We'll examine some of those tools later in this chapter.

Oscillating Circuits

Another testability problem is created when PLDs are used to create *oscillator* circuits, such as the one shown in Figure 10.17. An oscillator is a circuit that won't stabilize to a known logic level, and changes states at a speed determined by the delay paths of the circuit. Testers for PLDs generally are capable of testing only steady state values, and are not capable of testing for proper oscillator operation. Even if testers could test for oscillator operation, generating test vectors for these designs would be extremely time consuming. If you need to implement an oscillator or pulse generator circuit in a PLD, you should provide a signal to turn off the oscillator for testing purposes.

10.7 DESIGNING TESTABLE SEQUENTIAL CIRCUITS

PLDs, as we have seen in previous chapters, are very useful for implementing finite state machines. It's important to realize, however, that the design of a state machine is critical to ensuring that it is testable.

First, in order to isolate faults in a PLD it's necessary to get all of the flip-flops in that device to a known state. At that point the device can be evaluated as though it was a combinational circuit. To do this, each fault detecting test vector must be preceded by however many test vectors are required to initialize the device to the required state. This first set of test vectors is known as the *initialization sequence*. Determining the required initialization sequences for detecting all the faults associated with a particular state machine can be a task of mammoth proportions.

If the specific PLD chosen for production has a register preload feature, the initialization sequence is quite simple. Preloadable PLDs have built-in circuitry

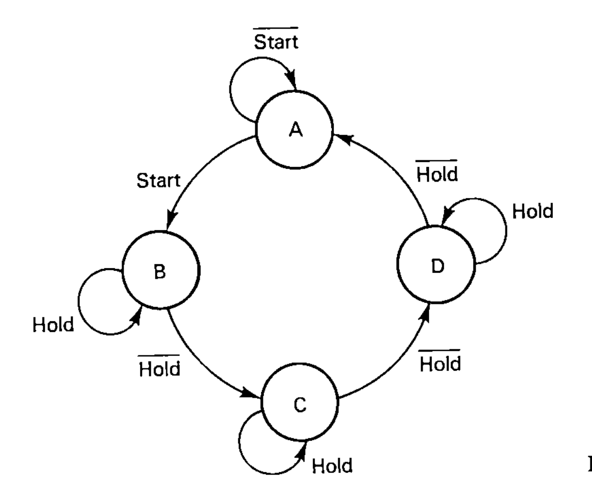

Figure 10.18 Simple state machine

to allow registers to be set to high or low as needed for testing. To set a register to a specific value, a higher than normal voltage is applied to the appropriate pins to force the registers into the desired states. This is sometimes referred to as *jam-loading*. Once the flip-flops of a circuit have been set to a known state, the testing process is similar to the process for combinational circuits.

Preloading is useful for testing devices individually (at the time they are programmed, for example), but often isn't possible during in-circuit testing since the high voltages used can destroy other devices on the board. Nonpreloadable devices introduce some serious complications to the test vector generation process. Without preload, it's necessary to analyze the design and determine what sequence of vectors will cause the device to step through all of its possible states.

An easier approach is to determine an initialization sequence that will get the device to a known state from any random state. For example, consider the state graph in Figure 10.18. This simple state machine has four possible states.

This state machine requires an initialization sequence that consists of the three identical test vectors shown in Figure 10.19. No matter what state the machine is in, the application of these three test vectors will result in a known state (state *A*) for the machine. The use of an initialization sequence such as this can have practical benefits. Because a given set of initialization and detection vectors can be used independent of previous sets, large test vector suits can be split up if they exceed the memory capabilities of device or board level testers.

No matter what approach is chosen for initializing a state machine circuit, if the state machine has states that aren't specified (illegal states) there may be power up or glitch conditions from which no sequence of vectors can recover. In extreme cases, uncontrolled oscillation of the state bits is possible. For this reason, it's important to ensure that all possible states in a PLD circuit have some method specified for returning to a known state.

There are many PLDs in common use that don't initialize to a known state after power-up. When designs with unspecified states are implemented in such

Clk	Start	Hold ⟶ State
⎍	0	0 ⟶ ?
⎍	0	0 ⟶ ?
⎍	0	0 ⟶ A

Figure 10.19 Initialization sequence

PLDs there are sure to be operational and testability problems. Devices with known power-up states can help to alleviate the problem, but relying on such device features limits the number of PLDs in which a given design can be implemented and places undue restrictions on those people who are responsible for device acquisition and qualification.

Determining whether a design can be initialized to a known state is a time consuming process, unless automated design analysis, fault grading, and test generation tools are used. Once it has been determined that a design cannot be initialized, there are a number of approaches that can be used to provide initialization. The easiest is to provide an additional input to the PLD which resets the device to a known state. On some devices (notably the FPLS devices from Signetics) a complement array feature is provided that can be used to force a transition out of any unknown state. In some cases it may be possible to renumber the state values until a combination is found that results in an initializable design.

Initializable state machines that have a large number of states can also cause testability problems, simply because of the large number of test vectors required to initialize them. Providing logic to load the state bits to known values will help to reduce the number of test vectors required to test such a design.

10.8 PLD TESTING IN PRODUCTION

Assuming that we've convinced you of the need for testability and testing, let's look at some tools and methods that can be used in production to provide the required levels of quality assurance.

Test Vector Generation

As we've seen, fully testing a PLD can require a large number of complex test vectors. From a designer's standpoint, it's tempting to just toss the master device into the hands of the production-test department and have them determine the set of test vectors that will result in adequate test coverage. Enlightened design engineers understand that testability problems begin early in the design phase, and will perform a testability analysis of the design before signing it off. The engineer responsible for the design may even provide the test engineer with a set of functional test vectors that were used during the design process.

Faced with the enormous task of analyzing the device and manually generating test vectors, many test engineering departments just punt on the testing issue and hope that badly programmed devices will be caught later on in the process. If volumes are relatively small, it may be cheaper to reject a percentage of boards due to faulty PLDs than to spend the testing resources developing adequate test suites. In most cases, however, it's significantly less costly in the long run to put the time in up-front and develop the needed test vector suites. The automatic test vector generation and fault grading products mentioned in Chapter 5 have alleviated much of the test vector generation burden for simpler designs and devices.

Device Testing

If device programming is done in-house and a complete set of test vectors is provided, device testing can be done as a part of the programming process. Most device programmers contain circuitry that will apply a set of test vectors to the device immediately after it's programmed. You should be aware, however, that different device testers and programmers apply test vectors in different ways. In particular, be aware that some device programmers will apply an input vector serially to the device's inputs, while others will apply the vector to all inputs simultaneously. If the design programmed into your PLD depends on a particular order of input stimulus, it may pass testing on one test platform but fail on another.

If the devices are programmed out-of-house, you have the choice of accepting the programming service's assurance of programming quality or testing the devices individually in receiving inspection. If you intend to test the devices yourself, you will need an IC tester that is capable of applying your test vectors to each device. This process requires custom sets of test vectors to be loaded into the IC tester and reconfiguration of the tester to accept the particular device under test. This process is no less involved that device programming, so it quickly becomes obvious that the devices should either be tested out-of-house or programmed and tested in-house.

If you are forced to sample-test programmed PLDs in receiving inspection because of rigid QA procedures, you will probably find it most cost-effective to do this testing on a device programmer since there are few IC testers that are capable of accepting a variety of PLDs.

Fusible link PLDs, since they aren't reprogrammable, do not lend themselves easily to incoming inspection testing. In particular, complex PLDs may have non-fuse related defects that are virtually impossible to detect until the device has been programmed and test vectors have been applied. If the goal is to detect defective devices before they are programmed, this is clearly unacceptable.

Many manufacturers have responded by adding additional circuitry to their PLDs that allow much of the internal circuitry to be tested non-destructively. The test circuitry may take the form of additional product terms and corresponding fuses that are preprogrammed at the factory. This test circuitry is used by the device manufacturer for final device testing, and can also be utilized by some device programmers to verify the device prior to actual programming. Some device manufacturers, such as Signetics, publish the test circuitry so the devices can be tested by the end-users' receiving inspection departments. Since the test circuitry can drive device elements such as output registers and other non-fuse-related circuitry, you can have relatively high confidence that these parts of the devices will not fail after a successful programming cycle.

If the device testing proves to be too formidable, there are testing services available from device manufacturers and distributors. These services are generally associated with high-volume programming services and include testability consulting as well as help with test suite development.

There are also many situations in which testing of unprogrammed PLDs is important. One such situation, as mentioned above, is then the PLDs are to be programmed in-circuit. In this situation it is critical that bad devices be identified before they are installed on a board. If erasable PLDs are used, receiving inspection is considerably easier. All that is needed is a test pattern that can be programmed into each PLD and a set of test vectors to verify the the device works as it should. Most vendors of erasable PLDs supply their devices with such test patterns already programmed in, further easing the testing process.

As you can see, one-time-programmable devices can present a challenge for incoming inspection. Before these devices are used in an in-circuit programming situation, you should carefully weigh the expected device yield against the number of devices utilized on each circuit board to determine whether the resulting in-circuit programming dropout rate is worth the marginally lower cost of the PLDs.

Checksum Verification

The JEDEC standard 3 format defines a checksum calculation that can be performed on each programming pattern. This checksum is appended to the JEDEC file and can be compared to the checksum calculated by the device programmer both before and after devices are programmed. The checksum can help to detect situations in which a device has been programmed with the incorrect pattern and can help verify the integrity of programming data. The checksum should not, however, be used to positively identify a programming pattern. Since the checksum is byte oriented, it is quite possible (and common) to get the same checksum for different fuse patterns. In many cases, an accidental complete reversal of two or more product terms can go undetected.

For these reasons, checksum verification should never be considered a completely safe way to prevent against an incorrect programming pattern being loaded into a device. Functional test vectors are still the best way to assure that a device has been programmed with the proper pattern.

Signature Analysis Testing

Signature analysis is a testing method that doesn't require test vector generation. It has been implemented in some device programmers. Basically, signature analysis is a method of comparing a programmed device with a known good master device. A signature analysis sequence begins with a known set of device inputs that are applied to the device. The resulting device outputs are then fed back to the device inputs through a shift register and the process is repeated. Each repetition results in a new set of values in the shift register. This output collection, shifting and applying of new inputs is repeated many thousands of times to complete a single analysis cycle.

When a specified number of cycles have been completed, the device's signature is reflected in the shift register. This value can be compared to the signature of the master device. If there is any difference in the signature value, a bad or improperly programmed device is indicated.

Signature analysis is a reasonably good method for testing, if the kinds of circuits being tested are not overly complex. Signature analysis can detect a large number of faulty devices and requires almost no work on the part of a test engineer. There is no guarantee, however, that signature analysis will detect an acceptable percentage of faulty devices. It's virtually impossible to calculate or predict the fault coverage that will be obtained as a result of signature analysis testing techniques. For this reason, signature analysis can't be considered as a replacement for structured test vectors.

Parametric Testing

The objective of parametric testing is to determine whether a device will operate correctly over all operating ranges as specified by the manufacturer. Parametric testing can catch errors that escape detection during device programming.

DC parametric tests include testing for electrical characteristics such as input and output voltage and current ranges, input leakage currents, input clamping diode voltages, output leakage currents, output short circuit current and power supply current. A complete DC parametric test will exercise the device with test vectors while all of the electrical parameters are tested.

AC parametric tests are used to verify the proper switching operation of devices over the published temperature and voltage ranges. These switching parameters include input to nonregistered output (t_{pd}), setup and hold times (t_s and t_h), clock period and width (t_p and t_w), and maximum frequency (f_{max}).

Testing for these parameters requires extremely complex and expensive test equipment, and is therefore rarely done for PLD devices. The expense of the test equipment coupled with the large amount of time required for each test makes such testing impractical for all but very low volume products. In practice, the number of parametric-related device failures is quite low (estimates are that these failures account for only .1 percent of all device failures—the vast majority of device failures are due to incorrect programming). The on-chip test circuitry mentioned earlier further reduces the need for in-house parametric testing, since the device manufactures can (and do) perform such testing themselves to verify published electrical and switching characteristics.

Board Level Testing of PLD-based Designs

Traditionally, PLD-based designs have caused problems for board-level testing. Board level testing hardware and software is typically not capable of accepting PLD test vectors directly. Even when test vectors for individual PLDs are successfully transferred to an ATE system, the development of ATE programs for large boards containing dozens of PLDs can be a mammoth task.

One common procedure is to socket all PLDs on a board and remove the devices prior to board test. After the board is tested, the PLDs are re-inserted into their sockets. Clearly this isn't the optimal solution, since IC sockets are a common area of failure. The increasing use of surface-mount technology makes this approach completely impractical for many applications.

ATE vendors recognize the need for more comprehensive test features for PLD-based designs, and are working with developers of test generation software to provide these features to their customers.

10.9 CONFIGURATION MANAGEMENT

The design and production process for PLD-based circuits is similar in many ways to the processes required for microprocessor-based systems. The final product consists not only of hardware, but of firmware as well. This firmware must be archived along with other design documentation, and must also be available for use in production.

When a PLD-based design is complete, you should archive not only your final JEDEC files, but also the source files from which the JEDEC files were derived and the software that was used to translate the source files into the programming patterns represented by the JEDEC files. PLD design tools, like their microprocessor counterparts, are subject to change from one revision to the next. There is no guarantee that a design file written in a four-year-old dialect of an obscure PLD design language can be reprocessed into a meaningful JEDEC file unless the appropriate version of that design tool has been kept in storage.

Never rely on a programmed master device alone for configuration purposes. Programmed devices are too easily destroyed, mislabeled and misplaced to be used for archival purposes, and a master device gives you no facility for archiving test vector data. Test vectors should be considered a critical item that must be provided for a complete product configuration.

10.10 SUMMARY

Designing PLD-based products that are testable, reliable, and economical can be a challenge. Design methods that provide for metastability and hazard protection often conflict with design for testability methods. While automated design tools certainly speed the design process, they can also introduce unforeseen timing or testability problems into PLD circuits. The designer must understand the causes of these problems and weigh the advantages and penalties of various design options in order to achieve an acceptable balance of design goals.

10.11 REFERENCES

Data I/O Corporation, *PLDtest Plus User Manual,* Data I/O Corp., Redmond, WA, 1989.

Durwood, Brian and O'Donnell, Gary. "Mastering PLD Test Aspects Eases PLD System Testing." *Electronic Design* (February 1988):103–106.

Kopec, Stan and Faria, Don. "Check List Helps You Avoid Trouble with PLD Designs." *Electronic Design News* (November 24, 1988):153–162.

Tavana, Danesh M. "Metastability—A Study of the Anomalous Behavior of Synchronizer Circuits." *MMI PAL/PLE Databook,* Monolithic Memories Inc., 1986.

Wang, Francis and Engstrom, Eric. "Designing PLD Circuits For Testability." *Electronic Design* (April 27, 1989):79–82.

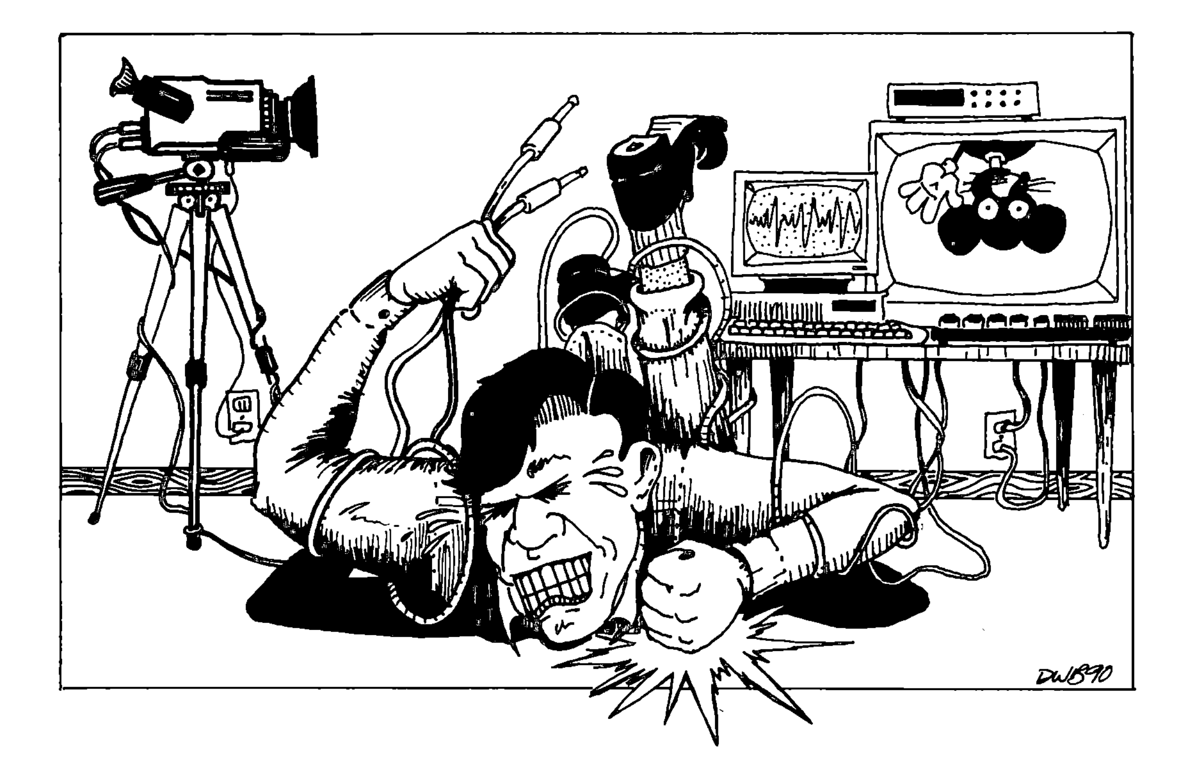

11

Putting It All Together

In this chapter we'll put together what we've discussed in previous chapters and show how a complex design can be implemented in PLDs. We'll also show how a similar design is implemented in an FPGA demonstrating some of the many differences found in these more advanced devices.

11.1 A VIDEO FRAME GRABBER

The design we'll present here is a *video frame grabber*. This design has been implemented in a number of forms, two of which we'll examine in detail. Our first video frame grabber consists of an IBM PC compatible board that digitizes a single field of an external video image (supplied from a camera or other video source) and allows it to be displayed on a video monitor, or accessed by MS-DOS software for storage and display purposes. The design utilizes PLDs for control and memory read/write functions. The frame grabber system includes a set of software control programs that run under MS-DOS.

The second implementation is a stand-alone system that is identical to the first, with the removal of the PC interface circuitry. This frame grabber design connects a video camera (or other video source) with a video monitor and allows the image from the video source to be frozen on the monitor with the push of a button. We have used different design techniques and hardware implementations for these two designs in order to demonstrate some of the many options available to designers.

11.2 GENERAL DESIGN REQUIREMENTS

Before launching into the design and construction of our two video frame grabbers, let's analyze the design requirements.

Video Overview

A standard NTSC video image (our frame grabbers are designed for NTSC video signals, but could quite easily be adapted for use with the European PAL and SECAM standards) is composed of 525 *scan lines*, as shown in Figure 11.1. Each video image, or *frame*, is actually divided into two interlaced *fields* of 262.5 lines. These fields are refreshed at a rate of 59.94 fields per second (the original RS-170 black and white standard specified a simpler 60 fields per second rate, but this was changed in the late 1950s when color was added). This results in a display rate of just under 30 frames each second.

To capture one frame of a live video image, high-speed circuits called *frame grabbers* are used to digitize and store hundreds of samples of video per scan line for the 525 lines of the image. Until recently, frame grabbers could only be built by using power-hungry and expensive ECL RAM devices. Recent advances in CMOS static RAM technology, however, have made it possible to construct a frame grabber at a reasonable cost.

In addition, video chips have become available that further simplify the design of low-cost frame grabbers. The Samsung KSV3100A video chip, for example, is a complete video system that includes an input preamp and clamping circuit, an 8-bit video A-D converter with reference and a 10-bit D-A converter. All of these functions are contained in one 40-pin package.

The video signals that we will be capturing are, of course, analog. An A-D converter is used to convert the incoming video signal to digital amplitude values. These sample values must be analyzed and stored in RAM as quickly as they arrive in order to capture a live image.

Complete Frame

Figure 11.1 Standard video image

Fortunately, the signals that identify the critical portions of a video image are relatively easy to detect. The start of each field is identified by a 20-line vertical blanking interval (the black bar seen at the top of the picture on a poorly adjusted television or monitor). Each scan line consists of ten microseconds of horizontal synchronization and color information (the *color burst*), followed by about 53 microseconds of video information (see Figure 11.2).

Choosing a Sample Rate

Sampling and digitizing an analog signal (of any type) can be done at a variety of rates depending on the quality required of the reconstructed signal. If the highest frequency in the sampled signal (the *Nyquist frequency*) is f_h; the minimum sampling rate at which the signal could theoretically be recovered is $2f_h$ (the *Nyquist rate*).

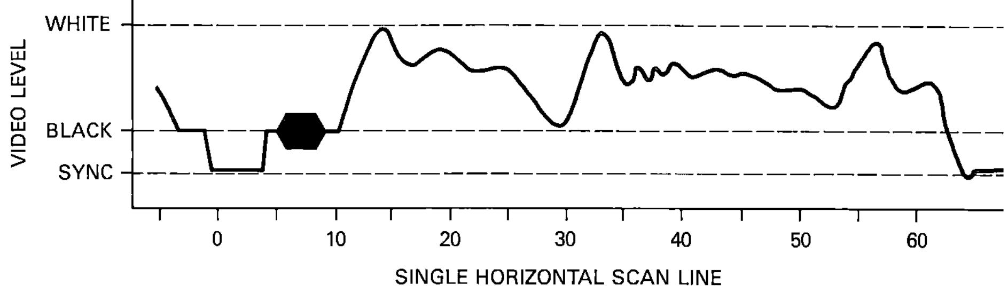

Figure 11.2 Video scan line

The NTSC color burst frequency is 3.579545 MHz. To reproduce a color image, the video signal must be sampled at more than twice the color burst frequency. In practice, sampling the video signal at exactly twice the color burst frequency produces a poor quality color picture, producing correct colors for, at best, half the image.

For our two frame grabber designs, we have chosen to capture only one field of video rather than the complete frame. This results in some loss of vertical resolution, but this loss isn't noticeable on a standard television screen. To sample at over twice the Nyquist rate, each field requires more that 119,437 samples ((2 × 3579545) ÷ 59.94).

Since the digitized image must be stored in high-speed RAM, one or more appropriately sized RAM devices must be selected. The nearest standard RAM size is 128K, which provides a total of 131,072 bytes of storage. The ideal sample rate to fully use this amount of RAM would be 7.856456 MHz (131,072 × 59.94). The nearest standard frequency oscillators available are 7.3728 MHz and 8.000 MHz. A 15.000 MHz oscillator divided by two will also produce a good sample rate. The higher speed 8.000 MHz oscillator will provide the best image, but will cause an overflow of the storage RAM and resultant loss of some image data.

If we sacrifice some image data and utilize the 8.000 MHz oscillator, the result will be that the image is replayed too fast, since it's short by about five scan lines. This doesn't seem to be a problem, since most televisions are capable of synchronizing to a wide variety of marginal quality signals. The advantage to the higher sample rate is better color definition; the reconstructed picture simply looks better.

The 8.000 MHz frame grabber design can be improved by "faking" the missing five scan lines. This could be done by replaying five of the existing scan lines. These lines wouldn't be seen if they were part of the vertical blanking interval. This could be done by adding an additional bit to the frame grabber's address counter and counting to a value higher than the highest RAM address. This would result in the RAM address wrapping back around to zero.

Another approach is to synchronize the saved image with a live image by using part of the live image's vertical blanking interval to fill out the missing field data. In video parlance, this is known as *genlocking*. To genlock the frozen image, the frame grabber circuit must monitor a live video signal and wait for its vertical blanking interval to appear. When this blanking interval is detected, the frame grabber must switch from a live display mode to a frozen image display mode for the remainder of the field. At the completion of the (partial) field, the circuit must revert to live display mode momentarily and then repeat the process.

Choosing a Sample Size

The number of bits in each sample determines the dynamic range (or levels of intensity) that can be reproduced in the reconstructed image. These levels of intensity are also known as the *gray scale*. A 4-bit sample can reproduce sixteen levels of intensity, while eight bits can reproduce 256 levels. Most home video systems don't benefit from more than eight bits, while studio quality equipment will reproduce better images if ten bits are used.

4-bit sample data is often used for special effects. Figures 11.3 and 11.4 illustrate the difference between images saved using four and eight bits of sample information, respectively. We have used eight bits of sample data in our frame grabber circuits since this is the size of the sample provided by the Samsung video chip.

Selecting RAM Chips

There are many options available to circuit designers who require high-speed memory. Specialized *frame memory* chips exist for video frame storage (the Hitachi HM53051P and Toshiba TC521000P chips, for example), but these chips are expensive and aren't typically found on the shelves of your local electronics supply store.

Dynamic RAMS could be used, but the multiplexed address lines add to the design's complexity. For a frame buffer with no read-out capability, the refresh circuitry of DRAMs isn't needed, somewhat simplifying their use. The constant redisplay of the saved image results in an automatic refresh, since each display sequence completely cycles through all the frame grabber's memory addresses.

For this design, CMOS static RAMs were selected because of their simplicity. Byte-wide static RAMs are low in cost and readily available. The non-multiplexed address inputs eliminate the need for the multi-phased system clock that would be required if DRAMs were used.

With an 8.000 MHz sample rate, 125 nanosecond (nsec) memory cycle times are required. While 120-nsec RAM chips would work, 100-nsec RAMs are commonly available and make the circuit a bit easier to design. The critical write cycle parameters that must be provided for in the design are *address setup and hold time* and *data setup and hold time*.

Figure 11.3 Image reconstructed on a television monitor from 4-bit sample data

Figure 11.4 Image reconstructed on a television monitor from eight-bit sample data

11.3 AN IBM PC-BASED VIDEO FRAME GRABBER

As we mentioned previously, our first video frame grabber design plugs into the motherboard of an IBM PC or AT and allows digitized images (from a camera, VCR, or any other NTSC video source) to be displayed on a standard video monitor or stored in MS-DOS files for later use. This board is controlled by software (written in C) that runs on the PC. The control circuitry for the video frame grabber is implemented exclusively in PLDs.

The video frame grabber, as shown in the block diagram in Figure 11.5, monitors the digitized video signal from the A-D converter and, on command, scans and stores a single field of the live image. The frame grabber board has two RCA-type phono jacks for the video input and output signals.

When instructed by the controlling software to record an image, the frame grabber circuit must first wait for and identify the beginning of a field (indicated by the vertical blanking interval). When the blanking interval of the field is detected, the frame grabber begins sampling digitized scan lines and writing them to the four 32K-byte onboard static RAMs. When the image has been completely read, the contents of the RAM chips can either be sent to the D-A converter to reconstruct a frozen video image or be accessed by the controlling software for storage in an MS-DOS file.

Selecting PLDs

The first implementation decision made about this design, after choosing the video and RAM chips, was that it would be implemented in simple 20-pin PAL-type devices. This meant that the design would have to be optimized for a limited amount of I/O resources. As the design evolved, many fundamental design changes and tradeoffs were made to fit the design into the constrained architectures of this class of devices.

The frame grabber circuit was developed using the PEEL 18CV8 electrically erasable PLDs produced by ICT and Gould. The biggest advantage to using these

Figure 11.5 Frame grabber block diagram

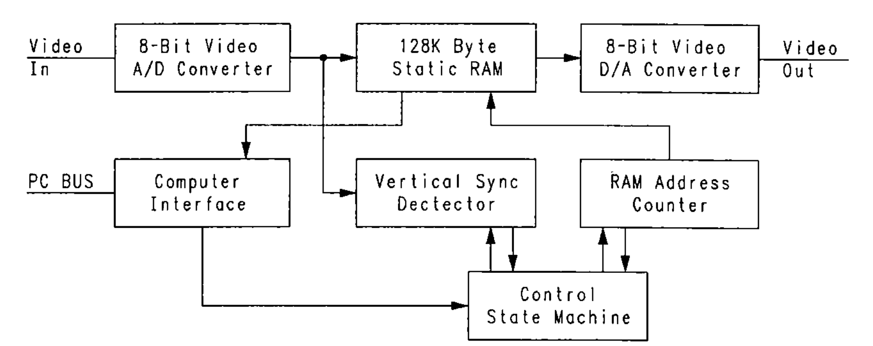

devices during development is that they can be quickly reprogrammed. Architectural features of the PEEL devices, such as their programmable output polarity and common reset terms, were also used to advantage in this design. With slight design modifications the design could fit into simpler 20-pin PAL devices: one reasonable alternative would be the GAL 16V8 from Lattice, National, and other manufacturers. The simpler fixed-polarity 16R8 could also be used if the active levels of some of the signals (in particular, the counter signals) were inverted.

Design Method

Using the block diagram as a guide and ABEL as a tool, the frame grabber circuit was developed. The design was functionally partitioned and the various modules were initially developed independently. Some of these modules required a single PLD, while others required multiple PLDs. As we mentioned previously, electrically erasable devices were used during development. This allowed design changes to be made and tried out with no waste of chips and no need to wait for UV erasure.

One of the many advantages of using erasable PLDs is that test circuitry can be added with little effort during debugging. This feature came in handy when an early version of the design was moved from the IBM XT computer to an IBM AT. When the board was installed in the AT and tried out, some of the scan lines that were read back from the frame grabber were offset to the left or right. Because of the many changes in the hardware and software that were made for the new system, the problem could have been anywhere. Software testing indicated that the RAM address would sometimes fail to increment or would double count.

The problem was successfully diagnosed by comparing the increment address signal with the least significant bit of the address counter chain. This comparator circuit was programmed into one of the erasable PLDs that had an available output and access to both signals. Examination of this test output with a simple logic probe confirmed the problem. The timing problem, due to the different bus clocking design of the AT, was corrected by synchronizing the control signals from the AT bus by means of a simple modification to one of the PLD circuits.

As design-level optimizations were identified, some of the circuit functions that were initially placed in one module were moved to other modules, or spread between modules. The final design is implemented in a total of six PLDs. Three of the devices are used for the RAM address counter and memory chip select logic, while the remaining three are used for bus interface, field detection, and frame grabber control logic.

Control Logic Module

The ABEL-HDL source file for the *CONTROL* module (shown in Figure 11.6) describes the state machine that runs the whole show. The state machine controller has nine possible states. While these nine states could have been expressed

```
Module CONTROL;
title 'Video Frame Grabber Control  Location U002
Michael Holley and Dave Pellerin   30 Aug 1990'

" Modes
" M2    0 = Read Memory     1 = Frame Grab and Playback Mode
"
" M1    0 = Live Video      1 = Still Video
"
" M0    0 = Hold Address    1 = Step Address

        control device  'P18CV8';

        Osc,M2,M1,M0,!DataEna              pin 1,2,3,4,5;
        !IOW,!IOR,VertSync,Write,OE2       pin 6,7,8,9,11;

        ClkAD,Inc,!RAMOE,!RAMWE            pin 14,13,12,19;
        AdrClr,EnaAD,S1,VertClr            pin 15,16,17,18;

        Inc,S1,VertClr,AdrClr,EnaAD        istype 'pos,reg_D,buffer';

        Sreg    = [S1,VertClr,AdrClr,EnaAD,Inc];
        Live    = [ 0,   0   ,   0  ,  0  , 0 ];
        GenLock = [ 0,   1   ,   0  ,  0  , 1 ];
        Reset   = [ 0,   1   ,   0  ,  1  , 0 ];
        Wait    = [ 0,   1   ,   1  ,  1  , 0 ];
        Play    = [ 0,   1   ,   1  ,  1  , 1 ];
        Scan    = [ 1,   1   ,   0  ,  0  , 1 ];
        Record  = [ 1,   1   ,   1  ,  0  , 1 ];
        Read    = [ 1,   1   ,   1  ,  1  , 0 ];
        Step    = [ 1,   1   ,   1  ,  1  , 1 ];

Test_Vectors
([Osc,M2,M1,M0,VertSync,IOR,IOW,DataEna] -> [Sreg    ,RAMOE,RAMWE]);
 [.C., 1, 0, 0,    0    , 0 , 0 ,   0    ] -> [Live    ,  0  ,  0  ];
 [.C., 1, 0, 0,    0    , 0 , 0 ,   0    ] -> [Live    ,  0  ,  0  ];
 [.C., 1, 1, 0,    0    , 0 , 0 ,   0    ] -> [Scan    ,  0  ,  0  ];
 [.C., 1, 1, 0,    0    , 0 , 0 ,   0    ] -> [Scan    ,  0  ,  0  ];
 [.C., 1, 1, 0,    1    , 0 , 0 ,   0    ] -> [Record  ,  0  ,  1  ];
 [ 0 , 1, 1, 0,    1    , 0 , 0 ,   0    ] -> [Record  ,  0  ,  1  ];
 [ 1 , 1, 1, 0,    1    , 0 , 0 ,   0    ] -> [Record  ,  0  ,  0  ];
 [.C., 1, 1, 0,    1    , 0 , 0 ,   0    ] -> [Record  ,  0  ,  1  ];
 [.C., 1, 1, 0,    0    , 0 , 0 ,   0    ] -> [Play    ,  1  ,  0  ];
 [.C., 1, 1, 0,    1    , 0 , 0 ,   0    ] -> [Play    ,  1  ,  0  ];
 [.C., 1, 1, 0,    1    , 0 , 0 ,   0    ] -> [Play    ,  1  ,  0  ];
 [.C., 1, 1, 0,    0    , 0 , 0 ,   0    ] -> [GenLock,   0  ,  0  ];
 [.C., 1, 1, 0,    0    , 0 , 0 ,   0    ] -> [GenLock,   0  ,  0  ];
 [.C., 1, 1, 0,    1    , 0 , 0 ,   0    ] -> [Play    ,  1  ,  0  ];
 [.C., 0, 1, 1,    0    , 0 , 0 ,   0    ] -> [Reset   ,  0  ,  0  ];
 [.C., 0, 1, 1,    0    , 0 , 0 ,   0    ] -> [Reset   ,  0  ,  0  ];
 [.C., 0, 1, 0,    0    , 0 , 0 ,   0    ] -> [Read    ,  0  ,  0  ];
 [.C., 0, 1, 0,    0    , 0 , 0 ,   1    ] -> [Read    ,  0  ,  0  ];
 [.C., 0, 1, 0,    0    , 1 , 0 ,   1    ] -> [Read    ,  1  ,  0  ];
 [.C., 0, 1, 0,    0    , 0 , 1 ,   1    ] -> [Read    ,  0  ,  1  ];
 [.C., 0, 1, 0,    0    , 1 , 0 ,   0    ] -> [Read    ,  0  ,  0  ];
 [.C., 0, 1, 1,    0    , 0 , 0 ,   0    ] -> [Step    ,  0  ,  0  ];
 [.C., 0, 1, 1,    0    , 0 , 0 ,   0    ] -> [Wait    ,  0  ,  0  ];
 [.C., 0, 1, 1,    0    , 0 , 0 ,   0    ] -> [Wait    ,  0  ,  0  ];
 [.C., 0, 1, 0,    0    , 0 , 0 ,   0    ] -> [Read    ,  0  ,  0  ];
 [.C., 1, 1, 0,    0    , 0 , 0 ,   0    ] -> [Play    ,  1  ,  0  ];
 [.C., 1, 0, 0,    0    , 0 , 0 ,   0    ] -> [Live    ,  0  ,  0  ];

@DCSET
Equations
        Sreg.RE = !M1;
        Sreg.C  = Osc;
        ClkAD   = Osc;
```

Figure 11.6 CONTROL.ABL

```
State_Diagram Sreg

State Live:                " Live video to monitor
        RAMWE = 0;
        RAMOE = 0;
        GOTO Scan;

State Scan:                " Look for start of field
        RAMWE = 0;
        RAMOE = 0;
        IF (VertSync) THEN Record ELSE Scan;

State Record:              " Capture one field in RAM
        RAMWE = !Osc;
        RAMOE = 0;
        IF (!VertSync) THEN Play ELSE Record;

State Play:                " Display field on monitor
        RAMWE = 0;
        RAMOE = 1;
        CASE    !M2              : Reset;
                 M2 &  VertSync : Play;
                 M2 & !VertSync : GenLock;
        ENDCASE;

State GenLock:             " Wait for next vertical sync
        RAMWE = 0;
        RAMOE = 0;
        IF (VertSync) THEN Play ELSE GenLock;

State Reset:               " Clear address counter
        RAMWE = 0;
        RAMOE = 0;
        IF (M0) THEN Reset ELSE Read;

State Read:                " PC bus can read or write RAM
        RAMWE = IOW & DataEna;
        RAMOE = IOR & DataEna;
        CASE     M2              : Play;
                !M2 & !M0        : Read;
                !M2 &  M0        : Step;
        ENDCASE;

State Step:                " Increment to next address
        RAMWE = 0;
        RAMOE = 0;
        GOTO Wait;

State Wait:                " Wait for M0 to go low
        RAMWE = 0;
        RAMOE = 0;
        IF (M0) THEN Wait ELSE Read;

End
```

Figure 11.6 *(continued)*

using only four state bits, this would have meant that an additional six outputs would have been required to control the other frame grabber modules for a total of ten outputs. Since 20-pin PAL devices have, at most, eight outputs, this wouldn't allow the state machine to be implemented in a single such device. Since four of the control outputs are registered (*VertClr, AdrClr, EnaAD,* and *Inc*), they can be combined with the state memory bits. The *RAMOE* and *RAMWE* outputs must remain asynchronous.

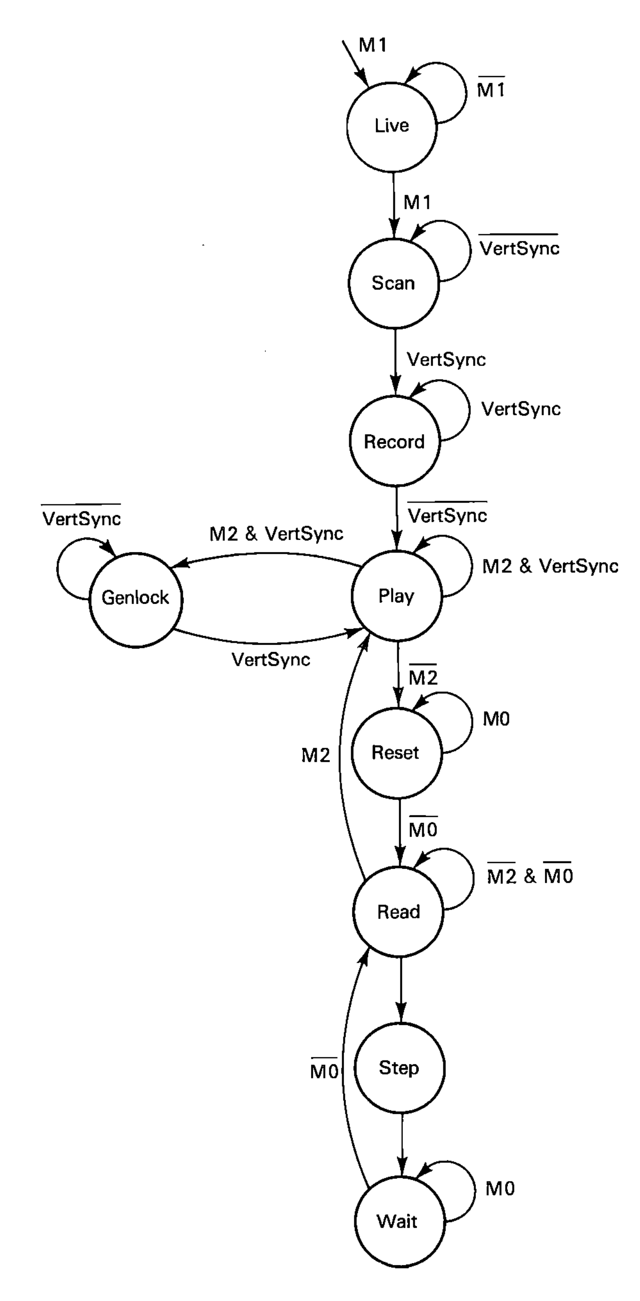

Figure 11.7 Frame grabber controller state machine

Using the four existing registered outputs for state bits wasn't sufficient to support the required state values (states *Play* and *Step* required the same specific values for their outputs, as did *Read* and *Wait*), so a fifth state bit output was added. The *S*1 bit was used to distinguish between the ambiguous states. The total number of outputs required then became seven, which fit into a 20-pin PLD with one output left over. This extra PLD output was later used as a delay element for the video chip's clock. The various states of the frame grabber controller are shown as a state graph in Figure 11.7.

The initial state of the machine is *Live*. During this state, the image is routed directly through the video chip, and the RAM and address counters are disabled. The *VertClr* is asserted in this state, resulting in the clearing of the field detector. The state machine is reset by an external signal, *M*1, which is controlled by the PC software. The use of an explicit reset signal simplifies the design of the state machine and allows the frame grabber to be reset at any time.

The machine remains in state *Live* as long as the *M*1 mode control bit remains low. When *M*1 goes high, the machine advances to the *Scan* state. In this state, the field detector becomes active, and the machine loops until the *VertSync* input is asserted by the field detector, indicating that the vertical blanking interval has been detected. To save inputs to the controller, the *address terminal count* (*ATC*) signal is not used directly, but is instead monitored through the *VertSync* input generated in the *Field* module.

When the start of the image is indicated, the machine advances to state *Record* to begin capturing and storing video samples. The *ATC* input from the *RAMCOM* module indicates that the RAM is full. When this occurs, the machine advances to the *Play* mode.

In the *Play* mode, the A-D is disabled and the RAM output is instead fed to the D-A section of the video chip. The address counter continues to operate, so the saved image (missing five scan lines) is written to the D-A. At the conclusion of the partial field, the state machine advances to the *GenLock* state, where it waits for a vertical blanking interval to be detected from the incoming synchronization signal. When the next blanking interval appears, the machine returns again to the *Play* state to repeat the process.

While the frame grabber controller is in the *GenLock* state (a period of time corresponding approximately to the missing scan lines), the live image is output directly to the D-A converter. Since the *GenLock* state occurs during the vertical blanking interval, the momentary change in video source isn't noticed. Since all saved images are genlocked with a live video signal, there is never a loss of vertical synchronization, even when new images are read and displayed.

The interface to the IBM PC is found in the remaining states of the frame grabber controller. The *M*2 bit, asserted while in state *Play,* results in the machine advancing to the *Reset* state, which clears the address counter, and then to the *Read* state, in which a single byte of RAM data is accessed.

*M*0 is the address increment control used in the *Read, Step,* and *Wait* states. During the *Step* state, the *Inc* signal is asserted, resulting in the advancing of the address counter. The *Wait* state pauses the machine until the PC software requests another address increment via the *M*0 bit. This synchronization of the PC software and the address counters is necessary to avoid missed or multiple counts.

As implemented, the *CONTROL* module uses two specialized features of the 18CV8: the clock input to the logic array, and the global reset term. For other PLDs, the clock can be connected to pin 9 in addition to pin 1 and the reset logic can be expressed as a part of the state diagram.

Field Detector Module

The *FIELD* field detector module (shown in Figure 11.8) provides a signal to the controller state machine that indicates the start and end of each video field. This signal, *VertSync,* goes high after 128 synchronization-level samples are counted. It goes low after the address counter reaches its terminal count (indicated by the signal *ATC*).

The number of samples used for vertical blanking interval detection is important. The horizontal synchronization pulse that begins each scan line has a duration of 4.7 microseconds. This corresponds to about 38 samples at our 8.00MHz speed. During the vertical blanking interval, the video signal is at zero for about half the duration of a scan line (31 microseconds or 248 samples) and the signal is below the black level for 17 to 20 scan lines. At 8.000MHz, then, the number of samples required to correctly identify the vertical blanking interval is between about 40 and 240 samples. One hundred twenty-eight samples was chosen for this design since a 128-state counter is trivial to design.

Address Counter And RAM Control

The frame grabber, as described earlier, must sample and store 131,072 bytes of data. This data must be written into the static RAM chips, of which there are four in this design. To do this, the frame grabber requires an address counter and chip select logic. An address counter that increments beyond 256 states is impossible to construct using only the eight outputs found in a 20-pin PAL (in fact, a counter with more than 128 states is impossible to implement in most 20-pin PALs because of product-term limitations), so it was necessary to split the frame grabber's large address counter into three smaller counters. Two of the counters are found in modules *COUNT*128 and *COUNT*64 while the third, and smallest, of the counters is combined with the chip select logic in the *RAMCON* module. The three ABEL source files for these modules are shown in Figures 11.9, 11.10, and 11.11.

The three counters are all described using a common format and all have three modes of operation—increment, hold, and clear. The *AdrClr* input, when asserted, causes all counters to reset. In addition, the chip select (consisting of the outputs *CE*3 through *CE*0) is reset to its initial state where *CE*0 is asserted.

```
module FIELD
title 'Vertical Sync Detector  Location U007
Michael Holley and Dave Pellerin   30 Aug 1990'

        field   device  'P18CV8';

        Osc,OE7                         pin 1,11;
        VertClr,Inc,zero,ATC            pin 2,3,4,5;
        AD7,AD6,AD5,AD4                 pin 6,7,8,9;
        Q6,Q5,Q4,Q3,Q2,Q1,Q0            pin 18,17,16,15,14,13,12;
        VertSync                        pin 19;

        VertSync,Q6,Q5,Q4,Q3,Q2,Q1,Q0   istype 'reg_D';

        CountZ = [Q6..Q0];
        Data   = [AD7..AD4,.X.,.X.,.X.,.X.];

        Zero   = (Data == 0);

Equations
        CountZ    := (CountZ.fb + 1) & Zero & VertClr & !ATC &  Inc
                   #  CountZ.fb      & Zero & VertClr & !ATC & !Inc;

        VertSync := (CountZ.fb == 127) & VertClr & !ATC
                   #  VertSync.fb       & VertClr & !ATC;

        CountZ.c  = Osc;
        CountZ.oe = !OE7;

        VertSync.c  = Osc;
        VertSync.oe = !OE7;

Test_Vectors
       ([Osc,OE7,VertClr,Inc,Data,ATC] -> [CountZ,VertSync])
        [.c., 0 ,   0   , 0 ,  0 , 0 ] -> [  0  ,   0    ];
        [.c., 0 ,   1   , 0 ,  0 , 0 ] -> [  0  ,   0    ];
        [.c., 0 ,   1   , 0 ,  0 , 0 ] -> [  0  ,   0    ];
        [.c., 0 ,   1   , 1 ,  0 , 0 ] -> [  1  ,   0    ];
        [.c., 0 ,   1   , 1 ,  0 , 0 ] -> [  2  ,   0    ];
        [.c., 0 ,   1   , 1 ,  0 , 0 ] -> [  3  ,   0    ];
        [.c., 0 ,   1   , 1 ,  0 , 0 ] -> [  4  ,   0    ];
        [.c., 0 ,   1   , 1 ,  0 , 0 ] -> [  5  ,   0    ];
        [.c., 0 ,   1   , 1 , 33 , 0 ] -> [  0  ,   0    ];
        [.c., 0 ,   1   , 1 ,  0 , 0 ] -> [  1  ,   0    ];
        [.c., 0 ,   1   , 1 ,  0 , 0 ] -> [  2  ,   0    ];
        [.c., 0 ,   1   , 1 ,  0 , 0 ] -> [  3  ,   0    ];
        [.c., 0 ,   1   , 1 ,  0 , 0 ] -> [  4  ,   0    ];
        [.c., 0 ,   1   , 1 ,  0 , 0 ] -> [  5  ,   0    ];
        [.c., 0 ,   1   , 1 ,  0 , 1 ] -> [  0  ,   0    ];

@const i=1; @repeat 125 {
        [.c., 0 ,   1   , 1 ,  0 , 0 ] -> [@expr i;, 0 ]; @const i=i+1;}

        [.c., 0 ,   1   , 1 ,  0 , 0 ] -> [  126,   0    ];
        [.c., 0 ,   1   , 1 ,  0 , 0 ] -> [  127,   0    ];
        [.c., 0 ,   1   , 1 ,  0 , 0 ] -> [  0  ,   1    ];
        [.c., 0 ,   1   , 1 ,  0 , 0 ] -> [  1  ,   1    ];
        [.c., 0 ,   1   , 1 ,  0 , 0 ] -> [  2  ,   1    ];
        [.c., 0 ,   1   , 1 ,  0 , 0 ] -> [  3  ,   1    ];
        [.c., 0 ,   1   , 1 ,  0 , 0 ] -> [  4  ,   1    ];
        [.c., 0 ,   0   , 1 ,  0 , 0 ] -> [  0  ,   0    ];
 End
```

Figure 11.8 FIELD.ABL

```
Module Count128
Title '7-bit counter with registered carry out   Location U006
Michael Holley and Dave Pellerin    30 Aug 1990'

        Count128 device 'P18CV8';

        Osc,Inc,AdrClr,OE6                pin 1,3,2,11;
        A6,A5,A4,A3,A2,A1,A0              pin 18,17,16,15,14,13,12;
        CarryA                            pin 19;

        H,L,Z,X,C       = 1, 0, .Z., .X., .C.;

        CountA  = [A6,A5,A4,A3,A2,A1,A0];

Equations

        CountA := (CountA.fb + 1)     &  AdrClr &  Inc       " Inc
                #  CountA.fb          &  AdrClr & !Inc       " Hold
                #  0                  & !AdrClr;             " Clear

        CarryA := (CountA.fb == 126)  &  AdrClr &  Inc       " Carry
                #  CarryA.fb          &  AdrClr & !Inc       " Hold
                #  0                  & !AdrClr;             " Clear

        [CountA,CarryA].c  = Osc;
        [CountA,CarryA].oe = !OE6;

test_vectors
       ([Osc,OE6,AdrClr,Inc] -> [CarryA,CountA])
        [ C , 0 ,  0   , 1 ] -> [  0   ,   0  ];
        [ C , 0 ,  1   , 1 ] -> [  0   ,   1  ];
        [ C , 0 ,  1   , 1 ] -> [  0   ,   2  ];
        [ C , 0 ,  1   , 1 ] -> [  0   ,   3  ];
        [ C , 0 ,  1   , 1 ] -> [  0   ,   4  ];
        [ C , 0 ,  1   , 1 ] -> [  0   ,   5  ];
        [ C , 0 ,  1   , 0 ] -> [  0   ,   5  ];
        [ C , 0 ,  1   , 1 ] -> [  0   ,   6  ];
        [ C , 0 ,  0   , 1 ] -> [  0   ,   0  ];

@const i=1; @repeat 123 {
        [ C , 0 ,  1   , 1 ] -> [  0   , @expr i; ]; @const i=i+1;}

        [ C , 0 ,  1   , 1 ] -> [  0   , 124  ];
        [ C , 0 ,  1   , 1 ] -> [  0   , 125  ];
        [ C , 0 ,  1   , 1 ] -> [  0   , 126  ];
        [ C , 0 ,  1   , 0 ] -> [  0   , 126  ];
        [ C , 0 ,  1   , 1 ] -> [  1   , 127  ];
        [ C , 0 ,  1   , 0 ] -> [  1   , 127  ];
        [ C , 0 ,  1   , 1 ] -> [  0   ,   0  ];
        [ C , 0 ,  1   , 1 ] -> [  0   ,   1  ];
end
```

Figure 11.9 COUNT128.ABL

```
Module Count64
Title '6 Bit Counter with Carry Out Location U005
Michael Holley and Dave Pellerin   30 Aug 1990'

        Count64 device 'P18CV8';

        Osc,Inc,AdrClr,CarryA,OE5       pin 1,3,2,9,11;
        A12,A11,A10,A9,A8,A7            pin 18,17,16,15,14,13;
        CarryB                          pin 19;

        A12,A11,A10,A9,A8,A7            istype 'pos,reg_D';

        H,L,Z,X,C       = 1, 0, .Z., .X., .C.;

        CountB  = [A12..A7];

Equations

        CountB := (CountB.fb + 1) &  AdrClr &  Inc &  CarryA  " Inc
                #  CountB.fb      &  AdrClr &  Inc & !CarryA  " Hold
                #  CountB.fb      &  AdrClr & !Inc            " Hold
                #  0              & !AdrClr;                  " Clear

        CarryB  = (CountB.fb == 63) &  AdrClr &  Inc &  CarryA;

        [CountB,CarryB].c  = Osc;
        [CountB,CarryB].oe = !OE5;

test_vectors
       ([Osc,OE5,AdrClr,Inc,CarryA] -> [CarryB,CountB])
        [ C , 0 ,  0   , 1 ,   1  ] -> [  0   ,   0  ];
        [ C , 0 ,  1   , 1 ,   1  ] -> [  0   ,   1  ];
        [ C , 0 ,  1   , 1 ,   1  ] -> [  0   ,   2  ];
        [ C , 0 ,  1   , 1 ,   1  ] -> [  0   ,   3  ];
        [ C , 0 ,  1   , 1 ,   1  ] -> [  0   ,   4  ];
        [ C , 0 ,  1   , 1 ,   1  ] -> [  0   ,   5  ];
        [ C , 0 ,  1   , 0 ,   1  ] -> [  0   ,   5  ];
        [ C , 0 ,  1   , 1 ,   1  ] -> [  0   ,   6  ];
        [ C , 0 ,  0   , 1 ,   1  ] -> [  0   ,   0  ];

@const i=1; @repeat 60 {
        [ C , 0 ,  1   , 1 ,   1  ] -> [  0   , @expr i; ]; @const i=i+1;}

        [ C , 0 ,  1   , 1 ,   1  ] -> [  0   ,  61  ];
        [ C , 0 ,  1   , 1 ,   1  ] -> [  0   ,  62  ];
        [ C , 0 ,  1   , 1 ,   1  ] -> [  1   ,  63  ];
        [ C , 0 ,  1   , 0 ,   1  ] -> [  0   ,  63  ];
        [ C , 0 ,  1   , 1 ,   1  ] -> [  0   ,   0  ];
        [ C , 0 ,  1   , 0 ,   1  ] -> [  0   ,   0  ];
        [ C , 0 ,  1   , 1 ,   1  ] -> [  0   ,   1  ];
        [ C , 0 ,  1   , 1 ,   1  ] -> [  0   ,   2  ];
end
```

Figure 11.10 COUNT64.ABL

```
Module RAMCON
title 'RAM Chip Enable Control    Location U004
Michael Holley and Dave Pellerin     30 Aug 1990'

        ramcon  device 'P18CV8';

        Osc,AdrClr,Inc,OE4        pin 1,2,3,11;
        CarryB                    pin 9;
        A15,A14,A13,ATC           pin 19,18,13,12;
        !CE3,!CE2,!CE1,!CE0       pin 17,16,15,14;

        A14,A13,CE3,CE2,CE1,CE0 istype 'reg_D';

        CE      = [CE3,CE2,CE1,CE0];
        CEshift = [CE2,CE1,CE0,CE3];

        CountC  = [A14,A13];

        CarryC   = (CountC.fb == 3) & CarryB;

Equations
        CountC := (CountC.fb + 1) &  AdrClr &  Inc &  CarryB      " Count
                #  CountC.fb      &  AdrClr &  Inc & !CarryB      " Hold
                #  CountC.fb      &  AdrClr & !Inc                " Hold
                #  0              & !AdrClr;                      " Clear

        CE     := CEshift.fb      &  AdrClr &  Inc &  CarryC      " Shift
                # CE.fb           &  AdrClr &  Inc & !CarryC      " Hold
                # CE.fb           &  AdrClr & !Inc                " Hold
                # [0,0,0,1]       & !AdrClr;                      " Clear

        ATC := (CE.fb == ^b1000) & (CountC.fb == 3) & Inc & AdrClr & CarryB;

        [CountC,CE,ATC].c  = Osc;
        [CountC,CE,ATC].oe = !OE4;

test_vectors
([Osc,OE4,AdrClr,Inc,CarryB] -> [CountC,CE3,CE2,CE1,CE0,ATC])
 [.C., 0 ,  0   , 1 ,   1  ] -> [   0  , 0 , 0 , 0 , 1 , 0 ];
 [.C., 0 ,  1   , 1 ,   1  ] -> [   1  , 0 , 0 , 0 , 1 , 0 ];
 [.C., 0 ,  1   , 1 ,   1  ] -> [   2  , 0 , 0 , 0 , 1 , 0 ];
 [.C., 0 ,  1   , 1 ,   1  ] -> [   3  , 0 , 0 , 0 , 1 , 0 ];
 [.C., 0 ,  1   , 1 ,   1  ] -> [   0  , 0 , 0 , 1 , 0 , 0 ];
 [.C., 0 ,  1   , 1 ,   1  ] -> [   1  , 0 , 0 , 1 , 0 , 0 ];
 [.C., 0 ,  1   , 1 ,   1  ] -> [   2  , 0 , 0 , 1 , 0 , 0 ];
 [.C., 0 ,  1   , 1 ,   1  ] -> [   3  , 0 , 0 , 1 , 0 , 0 ];
 [.C., 0 ,  1   , 1 ,   1  ] -> [   0  , 0 , 1 , 0 , 0 , 0 ];
 [.C., 0 ,  1   , 1 ,   1  ] -> [   1  , 0 , 1 , 0 , 0 , 0 ];
 [.C., 0 ,  1   , 0 ,   1  ] -> [   1  , 0 , 1 , 0 , 0 , 0 ];
 [.C., 0 ,  1   , 1 ,   1  ] -> [   2  , 0 , 1 , 0 , 0 , 0 ];
 [.C., 0 ,  1   , 1 ,   1  ] -> [   3  , 0 , 1 , 0 , 0 , 0 ];
 [.C., 0 ,  1   , 1 ,   1  ] -> [   0  , 1 , 0 , 0 , 0 , 0 ];
 [.C., 0 ,  1   , 1 ,   0  ] -> [   0  , 1 , 0 , 0 , 0 , 0 ];
 [.C., 0 ,  1   , 1 ,   1  ] -> [   1  , 1 , 0 , 0 , 0 , 0 ];
 [.C., 0 ,  1   , 1 ,   1  ] -> [   2  , 1 , 0 , 0 , 0 , 0 ];
 [.C., 0 ,  1   , 1 ,   1  ] -> [   3  , 1 , 0 , 0 , 0 , 0 ];
 [.C., 0 ,  1   , 1 ,   1  ] -> [   0  , 0 , 0 , 0 , 1 , 1 ];
 [.C., 0 ,  1   , 1 ,   1  ] -> [   1  , 0 , 0 , 0 , 1 , 0 ];
End
```

Figure 11.11 RAMCON.ABL

The counter segments will hold whenever their carry inputs (*CarryA, CarryB,* or *CarryC*) are asserted by the previous counter segment.

The *Inc* input is used during read and write operations, when the memory addressing is being controlled from the PC-resident software. At these times, the counter must operate in a single-step mode, since the software can't operate at the speed of the frame grabber circuitry.

Chip select is provided by using a 4-bit shifter in the *RAMCON* module. This eliminates the need for the most significant two bits of the address counter. If a single 128K-RAM chip were used, the ship select shifter logic would be eliminated by a larger (17-bit) counter chain.

PC Interface Module

The final module of the frame grabber circuit provides the interface to and from the PC-resident software. The *PCHOST* module decodes the hexadecimal addresses 300 to 31F from the PC's bus (this is the address range of the IBM prototyping card). Writing to addresses 310 to 317 (from the frame grabber control software) sets the state of the three mode control bits (*M*2, *M*1, through *M*0) as shown in Figure 11.12. Address 300 contains the byte-wide data that is read from or written to the current address of the frame grabber's RAM.

You might have noticed that there is no provision for writing to the frame grabber's memory randomly; all memory read and write operations must occur sequentially, regardless of whether the memory is being accessed by the frame grabber controller or from the PC interface module.

The *PCHOST* module uses pin 1 of the 18CV8 (*IOW*) both for clocking of the mode control bits (*M*2 and *M*0), and as a combinational input for controlling *DataEna*. Unlike most PLDs, the 18CV8 allows the global clock pin to be fed into the logic array for applications such as this. If a device such as GAL 16V8 were used, the *IOW* would have to be split between two input pins (in this case, pin 17 is unused and could therefore be utilized for this purpose). If a simpler device such as the 16R4 were used, the RAM write feature would have to be omitted. Figure 11.13 shows the ABEL source files for the *PCHOST* module.

Address	M2	M1	M0	Function
310	0	0	0	Live Video
316	1	1	0	Freeze Image
313	0	1	1	Step RAM Address
312	0	1	0	Hold RAM Address
300	-	-	-	Read Image RAM

Figure 11.12 Mode control bit values

```
Module PCHOST
title 'PC Interface and I/O address decode  Location U001
Michael Holley and Dave Pellerin    30 Aug 1990'

        pchost device 'P18CV8';

        A9,A8,A7,A6,A5,A4       pin 3,4,5,6,7,8;
        A2,A1,A0                pin 13,12,9;
        !AEN,!IOR,!IOW,OE1      pin 2,19,1,11;
        M2,M1,M0                pin 16,15,14;
        !DataEna                pin 18;

        M2,M1,M0                istype 'reg_D';

        Addr    = [A9..A4,.X.,.X.,.X.,.X.];
        Address = [A9..A4,.X.,A2..A0];

        CtrlPort = (AEN & (Addr == ^h310));
        DataPort = (AEN & (Addr == ^h300));

Equations

    [M2..M0]      :=  CtrlPort & [A2..A0]         "Set  M2..M0
                   # !CtrlPort & [M2..M0].fb;     "Hold M2..M0

    [M2..M0].c    = !IOW;
    [M2..M0].oe   = !OE1;

    DataEna       = IOW & DataPort & !M2.fb
                  # IOR & DataPort & !M2.fb;

test_vectors
       ([IOW,IOR,AEN,Address,OE1] -> [M2,M1,M0,DataEna])
        [.c., 0 , 1 ,  ^h310, 0 ] -> [ 0, 0, 0,   0   ];
        [.c., 0 , 0 ,  ^h311, 0 ] -> [ 0, 0, 0,   0   ];
        [.c., 0 , 1 ,  ^h311, 0 ] -> [ 0, 0, 1,   0   ];
        [.c., 0 , 1 ,  ^h312, 0 ] -> [ 0, 1, 0,   0   ];
        [.c., 0 , 1 ,  ^h313, 0 ] -> [ 0, 1, 1,   0   ];
        [.c., 0 , 1 ,  ^h324, 0 ] -> [ 0, 1, 1,   0   ];
        [.c., 0 , 1 ,  ^h314, 0 ] -> [ 1, 0, 0,   0   ];
        [.c., 0 , 1 ,  ^h315, 0 ] -> [ 1, 0, 1,   0   ];
        [.c., 0 , 1 ,  ^h316, 0 ] -> [ 1, 1, 0,   0   ];
        [.c., 0 , 1 ,  ^h317, 0 ] -> [ 1, 1, 1,   0   ];
        [ 0 , 1 , 1 ,  ^h300, 0 ] -> [ 1, 1, 1,   0   ];
        [ 1 , 0 , 1 ,  ^h300, 0 ] -> [ 1, 1, 1,   0   ];
        [.c., 0 , 1 ,  ^h310, 0 ] -> [ 0, 0, 0,   0   ];
        [ 0 , 1 , 1 ,  ^h300, 0 ] -> [ 0, 0, 0,   1   ];
        [ 1 , 0 , 1 ,  ^h300, 0 ] -> [ 0, 0, 0,   1   ];
        [ 0 , 0 , 1 ,  ^h300, 0 ] -> [ 0, 0, 0,   0   ];
        [ 0 , 1 , 0 ,  ^h300, 0 ] -> [ 0, 0, 0,   0   ];
End
```

Figure 11.13 PCHOST.ABL

The Complete Frame Grabber

The complete frame grabber board is pictured in Figure 11.14. A schematic of the board is illustrated in Figure 11.15.

As implemented, the circuit requires six PLDs, one fixed-function TTL chip, the four 32K static RAMs, and the video chip. If more complex PLDs had been used, the total number of packages could have been reduced, but for this design, it was decided that the six low-cost PLDs were the best compromise.

Frame Grabber Control Software

The PC-resident frame grabber software was developed using Microsoft C; complete source listings are found in Appendix C. The software accesses the video

Figure 11.14 Video frame grabber board

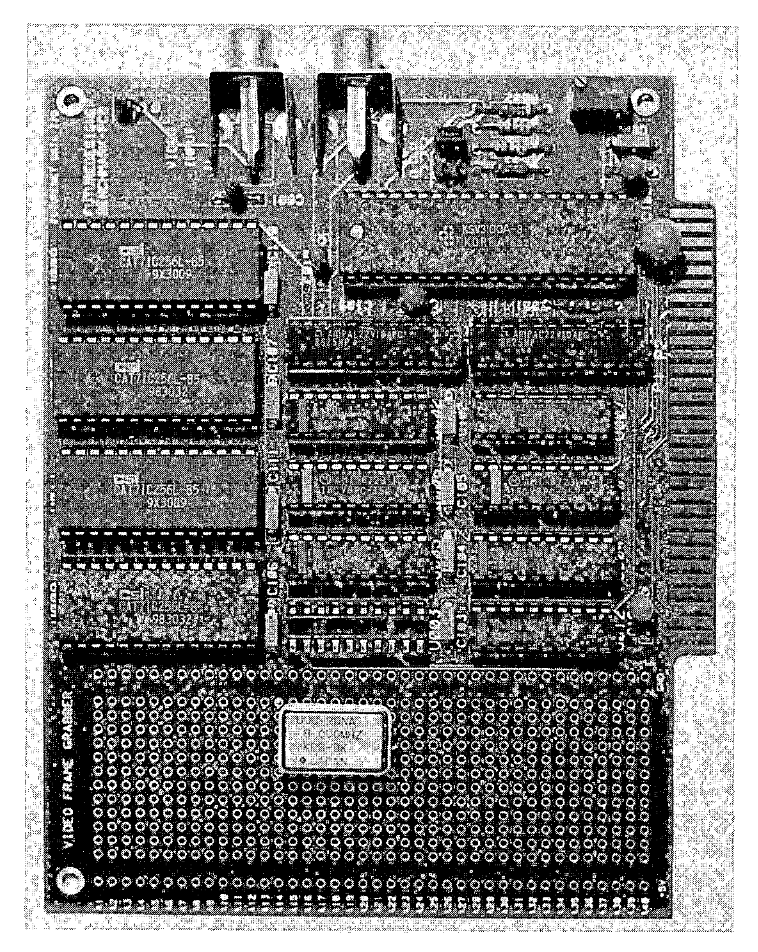

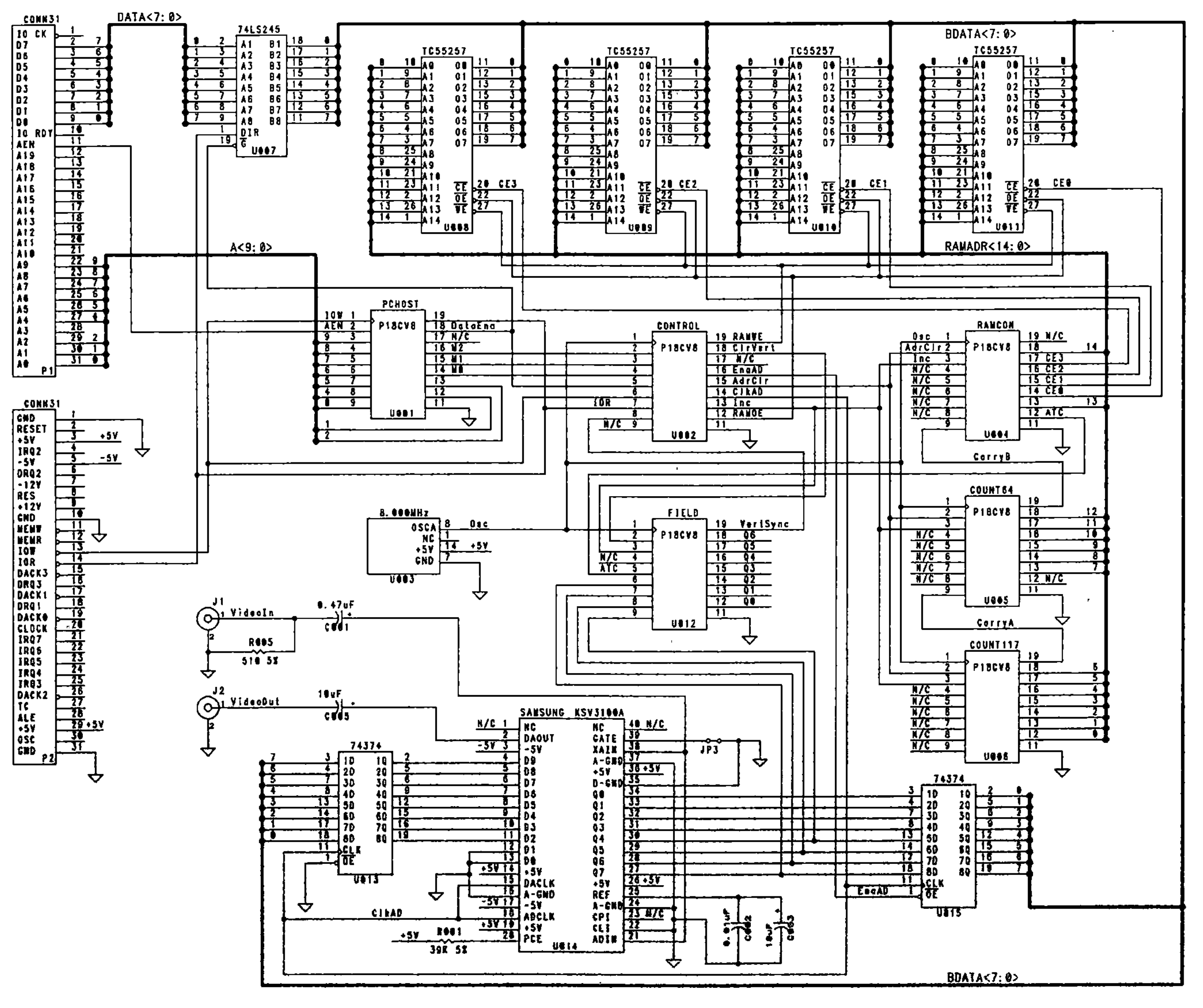

Figure 11.15 Schematic of video frame grabber

frame grabber at hexadecimal addresses 310 through 317. These are the addresses used to trigger the mode control bits $M0$, $M1$, and $M2$.

The video frame grabber control program reads and writes the frame grabber's onboard memory using PC memory address 300 as the data address for each image sample. The image data is stored directly in MS-DOS files. When this program is used, images (either live or frozen, depending on the operation) are sent to the frame grabber's video output jack, but are not displayed on the PC's monitor. Since all displayed images must be genlocked with a live image, the video input jack of the frame grabber board must at all times be connected to a valid video signal (the output of a video camera or other video source—reruns of Star Trek are perfectly acceptable). The program accepts the following single character commands:

L	Display live image
F	Capture (freeze) image
W	Write image to disk file
R	Read image from disk file
Q	Quit

Displaying and Printing Saved Images with an IBM PC

Since the frame grabber control software saves images in MS-DOS files, there is no reason why the images can't be displayed on the PC's monitor or output to a graphics printer. In order to do this, it's necessary to convert the image to a form that matches the particular attributes of the chosen display peripheral.

On video monitors in general, the image quality is determined by the amount of gray scale provided. On an analog VGA monitor, for example, monochrome picture quality is nearly as good as would be observed on a black and white TV set. (The color image can't be reconstructed on a computer monitor because, in NTSC video signals, the color information is encoded as a part of the composite video signal. On a computer monitor, on the other hand, a color image is separated into individual signals for red, green, and blue.)

On the more common EGA monitors, the screen image is composed of 350 lines of 640 pixels each. Since the saved images created by the frame grabber circuit consist of 262 lines of 504 samples each, we have a choice of either truncating the image and using four pixels per image sample or using only a portion of the EGA's screen area. We have chosen to truncate the picture since some of the lost information is simply the vertical and horizontal sync information, which isn't required. The truncated image reflects approximately the center seventy-five percent of the saved image.

To help overcome the gray scale limitations of the EGA monitor, each image sample can be decoded into five distinct gray levels by using four pixels displayed in the patterns shown in Figure 11.16. The image displayed in Figure 11.17 uses this gray scale decoding.

When images are printed or when the image is output to a monitor with more

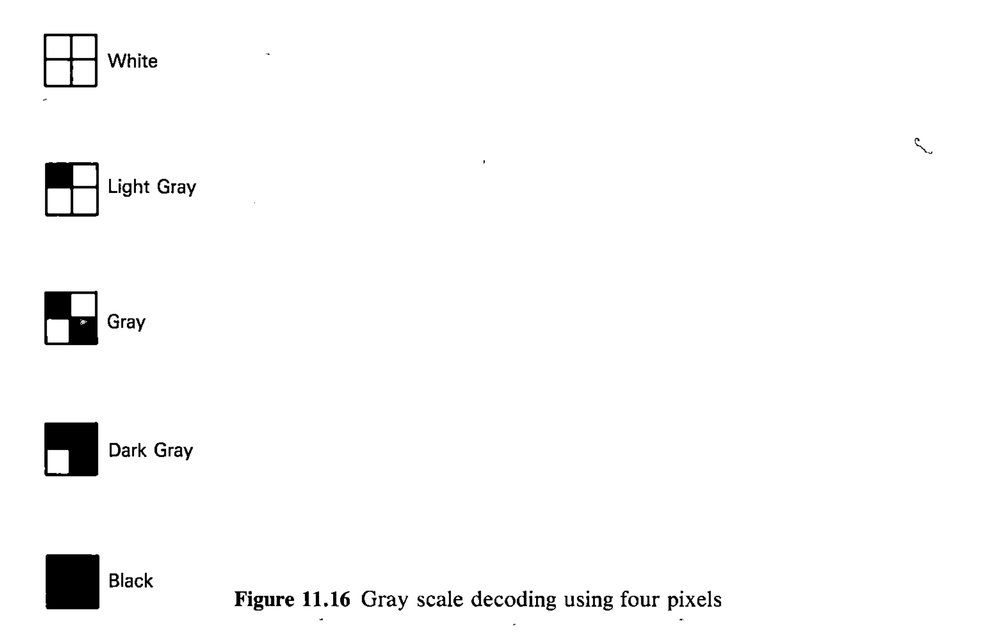

Figure 11.16 Gray scale decoding using four pixels

limited gray scale (for example, a Hercules or equivalent monochrome display) the image can be improved by employing some simple image processing techniques. At the extreme end of the gray scale limited devices, dot matrix printers tend to support only two levels of intensity—either black or white. With these limitations, some image processing must be done to get a good image on paper. With the addition of a simple dithering technique, a reasonably good image such as that displayed in Figure 11.18 can be obtained.

11.4 A STAND-ALONE VIDEO FRAME GRABBER

The stand-alone version of the video frame grabber was developed as a vehicle for experimenting with the FPGAs (in particular, the LCA devices from Xilinx). Functionally, this version of the frame grabber is quite similar to the earlier frame grabber design but has no interface to the IBM PC.

The operation of this frame grabber is quite simple; it has a user interface that consists of one button. When the button is depressed, the frame grabber routes the live video input signal directly to the output and when the button is released, the image freezes. The overall schematic for this design is shown in Figure 11.19.

The stand-alone frame grabber circuit is quite compact, requiring a total of only four ICs. The Samsung video chip was again utilized along with a 128K-byte static RAM to keep the chip count to an absolute minimum. The board includes options for loading the LCA device from either a serial PROM, an EPROM, or

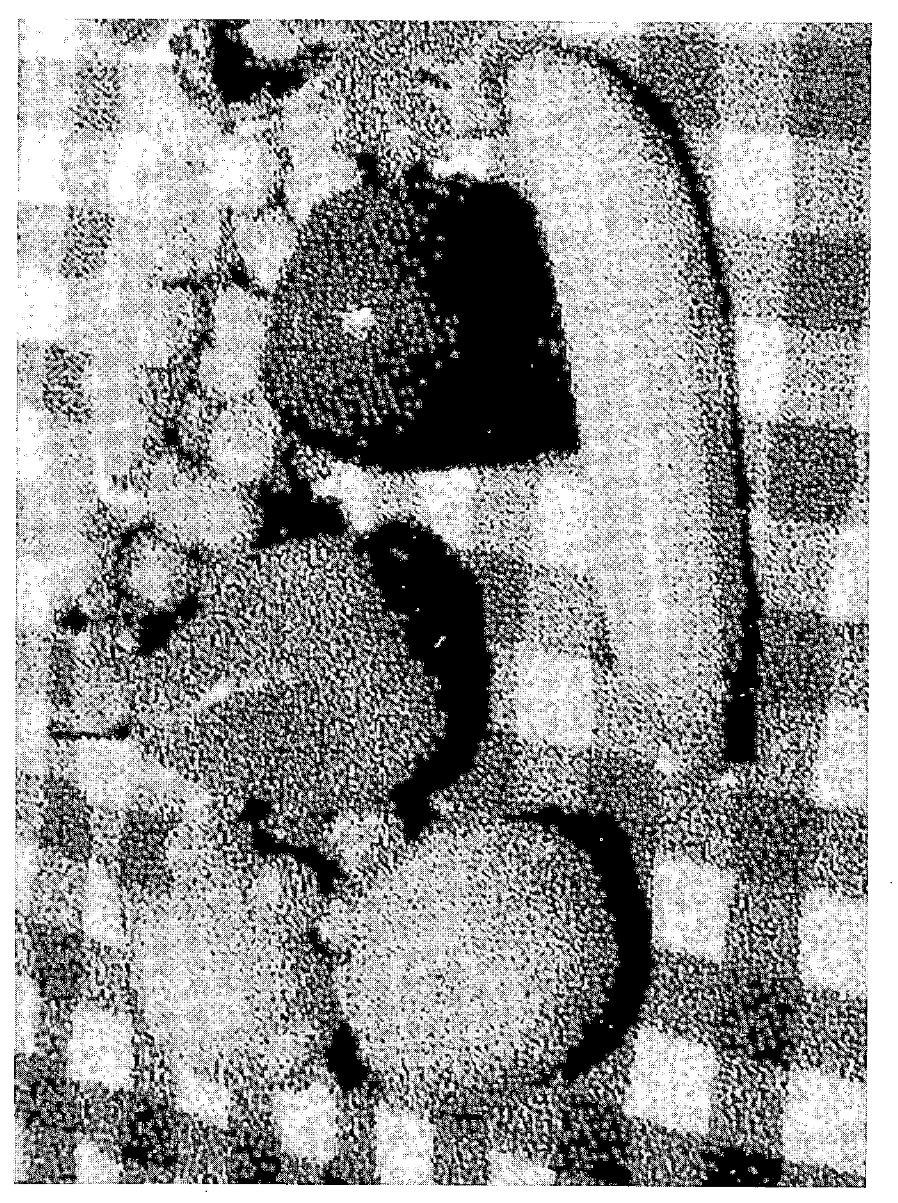

Figure 11.17 Image constructed with simple gray scale decoding on EGA display

Figure 11.18 Printed image constructed using dithering technique

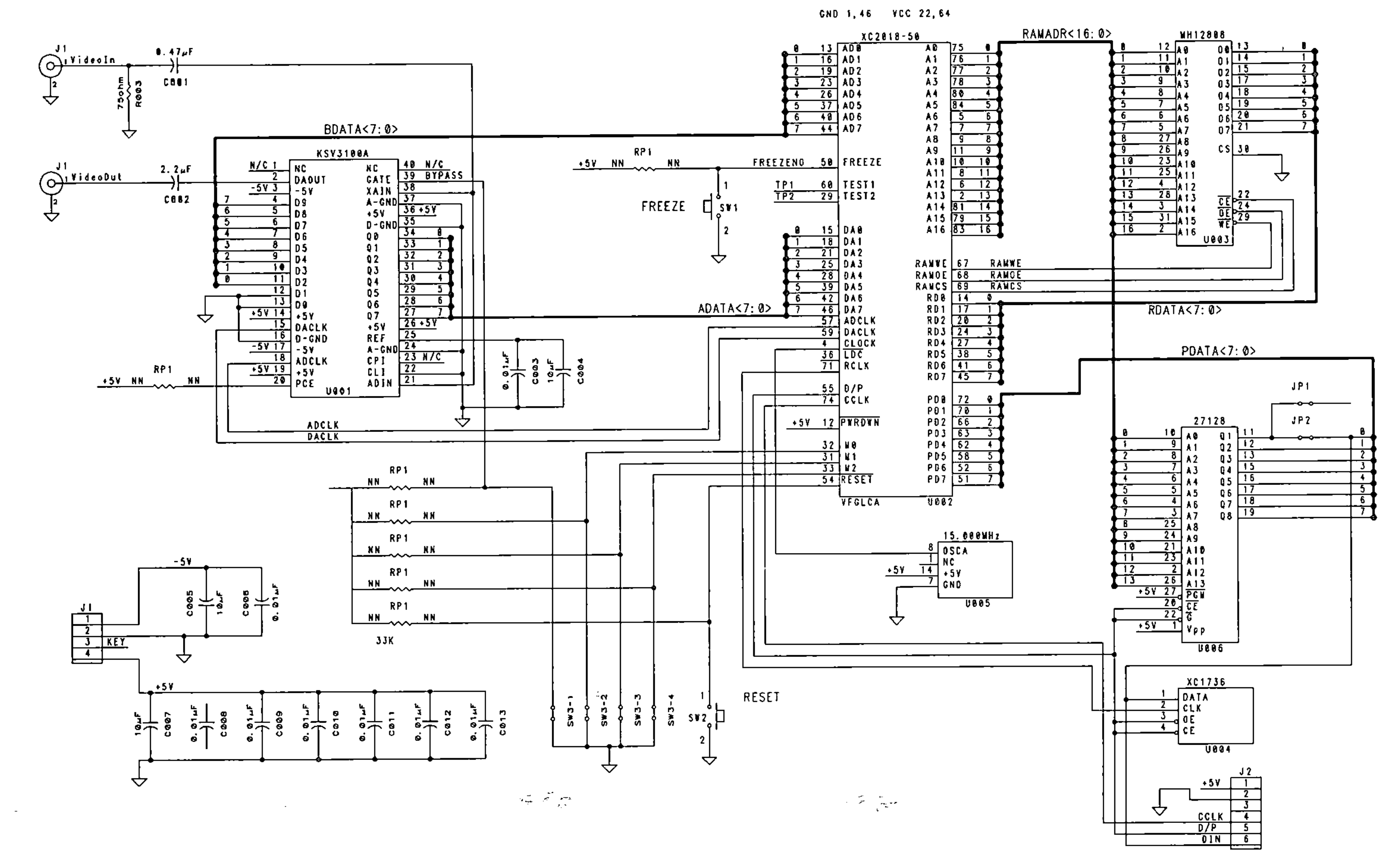

Figure 11.19 Frame grabber schematic

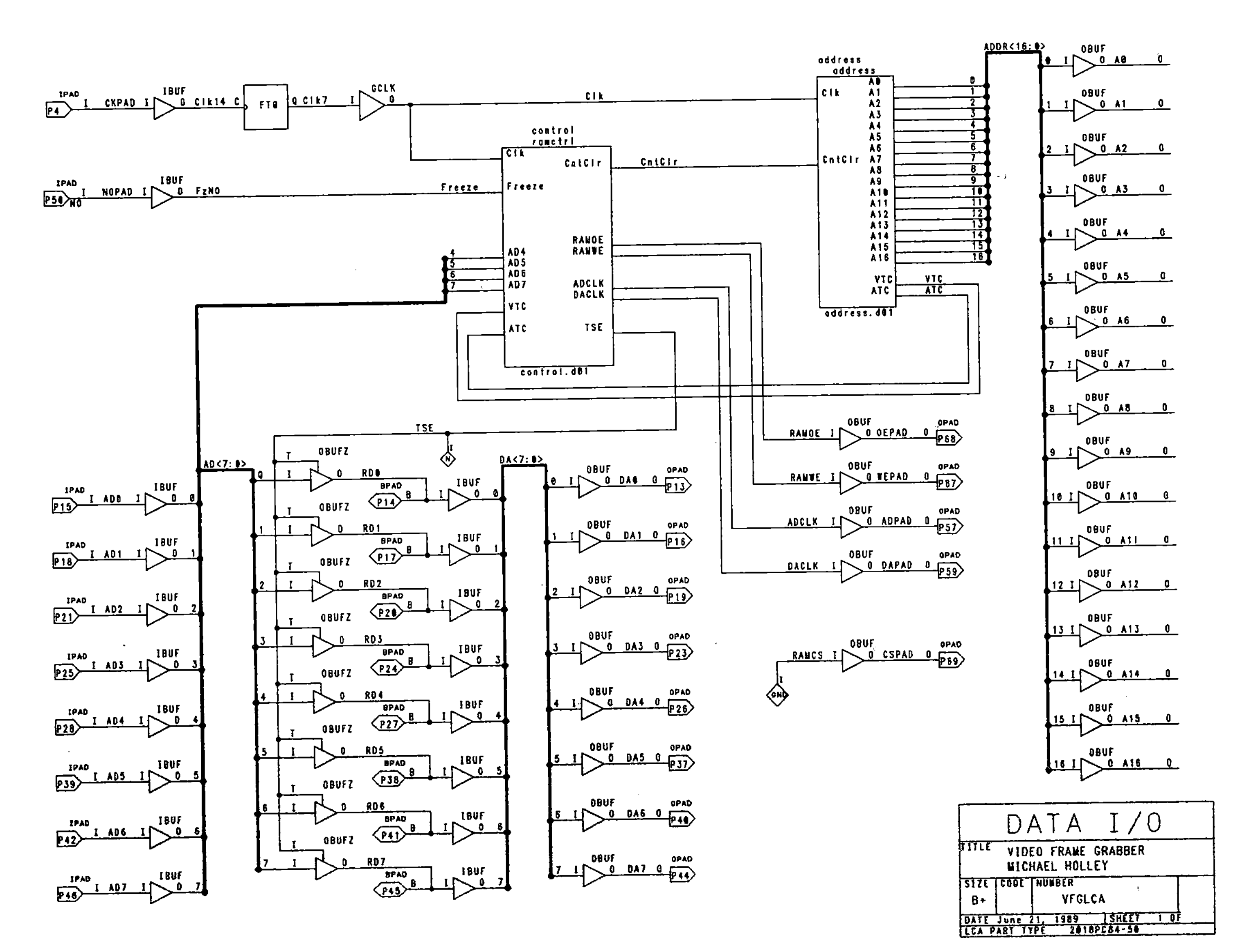

Figure 11.20 LCA pin block diagram

from the download cable supplied with the Xilinx PGA development system. The configuration source is selected with a jumper block that is connected to the LCA's mode control inputs. The following table shows the jumper values for each configuration option:

Source	*M2*	*M1*	*M0*
Serial PROM	0	0	0
EPROM	1	0	0
Cable	1	1	1

Developing the LCA Circuitry

For this version of the design, the frame grabber was split into two functional modules. The first module, *CONTROL,* contains the controller state machine and the field detector circuitry. The second module, *ADDRESS,* contains the memory address counter circuitry. Both of these modules are implemented in a single LCA device. The state machine is simplified over the earlier design by the elimination of the computer interface logic. The counter is also simplified since there is no need to chain smaller counters to create a large address counter.

Figure 11.20 shows the LCA pin assignments and the two functional blocks of the frame grabber LCA. The schematic shown, and the behavioral descriptions for the lower-level logic of the functional blocks, were developed using Data I/O's FutureDesigner design entry system.

CONTROL, which contains the field detector and mode control state machine, was described using high-level equations and FutureDesigner's Gates state description language. *ADDRESS* contains the address counter, and was designed using high-level equations. This address counter differs from the counter used in the PLD implementation in that it didn't need to be split into smaller units. In addition, the counter used in the LCA uses T-type flip-flops. These changes reflect the different architectures of PLDs and LCAs.

Figure 11.21 shows the Gates declaration, state description and equation forms describing the field detection and mode control segments of the design (complete listings are found in Appendix C). Figure 11.22 shows the declaration and equation forms for the address counter.

Gates' logic reduction algorithm (Espresso) is used to minimize the design's logic. For an LCA device, the reduced equations are not in a form appropriate for direct implementation since the AND and OR gates required are much too wide. An LCA device is composed of many individual logic cells (CLBs), each of which has only a small number of inputs feeding a sum-of-products PROM array and one of two registers.

As we mentioned in earlier chapters, FPGA design tools (such as XACT from Xilinx) include automated placement and routing routines that take care of squeezing designs into the LCA cells. These APR routines work most effectively if the designs fed into them are decomposed into multiple levels of logic that don't exceed the gate width constraints of the cell.

```
      Form type:  declaration     Form name:  control

Name                                Type        Definition/Comment
Clk                                 input       "System clock
Freeze                              input       "Freeze Frame Signal
VTC                                 input
RAMOE,RAMWE                         bidir       comb
DACLK,ADCLK                         bidir       comb
AD7..AD0                            input       "Video AD signal
AD                                  set         [AD7..AD0]
CntClr                              output      d
SyncLevel                           bidir       comb
TSE                                 bidir       comb
ATC                                 input
S1..S0                              bidir       d
statereg                            set         [S1..S0]


      Form type:  equation        Form name:  vertsync

SyncLevel = (AD == [0,0,0,0,~x,~x,~x,~x])

statereg.clk = Clk

statereg.ar = !Freeze

TSE = !RAMOE

ADCLK = !Clk

DACLK = !Clk

CntClr.clk = Clk
```

Figure 11.21 Gates design forms for controller

Gates includes a factoring feature that allows design equations to be converted in such a way. Figure 11.23 shows a factor form for this design. Notice that the AND maximum and OR maximum fields are specified with a maximum gate width of three. Although the LCA cells have a typical gate width of four or five inputs, the XACT optimizer is better able to utilize CLBs if the design is provided to it in smaller chunks of logic. For example, two three-input equation factors that have two common inputs can be mapped into a single two-output LCA cell.

After the design equations were factored, Gates was used to automatically generate a schematic for the design. This schematic is composed of logic gates and symbols that are understood by the Xilinx XACT tools. The netlist created from this schematic was then fed into the XACT system and, after being processed by the XACT tools, was downloaded as a bit pattern to the LCA and placed in the frame grabber circuit board. This process was detailed in Chapter 9.

```
        Form type:  state           Form name:  control

           State Register Set Name: statereg

        State Name:                      State Value:
        Live                             0
        Scan                             1
        Record                           3
        PlayBack                         2

  State: Live
    Equations:
      : RAMOE = 1
      : RAMWE = 1
      : CntClr.d = 1
    Transitions:          Type: dependent
      Case: default
            Goto: Scan

  State: Scan
    Equations:
      : RAMOE = 1
      : RAMWE = !Clk
      : CntClr.d = !SyncLevel
    Transitions:          Type: dependent
      Case: VTC
            Goto: Record
      Case: default
            Goto: Scan

  State: Record
    Equations:
      : RAMOE = 1
      : RAMWE = !Clk
      : CntClr.d = ATC
    Transitions:          Type: dependent
      Case: CntClr.q
            Goto: PlayBack
      Case: default
            Goto: Record

  State: PlayBack
    Equations:
      : RAMOE = 0
      : RAMWE = 1
      : CntClr.d = ATC
    Transitions:          Type: dependent
      Case: default
            Goto: PlayBack
```

Figure 11.21 *(continued)*

As we mentioned in earlier chapters, the design of an LCA circuit requires a somewhat different approach that the design of a circuit to be implemented in PLDs. The approach that worked best for this design was to use a modular approach, where the design was broken up into smaller segments that were designed independently, then downloaded to the LCA and tested. (This approach can also

```
        Form type:  declaration     Form name:  address

   Name                                  Type        Definition/Comment
   A17..A0                               bidir       t
   Clk,CntClr                            input
   Address                               set         [A17..A0]
   ATC                                   output      d
   VTC                                   output      d
  " Terminal Count = Clock / 59.94 = 8.0 MHz / 59.94
   Terminal_Count                        macro       133467
   Sync_Count                            macro       127


        Form type:  equation        Form name:  address

   ATC.d = (Address == Terminal_Count)

   VTC.d = (Address == Sync_Count)

   Address.t  = (Address.q + 1)
              $  Address.q

   Address.ar = CntClr

   Address.clk = Clk

   ATC.clk = Clk

   VTC.clk = Clk
```

Figure 11.22 Address counter forms

be used for large multiple-PLD designs when electrically erasable devices are used.)

Additional LCA circuitry, unrelated to the actual design, was frequently developed to aid in routing internal signals to output pins during testing. This additional circuitry could be discarded when the subcircuit was functioning correctly. Since the LCA is RAM based, there is no penalty for repeatedly loading different patterns into the chip.

The complete stand-alone frame grabber system is shown in Figure 11.24. While this frame grabber is of little value in itself, it could quite easily be expanded in function. The XC2018 device chosen for this design is only partially utilized by the frame grabber circuit we have presented. It would be relatively easy to add more capabilities to the design without greatly increasing the size or complexity of the circuit board.

For example, a serial port could be added to the board to transfer image data to a host computer. The associated UART circuitry could be controlled from (and possibly even implemented in) the LCA device. Another option might be to add some simple special effects image processing circuitry such as byte-wide

```
        Form type:  factor          Form name:  Factor

  Group: ad          Def: VTC.d,ATC.d,Address.t

                                Factored Output
  Group: ad          AND Min: 3   AND Max: 3   OR Min: 3   OR Max: 3
                     Stage Count: 99  Factor Target: g  Polarity: +

Stages  Factored Equations:
[1]     A0.t = 1
[1]     A1.t = A0.q
[1]     A2.t = A1.q & A0.q
[1]     A3.t = ad@0
[2]     A4.t = ad@0 & A3.q
[2]     A5.t = ad@1
[3]     A6.t = ad@1 & A5.q
[3]     A7.t = ad@2
[4]     A8.t = ad@2 & A7.q
[4]     A9.t = ad@3
[5]     A10.t = ad@3 & A9.q
[5]     A11.t = ad@4
[6]     A12.t = ad@4 & A11.q
[6]     A13.t = ad@5
[7]     A14.t = ad@5 & A13.q
[7]     A15.t = ad@6
[8]     A16.t = ad@6 & A15.q
[8]     A17.t = ad@6 & A16.q & A15.q
[8]     ATC.d = ad@14 & ad@13
[8]     VTC.d = ad@16 & ad@15

Stages  Intermediate Equations:
[1]     ad@0 = A2.q & A1.q & A0.q
[2]     ad@1 = ad@0 & A4.q & A3.q
[3]     ad@2 = ad@1 & A6.q & A5.q
[4]     ad@3 = ad@2 & A8.q & A7.q
[5]     ad@4 = ad@3 & A10.q & A9.q
[6]     ad@5 = ad@4 & A12.q & A11.q
[7]     ad@6 = ad@5 & A14.q & A13.q
[1]     ad@7 = A0 & A1 & A3
[2]     ad@8 = ad@7 & A4 & A6
[3]     ad@9 = ad@8 & !A7 & !A9
[4]     ad@10 = ad@9 & !A10 & !A12
[5]     ad@11 = ad@10 & !A13 & !A14
[6]     ad@12 = ad@11 & !A15 & !A16
[7]     ad@13 = ad@12 & !A2 & !A5
[1]     ad@14 = A8 & A11 & A17
[7]     ad@15 = ad@12 & A2 & A5
[1]     ad@16 = !A8 & !A11 & !A17

Output Statistics:
    Products: Reduced 188, Factored 74
    Sums:     Reduced 0, Factored 0
```

Figure 11.23 Gates factor form

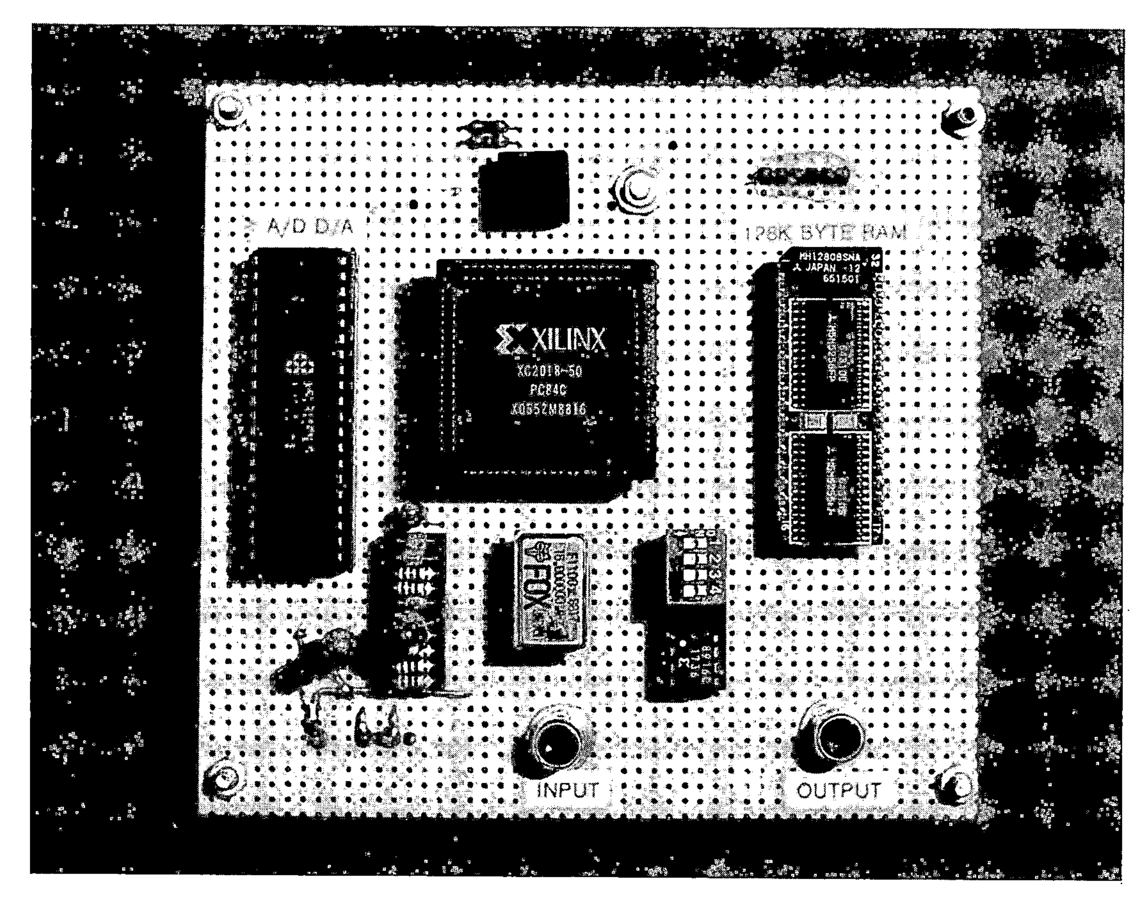

Figure 11.24 Stand-alone video frame grabber system

shifters or comparators. These options could be controlled by one or more additional switches on the board, to provide (in hardware) such things as edge enhancement, tiling effects, or partial-frame capture and playback.

11.5 SUMMARY

In this chapter, we have shown how an actual design is implemented using high-level design tools. In the process, we have illustrated how design decisions are frequently affected by the target device technology. The video frame grabber is typical of applications being implemented in PLDs. It's now normal to see complete systems, requiring perhaps dozens of PLDs, being created for production. FPGAs can be used to implement similarly sized applications, but the designer must be aware that there are many differences between FPGAs and simple PLDs that affect the fundamental design of a circuit.

As we pointed out in Chapter 4, an application that maps quite easily into one or more PLDs may not map quite so easily into an FPGA without major design changes. High-level design tools make it easier to identify and implement such changes.

11.6 REFERENCES

Gilbert, Alfie. "Video Frame Grabber." *Programmable Logic Handbook, 4th Edition,* Monolithic Memories, Inc., Santa Clara, CA, 1985.

Stanley, William D. *Digital Signal Processing,* Reston Publishing Company (a Division of Prentice-Hall), Reston, VA, 1975.

Appendix A

Glossary of PLD-Related Acronyms

One of the most striking—and often amusing—by-products of the PLD marketing frenzy of recent years is the large number of acronyms coined by device manufacturers and other industry participants. In the following list, we have tried to include as many of these acronyms as possible, and have indicated any trademarks that we are aware of.

ABEL *Advanced Boolean Expression Language (Data I/O trademark). Data I/O's universal PLD design language and supporting software.*

ACB *Architecture Control Block. Altera's name for the EP600/EP900 configurable I/O macrocell.*

ACT *Application Configurable Technology (Actel trademark). A family of field programmable gate arrays (FPGAs) offered by Actel Corporation.*

AIM *Avalanche-Induced Migration. A programming technique developed by Intersil that employs a single transistor as the programming element. Programming is accomplished by shorting the emitter to the base with a high current resulting in a diode connection between two signal paths.*

AMAZE *Automatic Map And Zap of Equations (Signetics trademark). Signetics' first proprietary design tool for FPLA devices. See* FLPA.

AMD *Advanced Micro Devices. World's largest manufacturer of PLDs.*

AmPAL *AMD Programmable Array Logic (AMD trademark). See* PAL.

APEEL *Assembler for PEEL (International CMOS Technology trademark). PEEL-specific PLD design tool developed by John Birkner for ICT. See* PEEL, ICT.

A+PLUS *Altera Programmable Logic User System (Altera trademark). High-level design system for Altera's PLDs.*

ASIC *Application Specific Integrated Circuit. An integrated circuit designed or tailored by an end-user for a particular application.*

ASPLD *Application Specific PLD. A PLD containing fixed-function circuitry targeted for a particular class of applications.*

ATG or ATVG *Automated Test Vector Generation. Computer generation of production test vectors with the goal of providing maximum fault coverage.*

BIC *Bus Interface Controller (Intel trademark). An application-specific PLD intended for bus interface applications.*

CHMOS *Complementary High-speed Metal Oxide Semiconductor (Intel trademark). A high-speed CMOS process. See* CMOS.

CLB *Configurable Logic Block. A logic module found in logic cell array devices. See* LCA.

CMOS *Complementary Metal Oxide Semiconductor. A Device technology characterized by minimal standby power requirements.*

CPL *CMOS Programmable Logic (Samsung trademark). Trade name for Samsung's family of PAL-type devices.*

CUPL *Universal Compiler for Programmable Logic (Logical Devices trademark). The first universal PLD design tool; originally developed by Assisted Technology.*

DIP *Dual-Inline Package. A rectangular IC package, either ceramic or plastic, with vertical leads symmetrically arranged on its long sides.*

ECL *Emitter Coupled Logic. A bipolar device technology characterized by high speeds, high power use and high cost.*

EDIF *Electronic Design Interchange Format. A standard format that is used to transfer design data in the form of schematics and netlists.*

E^2 *Electrically Erasable.*

EEPLD *Electrically Erasable PLD.*

EEPROM *Electrically Erasable PROM. See* PROM.

EPLD *Erasable PLD. Assumed to be ultra-violet erasable.*

EPROM *Erasable PROM. Assumed to be ultra-violet erasable.*

ERA *Electrically Reconfigurable Array. Plessey's trade name for their RAM-based field programmable gate arrays. See* FPGA.

ERASIC *Erasable, Reprogrammable ASIC (Exel trademark). A CMOS PLD featuring a folded NOR architecture.*

FPC *Fuse Programmable Controller (AMD trademark). A microcode-based sequencer device.*

FPGA *Field Programmable Gate Array. A large programmable device composed of independent configurable logic modules that are interconnected with programmable routing channels.*

FPIC *Field Programmable Integrated Circuit. Broad category of user-programmable integrated circuits, including ICs not intended for logic applications.*

FPLA *Field Programmable Logic Array. Acronym coined by Signetics for devices with a programmable AND array and a programmable OR array. See* PLA.

FPLD *Field Programmable Logic Device. See* PLD.

FPLS *Field Programmable Logic Sequencer. FPLA devices containing memory elements.*

FPRP *Field Programmable ROM Patch. An early Signetics programmable device.*

GAL *Generic Array Logic (Lattice Semiconductor trademark). Electrically erasable devices that are designed to replace most PAL-type devices. See* PAL.

HAL *Hard Array Logic (AMD trademark). Mask-programmed versions of the popular MMI PAL devices. See* PAL.

HDL *Hardware Description Language. A textual design description format used for design entry and synthesis or for circuit modeling and simulation.*

HiPAC *High-performance Programmable Array CMOS (MMI trademark). MMI's trade name for their CMOS PLD process.*

HPL *Harris Programmable Logic (Harris trademark). Harris' trade name for their family of PAL-type devices. See* PAL.

ICT *International CMOS Technology. Makers of the popular PEEL 18CV8 electrically erasable device. See* PEEL.

IFL *Integrated Fuse Logic (Signetics trademark). Archaic term for Signetics FPLA devices. See* FPLA, FPLS.

iPLDS *Intel Programmable Logic Design System (Intel trademark). IBM PC-based PLD design software for Intel PLDs.*

JEDEC *Joint Electron Device Engineering Council. A committee formed as a part of the Electronics Industry Association that provides and arbitrates standards for electronic devices and data formats.*

JLCC *J-Leaded Chip Carrier. See* PLCC.

LCA *Logic Cell Array (Xilinx trademark). A RAM-based field programmable gate array.*

LCC *Leadless Chip Carrier. A socketed ceramic IC package with flush leads.*

LSI *Large-Scale Integration. Generally accepted to be an integrated circuit containing 100 or more equivalent gates.*

MACH *Macro Array CMOS High Speed (AMD trademark). AMD's family of complex CMOS PLDs.*

MAX *Multiple Array Matrix (Altera trademark). A complex PLD offered by Altera.*

MMI *Monolithic Memories, Incorporated. Originators of the PAL architecture; now a part of AMD. See* PAL.

MOS *Metal Oxide Semiconductor. One of two basic types of transistor (the other being bipolar).*

MSI *Medium-Scale Integration. An integrated circuit containing between twelve and 100 equivalent gates.*

OTP *One-Time Programmable. A programmable device that can't be erased and reprogrammed.*

PAD *Programmable Address Decoder (TI trademark). A high-speed PLD featuring AND gates only.*

PALA *Programmable Associative Logic Array. An early PLD designed at General Electric and fabricated by MMI.*

PAL *Programmable Array Logic (AMD trademark). A device architecture featuring a programmable AND array and a fixed OR array.*

PALASM *PAL Assembler (AMD trademark). Design software developed by John Birkner for MMI's PAL devices.*

PCE *Polarity Control Element. A fuse-selectable inverter found in EXEL's ERASIC device. See* ERASIC.

PCSD *Programmable Chip Select Decoder (Harris trademark).*

PDS *PAL Design Specification. The design language used in PALASM 2.*

PEEL *Programmable Electrically Erasable Logic (ICT trademark). A family of EEPLD devices offered by ICT.*

PEG *Programmable Event Generator (AMD trademark). A programmable waveform generation device produced by AMD.*

PGA *Pin Grid Array. A square ceramic IC package with a matrix of vertical pins.*

PIC *Programmable Integrated Circuit. See* FPIC.

PLA *Programmable Logic Array. A structure for implementing sum-of-products logic functions. Characterized by a programmable AND array feeding a programmable OR array.*

PLAN *Programmable Logic Analysis by National (National Semiconductor trademark). Proprietary PLD design tool developed by National Semiconductor for their devices.*

PLCC *Plastic Leaded Chip Carrier. A plastic IC package intended for surface-mount applications.*

PLD *Programmable Logic Device. A user-configurable integrated circuit intended for digital logic applications.*

PLDE *Programmable Logic Design Environment (ELDEC trademark). Design language and equation entry tool supplied with the SUSIE simulator.*

PLDS *Programmable Logic Development System (Data I/O trademark). An early PLD programming system that bundled a programmer with a built-in PALASM assembler.*

PLE *Programmable Logic Element (Both Altera and MMI have claimed this trademark). A PROM device that is intended for logic applications. See* PROM.

PLEASM *PLE Assembler (AMD trademark).*

PLICE *Programmable Low Impedance Circuit Element (Actel trademark). A compact, vertical programmable element ("antifuse") that is similar to the AIM technology, but uses a simple dielectric insulator sandwiched between two conducting layers. A programming voltage applied across the insulator results in a permanent connection.*

PLPL *Programming Language for Programmable Logic. Proprietary design language and system developed by AMD.*

PLS *Programmable Logic Sequencer (Signetics trademark). New name for the Signetics registered FPLAs formerly known as IFLs.*

PML *Programmable Macro Logic (Signetics trademark). Encompasses the Signetics folded NAND devices (PLHS501 and PLHS502).*

PMSI *PAL Medium-Scale Integration. Small-scale PALs intended as replacement parts for discrete TTL devices. See* TTL.

PROM *Programmable Read-Only Memory. A programmable IC containing a sum-of-products logic array with a fixed AND array and programmable OR array.*

PROSE *Programmable Sequencer (AMD trademark). A programmable device intended for state machine applications. The first device to combine a PROM and PAL in a single programmable IC.*

PSG *Programmable Sequence Generator (TI trademark). A PLA-based PLD with a built-in counter intended for waveform generation applications.*

RAL *Reconfigurable Array Logic. Lattice's trade name for GAL devices that are preconfigured to emulate specific PLDs. A RAL device is JEDEC file compatible with the equivalent PAL device.*

SAM *Stand-Alone Microsequencer (Altera trademark). A large, erasable, microcode-based sequencer device produced by Altera.*

SPL *Sprague Programmable Logic (Sprague trademark). Sprague's family of PAL-type devices.*

TTL *Transistor-Transistor Logic. The family of bipolar logic devices typified by the 7400-series of fixed-function ICs.*

UCIC *User-Configurable Integrated Circuit. Acronym coined by Altera to describe all programmable devices.*

VAMP *Vertical Avalanche Migration Programming (Signetics trademark). See* AIM.

VHDL *VHSIC Hardware Description Language. Circuit description language developed as a part of the Department of Defense-funded VHSIC program. See* VHSIC.

VHSIC *Very High-Speed Integrated Circuit. A high-speed VLSI research program funded by the Department of Defense. See* VLSI.

VLSI *Very Large Scale Integration. An integrated circuit containing over 3000 equivalent gates.*

VTI *VLSI Technology, Incorporated.*

XACT *Xilinx Advanced CAD Technology (Xilinx trademark). Design software for users of Xilinx LCA devices.*

XNF *External Netlist Format. A common format for conversion of schematic data into the Xilinx LCA design system.*

ZHAL *Zero-power HAL (AMD trademark). A mask-programmed version of MMI's CMOS PAL products.*

Appendix B

Who's Who In The PLD Business

B.1 DEVICE MANUFACTURERS

The following companies offer programmable logic devices:

Actel Corporation
955 E. Arques Ave.
Sunnyvale, CA 94086
(408) 739-1010

Advanced Micro Devices, Incorporated
901 Thompson Place
P.O. Box 3453
Sunnyvale, CA 94088-3453
(408) 732-2400
(800) 538-8450

Altera Corporation
3525 Monroe Street
Santa Clara, CA 95051
(408) 984-2800

Atmel Corporation
2095 Ringwood Avenue
San Jose, CA 95131
(408) 434-9201

Cypress Semiconductor Corporation
3901 North First Street
San Jose, CA 95134
(408) 943-2600

Exel Microelectronics, Incorporated
2150 Commerce Drive
P.O. Box 49038
San Jose, CA 95131-9038
(408) 432-0500

Gazelle Microcircuits, Incorporated
2300 Owen Street
Santa Clara, CA 95054
(408) 982-0900

Gould Semiconductor Corporation
2300 Buckskin Road
Pocatello, ID 83201
(208) 233-4690

Harris Semiconductor Corporation
1301 Woody Burke Road
M/S CB1-25
Melbourne, FL 32901
(407) 724-3739
(800) 442-7747

Hyundai Semiconductor Corporation
2191 Laurelwood Road
Santa Clara, CA 95054
(408) 473-9200

Intel Corporation
1900 Prairie City Road
Folsom, CA 95630
(916) 351-8080
(800) 468-3548

International CMOS Technology, Incorporated
2125 Lundy Avenue
San Jose, CA 95131
(408) 434-0678

Lattice Semiconductor Corporation
P.O. Box 2500
Portland, OR 97208
(503) 681-0118

National Semiconductor Corporation
2900 Semiconductor Drive
Santa Clara, CA 95052
(408) 721-5000

Plessey Semiconductors
1500 Green Hills Road
Scotts Valley, CA 95066
(408) 438-2900

PLX Technology Corporation
625 Clyde Avenue
Mountain View, CA 94043
(415) 960-0448
(800) 759-3753

Samsung Semiconductor, Incorporated
5150 Great America Parkway
Santa Clara, CA 95054
(408) 727-7433

Seeq Technology, Incorporated
1849 Fortune Drive
San Jose, CA 95131
(408) 432-1550
(800) 333-7766

SGS Semiconductor Corporation
1000 East Bell Road
Phoenix, AZ 85022
(602) 867-6100

Signetics Corporation
811 E. Arques Avenue
P.O. Box 3409
Sunnyvale, CA 94088-3409
(408) 991-2000

Texas Instruments, Incorporated
P.O. Box 655303
Dallas, TX 75265
(214) 995-6611

VLSI Technology, Incorporated
1109 McKay Drive
San Jose, CA 95131
(408) 434-3100

Waferscale Integration, Incorporated
47280 Kato Road
Fremont, CA 94538
(415) 656-5400

Xilinx, Incorporated
2100 Logic Drive
San Jose, CA 95124
(408) 559-7778

B.2 PROGRAMMER MANUFACTURERS

The following companies offer device programming equipment:

Adams-MacDonald Enterprises
800 Airport Road
Monterey, CA 93940
(408) 373-3607

Advanced Microcomputer Systems
1321 NW 65th Place
Ft. Lauderdale, FL 33309
(305) 975-9515

Advin Systems, Incorporated
1050-L East Duane Avenue
Sunnyvale, CA 94086
(408) 984-8600
(800) 627-2456

BP Microsystems
10681 Haddington #190
Houston, TX 77043
(713) 461-9430
(800) 225-2102

Bytek Corporation
1021 South Rogers Circle
Boca Raton, FL 33437
(305) 994-3520
(800) 523-1565

Data I/O Corporation
10525 Willows Road NE
P.O. Box 97046
Redmond, WA 98073-9746
(206) 881-6444
(800) 247-5700

Digilec, Incorporated
1602 Lawrence Avenue
Ocean, NJ 07712
(201) 493-2420

Digital Media, Incorporated
11770 Warner Avenue, Suite 225
Fountain Valley, CA 92708
(714) 751-1373

Dynatec International
3594 West 1820 South
Salt Lake City, UT 84104
(801) 973-9500

Eden Engineering
P.O. Box 2200
Grass Valley, CA 95945
(916) 272-2770

Elan Digital Systems, Incorporated
516 Marin Drive
Burlingame, CA 94010
(415) 964-5338

GTEK, Incorporated
399 Highway 90, Drawer 1346
Bay St. Louis, MS 39520
(601) 467-8048

Inlab, Incorporated
2150-I West 6th Avenue
Broomfield, CO 80020
(303) 460-0103

International Microsystems, Incorporated
790 East Arques Avenue
Sunnyvale, CA 94086
(916) 885-7262

Jameco Electronics
1355 Shoreway Road
Belmont, CA 94002
(415) 592-8097

Kontron Electronics
1230 Charleston Road
Mountain View, CA 94039-7230
(415) 965-7020

Logical Devices, Incorporated
1201 NW 65th Place
Ft. Lauderdale, FL 33309
(305) 974-0967
(800) 331-7766

Nicolet Instrument Corporation
215 Fourier Avenue
Fremont, CA 94539
(415) 490-8870

Oliver Advanced Engineering, Incorporated
320 Arden Street
Glendale, CA 91203
(818) 240-0080

Pistohl Electronic Tool Company
22560 Alcalde Road
Cupertino, CA 95014
(408) 255-2422

Programmable Logic Technologies, Incorporated
P.O. Box 1567
Longmont, CO 80501
(303) 772-9059

Qwerty, Incorporated
5346 Bragg Street
San Diego, CA 92122
(619) 455-0500

Retnel Systems
P.O. Box 1348
Lawrence, MA 01842
(508) 683-4659

Royal Electronics
1314 Kilborn Avenue
Ottawa, Ontario K1H 6L3
CANADA
(613) 738-1202

Stag Microsystems, Incorporated
1600 Wyatt Drive
Santa Clara, CA 95054
(408) 988-1118

Storey Systems
3201 North Highway 67, Suite E
Mesquite, TX 75150
(214) 270-4135
(800) 852-2022

Structured Design
333 Cobalt Way, Unit 107
Sunnyvale, CA 94086
(408) 988-0725

Sunrise Electronics, Incorporated
524 South Vermont Avenue
Glendora, CA 91740
(818) 914-1926

Varix
P.O. Box 850605
Richardson, TX 75085
(214) 437-0777

Xeltek
473 Sapena Court, Unit 24
Santa Clara, CA 95054
(408) 727-6995

B.3 PLD DESIGN SOFTWARE VENDORS

The vendors in the following list supply computer-aided engineering (CAE) tools useful for PLD design. Note that, in addition to the companies listed here, many of the device manufacturers previously listed offer design tools for their specific devices. Refer to Chapter 5 for more information about these vendor-specific design tools.

Adams-MacDonald Enterprises
800 Airport Road
Monterey, CA 93940
(408) 373-3607

ALDEC Company
3525 Old Conejo Road, #111
Newbury Park, CA 91320
(805) 499-6867

Anvil Software
427-3 Amherst Street Suite 341
Nashua, NH 03063
(603) 891-1995

Bytek Corporation
1021 South Rogers Circle
Boca Raton, FL 33431
(305) 994-3520

Capilano Computing
P.O. Box 86971
North Vancouver, BC V7L 4P6
CANADA
(604) 669-6343

Dasix Corporation (Daisy Systems)
700 E. Middlefield Road
Mountain View, CA 94043
(415) 960-6593

Data I/O Corporation
10525 Willows Road, NE
Redmond, WA 98073-9746
(206) 881-6444
(800) 247-5700

Hewlett-Packard Company
1820 Embarcadero Road
Palo Alto, CA 94303
(415) 857-1501

Kontron Electronics
1230 Charleston Road
Mountain View, CA 94039-7230
(415) 965-7020

Logical Devices, Incorporated
1201 NW 65th Place
Fort Lauderdale, FL 33309
(305) 974-0967
(800) 331-7766

Minc, Incorporated
1575 York Road
Colorado Springs, CO 80918
(719) 590-1155

Pistohl Electronic Tool Company
22560 Alcade Road
Cupertino, CA 95014
(408) 255-2422

ViewLogic Systems, Incorporated
313 Boston Post Road W.
Marlboro, MA 01752
(508) 480-0881

Appendix C

Supplementary Listings for Chapter 11

```
/* Program 1 - Video Frame Grabber Control Program */
/* Copyright 1990   Michael Holley and Dave Pellerin */

/* Written for MicroSoft C version 5.1 */

#include <conio.h>
#include <stdio.h>

#define FILESIZE 131071L

char    name_buf[45];
```

```
main()
{
    int ch;

    print_menu();
    while(1)
    {
        ch = getch();
        switch(toupper(ch))
        {
            case 'F':   outp(0x316,0);  /*Freeze Picture */
                        break;

            case 'L':   outp(0x314,0);  /*Live Picture */
                        break;

            case 'W':   save_RAM();
                        break;

            case 'R':   load_RAM();
                        break;

            case 'Q':   exit(0);
                        break;

            default:    print_menu();
                        break;
        }
    }
}

print_menu()
{
    printf("\nFreeze video frame grabber\n");
    printf("Press 'L' for live image\n");
    printf("Press 'F' to freeze image\n");
    printf("Press 'W' to write image RAM to disk file\n");
    printf("Press 'R' to read disk file into image RAM\n");
    printf("Press 'Q' to return to DOS\n");
}



/* Copy image in VFG RAM to disk file */
save_RAM()
{
    unsigned    char data;
    long        i;
    FILE        *fp;

    name_buf[0] = '\0';
    printf("\nEnter file name: ");
    scanf("%40s",name_buf);
    if( (fp = fopen(name_buf,"wb") ) == NULL)
    {
        fprintf(stderr,"can't open output file '%s'\n",name_buf);
        return;
    }

    outp(0x313,0);      /*Clear address counter */
    for (i=1L; i < FILESIZE; i++)
```

```
    {
        outp(0x313,0);  /*Next Address */
        outp(0x312,0);  /*Hold Address */
        data = inp(0x300);
        putc(data,fp);
    }
    outp(0x316,0);      /*Freeze Picture */
    close(fp);
}

/* Copy a saved image from disk into VFG RAM for TV display */
load_RAM()
{
    unsigned    char data;
    long        i;
    FILE        *fp;

    name_buf[0] = '\0';
    printf("\nEnter file name: ");
    scanf("%40s",name_buf);
    if( (fp = fopen(name_buf,"rb") ) == NULL)
    {
        fprintf(stderr,"can't open input file '%s'\n",name_buf);
        return;
    }

    outp(0x313,0);      /*Clear address counter */
    for (i=1L; i < FILESIZE; i++)
    {
        outp(0x313,0);  /*Next Address */
        outp(0x312,0);  /*Hold Address */
        data = getc(fp);
        outp(0x300,data);
    }
    outp(0x316,0);      /*Freeze Picture */
    close(fp);
}
```

```
/* Program 2 - Format Video Image for HP LaserJet or EGA video display*/
/* Copyright 1990  Michael Holley and Dave Pellerin */

/* Written for MicroSoft C Version 5.1 */

#include <stdio.h>
#include <conio.h>

#ifdef HPLASER

#define XMAX      280   /* Number of pixel in X direction (width) */
#define YMAX      220   /* Number of pixel in Y direction (lines) */
#define TOP_SKIP   40   /* Scan lines skipped at top of frame */
#define SIDE_SKIP 140   /* Samples skipped at start of scan line */
#define CONTRAST   12

        FILE    *outfile;
        int     outline1[2*XMAX+4];     /* printer line buffer */
        int     outline2[2*XMAX+4];     /* printer line buffer */

#else   /* EGA video display */
```

```
#define XMAX      320  /* Number of pixel in X direction (width) */
#define YMAX      175  /* Number of pixel in Y direction (lines) */
#define TOP_SKIP   60  /* Scan lines skipped at top of frame */
#define SIDE_SKIP 100  /* Samples skipped at start of scan line */
#define CONTRAST    9

#include <graph.h>

#endif

/* Global variables */
        FILE    *infile;

        int     min_gray,max_gray,avg_gray;
        int     BlackValue,DGrayValue,GrayValue,LGrayValue,WhiteValue;
        int     Threshold1,Threshold2,Threshold3,Threshold4;
        int     UpperClip,LowerClip;

unsigned char   buffer[XMAX][YMAX];     /* Lines buffer */
        int     error[XMAX+3];          /* Line error buffer */
        int     linebuf[XMAX+3];        /* Current line buffer */

#define BLACK           0
#define DARKGRAY        1
#define GRAY            2
#define LIGHTGRAY       3
#define WHITE           4


/************************************************/
main(argc,argv)
int     argc;
char *argv[];
{
    int x, y;

    printf("Process video frame grabber disk file\n");

    /* Open data file */
    if(argc > 1)
    {
        printf("Opening input file '%s'\n",argv[1]);
        if( (infile = fopen(argv[1],"rb") ) == NULL)
        {
            fprintf(stderr,"can't open input file '%s'",argv[1]);
            exit(1);
        }
    }
    else
    {
        printf("Opening input file 'temp'\n");
        if( (infile = fopen("temp","rb") ) == NULL)
        {
            fprintf(stderr,"can't open input file '%s'","temp");
            exit(1);
        }
    }

    read_picture();
    close(infile);

    set_thresholds(min_gray,max_gray,avg_gray);
```

```
    init_plot();

    for(x = 0; x <= XMAX; x++)
    {
        error[x] = 0;    /* Clear error for 1st line */
    }

    for(y = 0; y < YMAX; y++)
    {
        fs_filter(y);    /* Process scan line */
        print_line(y);   /* Required for Laser printer */
    }

    done_plot();
    printf("min = %d, max = %d, avg = %d\n",min_gray,max_gray,avg_gray);
    exit(0);
}



/***********************************************/
/* Read picture from disk file into buffer */
read_picture()
{
    int i,y,x;
    unsigned char sample;
    unsigned long total;

    /* Skip first scan lines to center image */
    for(i=0; i < TOP_SKIP; i++)
    {
        find_hsync();
    }

    min_gray = 255;
    max_gray = 0;
    avg_gray = 0;
    total = 0L;

    /* Read scan lines */
    for(y=0; y < YMAX; y++)
    {
        for(x=0; x < XMAX; x++)
        {
            sample = getc(infile);
            buffer[x][y] = sample;
            total += sample;
            if(sample > max_gray)
            {
                max_gray = sample;
            }
            if(sample < min_gray)
            {
                min_gray = sample;
            }
        }
        find_hsync();    /* Find start of next scan line */
    }
    avg_gray = total / (YMAX * XMAX);
}
```

```
/*********************************************/
/* Find Horizontal sync pulse */
find_hsync()
{
    int sync_cnt;
    unsigned char sample;

    sync_cnt = 0;
    /* Look for 20 samples of zero (synclevel) */
    while(sync_cnt < 20)
    {
        sync_cnt++;
        sample = getc(infile);
        if(sample != 0)
        {
            sync_cnt = 0;
        }
    }
    /* Skip color burst samples */
    while(sync_cnt < SIDE_SKIP)
    {
        sample = getc(infile);
        sync_cnt++;
    }
}
/*********************************************/
set_thresholds(min,max,avg)
int     min,max,avg;
{
    /* Setting the Black and White values between the min and max
     * by about 10% give the image more contrast.  There will be
     * more areas of all black and all white.
     * Dividing (max-min) by values of 7 to 12 give good results.
     */

    int clip;

    clip = (max-min) / CONTRAST;

    BlackValue = min + clip;
    WhiteValue = max - clip;

    /* All the other values are scaled from Black and White */
    GrayValue  = (BlackValue + WhiteValue) / 2;
    DGrayValue = (BlackValue + GrayValue)  / 2;
    LGrayValue = (GrayValue + WhiteValue)  / 2;

    Threshold1 = (BlackValue + DGrayValue) / 2;
    Threshold2 = (DGrayValue + GrayValue)  / 2;
    Threshold3 = (GrayValue  + LGrayValue) / 2;
    Threshold4 = (LGrayValue + WhiteValue) / 2;

    LowerClip  = BlackValue;
    UpperClip  = WhiteValue;
}


/*********************************************/
/* Floyd-Steinberg error-diffusion filter is a method to smooth
 * transitions when reducing the number of gray scale levels.
 *
```

```
 * The Floyd and Steinberg filter divides the gray level for a pixel
 * by 16 and distributes it to the adjoining pixel in the ratios below:
 *
 *         *  7
 *      3  5  1
 *
 * Floyd, R. W. and L. Steinberg, "An Adaptive Algorithm for Spatial
 * Gray Scale.", SID International Symposium Digest of Technical
 * Papers, vol 1975m, pp. 36-37.
 *
 */
fs_filter(y)
int   y;
{
    unsigned char data;
    int x;
    int error_val,err;

    /* Clip data to increase areas of all white and all black */
    for(x = 0; x < XMAX; x++)
    {
        data = buffer[x][y] - error[x+1];

        /* Clip data between LowerClip and UpperClip */
        data = (data > LowerClip) ? data : LowerClip;
        data = (data < UpperClip) ? data : UpperClip;

        linebuf[x+1] = data;
        error[x+1] = 0;
    }

    /* Convert 256 levels of gray to 5 levels of gray */
    for(x = 0; x < XMAX; x++)
    {
        data = linebuf[x+1];
        if(data < Threshold1)
        {
            plot_gray(x,y,BLACK);
            error_val = BlackValue - data;
        }
        else if(data < Threshold2)
        {
            plot_gray(x,y,DARKGRAY);
            error_val = DGrayValue - data;
        }
        else if(data < Threshold3)
        {
            plot_gray(x,y,GRAY);
            error_val = GrayValue - data;
        }
        else if(data < Threshold4)
        {
            plot_gray(x,y,LIGHTGRAY);
            error_val = LGrayValue - data;
        }


        else
        {
            plot_gray(x,y,WHITE);
            error_val = WhiteValue - data;
        }
```

```
        /* Filter data to smooth transitions */
        err = error_val / 16;
        error[x]      += err * 3;        /* Down and left */
        error[x+1]    += err * 5;        /* Down */
        error[x+2]    += err;            /* Down and right */
        linebuf[x+1]  += err * 7;        /* Right */
    }
}


/**********************************************/
plot_gray(x,y,level)
int     x,y,level;
{
    /* Simple gray scale pattern alternates by a random number
     * to reduce the 'rivers' of white or black.
     *
     *          Black   D-Gray  Gray    L-Gray  White
     * dither 0  --      --      -#      ##      ##
     *           --      #-      #-      -#      ##
     *
     * dither 1  --      --      #-      ##      ##
     *           --      -#      -#      #-      ##
     *
     * dither 2  --      #-      #-      -#      ##
     *           --      --      -#      ##      ##
     *
     * dither 3  --      -#      -#      #-      ##
     *           --      --      #-      ##      ##
     */

    int dither,xvar,yvar;

    yvar = 2 * y;
    xvar = 2 * x;

    dither = rand() % 4;

    white_dot(xvar,0, yvar,0);
    white_dot(xvar,1, yvar,1);
    white_dot(xvar,0, yvar,1);
    white_dot(xvar,1, yvar,0);

    switch(level)
    {
        case BLACK:
                black_dot(xvar,0, yvar,0);
                black_dot(xvar,1, yvar,1);
                black_dot(xvar,0, yvar,1);
                black_dot(xvar,1, yvar,0);
                break;

        case DARKGRAY:
                switch(dither)
                {
                    case 0:     black_dot(xvar,0, yvar,0);
                                black_dot(xvar,1, yvar,0);
                                black_dot(xvar,1, yvar,1);
                                break;
```

```
                case 1:     black_dot(xvar,0, yvar,0);
                            black_dot(xvar,1, yvar,0);
                            black_dot(xvar,0, yvar,1);
                            break;

                case 2:     black_dot(xvar,1, yvar,0);
                            black_dot(xvar,0, yvar,1);
                            black_dot(xvar,1, yvar,1);
                            break;
                case 3:     black_dot(xvar,0, yvar,0);
                            black_dot(xvar,0, yvar,1);
                            black_dot(xvar,1, yvar,1);
                            break;

                default:    break;
            }
            break;

        case GRAY:
            switch(dither)
            {
                case 0:
                case 3:     black_dot(xvar,1, yvar,0);
                            black_dot(xvar,0, yvar,1);
                            break;

                case 1:
                case 2:     black_dot(xvar,0, yvar,0);
                            black_dot(xvar,1, yvar,1);
                            break;

                default:    break;
            }
            break;

        case LIGHTGRAY:
            switch(dither)
            {
                case 0:     black_dot(xvar,0, yvar,1);
                            break;

                case 1:     black_dot(xvar,1, yvar,1);
                            break;

                case 2:     black_dot(xvar,0, yvar,0);
                            break;

                case 3:     black_dot(xvar,1, yvar,0);
                            break;

                default:    break;
            }
            break;

        case WHITE:
        default:
            break;
    }
}
```

```
/**********************************************/
/* Initialize plot */
init_plot()
{
#ifdef HPLASER
    /* Open output file and set raster resolution */
    printf("Opening output file 'prtfile'\n");
    if( (outfile = fopen("prtfile","wb") ) == NULL)
    {
        fprintf(stderr,"can't open output file '%s'","prtfile");
        exit(1);
    }

    fprintf(outfile,"%c*t75R",27);      /* 75 pixels per inch */
    fprintf(outfile,"%c*rA",27);        /* Start raster graphics */

#else   /* EGA video display */
    /* Set graphics mode and fill screen to all white */
    _setvideomode(_ERESCOLOR);
    _rectangle(_GFILLINTERIOR, 0, 0, 2*XMAX, 2*YMAX );
    _setcolor(0);                       /* Plot with black dots */

#endif
}

/**********************************************/
/* Terminate plot */
done_plot()
{
#ifdef HPLASER
    fprintf(outfile,"%c*rB",27);        /* End raster graphics */
    close(outfile);

#else   /* EGA video display */
    while(!kbhit())                     /* Wait for keypress to exit */
    { }
    getch();
    _setvideomode(_DEFAULTMODE);        /* Return screen to normal mode */

#endif
}


/**********************************************/
/* Display or print a single pixel */
black_dot(xpos,xadd,ypos,yadd)
int xpos,xadd,ypos,yadd;
{
#ifdef HPLASER
    if(yadd == 0)
    {
        outline1[xpos+xadd] = 1;
    }
    else
    {
        outline2[xpos+xadd] = 1;
    }
#else
    _setpixel(xpos+xadd,ypos+yadd);
#endif

}
```

```
/***********************************************/
/* Display or print a single pixel */
white_dot(xpos,xadd,ypos,yadd)
int xpos,xadd,ypos,yadd;
{
#ifdef HPLASER
    if(yadd == 0)
    {
        outline1[xpos+xadd] = 0;
    }
    else
    {
        outline2[xpos+xadd] = 0;
    }
#else
    /* Nothing required for video */
#endif
}


/***********************************************/
print_line(y)
int   y;
{
#ifdef HPLASER
int     i,x,width,word;

    /* The sequence   ESC * B # W <raster data>   prepares the printer  */
    /* to receive # number of bytes of graphics data.                   */

    /* Each video scan line is printed as 2 or 3 lines on the laser     */
    /* printer to give the correct aspect ratio.                        */

    width = XMAX/4;

    /* Print first line */
    fprintf(outfile,"%c*b%dW",27,width);        /* Transfer graphics data */
    for(i = 0, x = 0; i < width; i++)
    {
        word  = outline1[x++] << 7;
        word += outline1[x++] << 6;
        word += outline1[x++] << 5;
        word += outline1[x++] << 4;
        word += outline1[x++] << 3;
        word += outline1[x++] << 2;
        word += outline1[x++] << 1;
        word += outline1[x++];

        fprintf(outfile,"%c",word);
    }

    /* Print second line */
    fprintf(outfile,"%c*b%dW",27,width);        /* Transfer graphics data */
    for(i = 0, x = 0; i < width; i++)
    {
        word  = outline2[x++] << 7;
        word += outline2[x++] << 6;
        word += outline2[x++] << 5;
        word += outline2[x++] << 4;
        word += outline2[x++] << 3;
        word += outline2[x++] << 2;
```

```
            word += outline2[x++] << 1;
            word += outline2[x++];

            fprintf(outfile,"%c",word);
        }

        /* Every third line, print an extra line to correct aspect ratio */
        if(y%3 == 0)
        {
            fprintf(outfile,"%c*b%dW",27,width);    /* Transfer graphics data */
            for(i = 0, x = 0; i < width; i++)
            {
                word  = outline1[x++] << 7;
                word += outline1[x++] << 6;
                word += outline1[x++] << 5;
                word += outline1[x++] << 4;
                word += outline1[x++] << 3;
                word += outline1[x++] << 2;
                word += outline1[x++] << 1;
                word += outline1[x++];

                fprintf(outfile,"%c",word);
            }
        }

#else   /* EGA video display */
    /* Nothing required for video */
#endif
}
```

```
 13-Jun-90 08:27 PM     address.doc           DASH-GATES Rev 4.11b

      Form type:  title          Form name:  Title

GATES  Rev      4.11b

Design name:  address
Date Last Saved:  13-Jun-90 08:10 PM          By GATES Rev  4.11b

Design Description:
Address counter for video frame grabber.

This counter reaches the terminal count of 119437 in 1/59.94 of a
second with a 7,159,090 clock. This allows one field of NTSC video
to be stored in 128K bytes of RAM.

There are two control inputs, a asynchronous reset named AdrClr, and
synchronous count/hold named Inc. In addition to the counter outputs,
A16..A0, a terminal count output is proviced, ATC.


      Form type:  declaration    Form name:  address

 Name                                  Type        Definition/Comment
 A17..A0                               bidir       t
 Clk,CntClr                            input
 Address                               set         [A17..A0]
 ATC                                   output      d
 VTC                                   output      d
" Terminal Count = Clock / 59.94 = 8.0 MHz / 59.94
 Terminal_Count                        macro       131071
 Sync_Count                            macro       127


      Form type:  equation       Form name:  address

 ATC.d = (Address == Terminal_Count)

 VTC.d = (Address == Sync_Count)

 Address.t   = (Address.q + 1)
             $  Address.q

 Address.ar = CntClr

 Address.clk = Clk

 ATC.clk = Clk

 VTC.clk = Clk
```

```
 13-Jun-90 08:27 PM     address.doc           DASH-GATES Rev 4.11b

      Form type:  factor         Form name:  Factor

 Group: ad          Def: VTC.d,ATC.d,A17.t,A16.t,A15.t,A14.t,A13.t,
                         A12.t,A11.t,A10.t,A9.t,A8.t,A7.t,A6.t,A5.t,
                         A4.t,A3.t,A2.t,A1.t,A0.t
```

```
                                   Factored Output
 Group: ad            AND Min: 3   AND Max: 3   OR Min: 3   OR Max: 3
                      Stage Count: 99  Factor Target: g  Polarity: +

Stages  Factored Equations:
[1]     A0.t = 1
[1]     A1.t = A0.q
[1]     A2.t = A1.q & A0.q
[1]     A3.t = ad@0
[2]     A4.t = ad@0 & A3.q
[2]     A5.t = ad@1
[3]     A6.t = ad@1 & A5.q
[3]     A7.t = ad@2
[4]     A8.t = ad@2 & A7.q
[4]     A9.t = ad@3
[5]     A10.t = ad@3 & A9.q
[5]     A11.t = ad@4
[6]     A12.t = ad@4 & A11.q
[6]     A13.t = ad@5
[7]     A14.t = ad@5 & A13.q
[7]     A15.t = ad@6
[8]     A16.t = ad@6 & A15.q
[8]     A17.t = ad@6 & A16.q & A15.q
[5]     ATC.d = ad@14 & ad@10
[5]     VTC.d = ad@19 & ad@15

Stages  Intermediate Equations:
[1]     ad@0 = A2.q & A1.q & A0.q
[2]     ad@1 = ad@0 & A4.q & A3.q
[3]     ad@2 = ad@1 & A6.q & A5.q
[4]     ad@3 = ad@2 & A8.q & A7.q
[5]     ad@4 = ad@3 & A10.q & A9.q
[6]     ad@5 = ad@4 & A12.q & A11.q
[7]     ad@6 = ad@5 & A14.q & A13.q
[1]     ad@7 = A0 & A1 & A2
[2]     ad@8 = ad@7 & A3 & A4
[3]     ad@9 = ad@8 & A5 & A6
[4]     ad@10 = ad@9 & A7 & A8
[1]     ad@11 = A9 & A10 & A11
[1]     ad@12 = A12 & A13 & A14
[1]     ad@13 = A15 & A16 & !A17
[2]     ad@14 = ad@13 & ad@12 & ad@11
[4]     ad@15 = ad@9 & !A7 & !A8
[1]     ad@16 = !A9 & !A10 & !A11
[1]     ad@17 = !A12 & !A13 & !A14
[1]     ad@18 = !A15 & !A16 & !A17
[2]     ad@19 = ad@18 & ad@17 & ad@16

Output Statistics:
```

```
13-Jun-90 08:27 PM     address.doc            DASH-GATES Rev 4.11b

  Products: Reduced 188, Factored 83
  Sums:     Reduced 0, Factored 0


     Form type:  simulation     Form name:  address
```

```
    Clk CntClr | Address ATC VTC
    ________________________________
1   ~c  1        0       0   0
2   ~c  0        1       0   0
3   ~c  0        2       0   0
4   ~c  0        3       0   0
5   ~c  0        4       0   0
6   ~c  0        5       0   0
7   ~c  0        6       0   0
8   ~c  0        7       0   0
9   ~c  0        8       0   0
10  ~p  0        125     0   0
11  ~c  0        126     0   0
12  ~c  0        127     0   0
13  ~c  0        128     0   1
14  ~c  0        129     0   0
15  ~p  0        131069  0   0
16  ~c  0        131070  0   0
17  ~c  0        131071  0   0
18  ~c  0        131072  1   0
19  ~c  1        0       0   0
20  ~c  0        1       0   0
21  ~c  0        2       0   0

  Input signals                      width   orientation
 Clk                                   3         h
 CntClr                                6         h

  Output signals                     width   orientation
 Address                               7         h
 ATC                                   3         h
 VTC                                   3         h

  Run Mode: until-error           Equations: reduced
```

```
 22-Jun-89 08:42 AM    control.doc         DASH-GATES Rev 4.11b


      Form type:  declaration    Form name:  control

Name                                Type       Definition/Comment
Clk                                 input      "System clock
Freeze                              input      "Freeze Frame Signal
VTC                                 input
RAMOE,RAMWE                         bidir      comb
DACLK,ADCLK                         bidir      comb
AD7..AD0                            input      "Video AD signal
AD                                  set        [AD7..AD0]
CntClr                              output     d
SyncLevel                           bidir      comb
TSE                                 bidir      comb
ATC                                 input
S1..S0                              bidir      d
statereg                            set        [S1..S0]

 22-Jun-89 08:42 AM    control.doc         DASH-GATES Rev 4.11b


      Form type:  equation       Form name:  vertsync
```

```
SyncLevel = (AD == [0,0,0,0,~x,~x,~x,~x])

statereg.clk = Clk

statereg.ar = !Freeze

TSE = !RAMOE

ADCLK = !Clk

DACLK = !Clk

CntClr.clk = Clk
```

```
  22-Jun-89 08:42 AM    control.doc          DASH-GATES Rev 4.11b
       Form type:  state           Form name:  control
          State Register Set Name: statereg
       State Name:                         State Value:
       Live                                0
       Scan                                1
       Record                              3
       PlayBack                            2
State: Live
  Equations:
    : RAMOE = 1
    : RAMWE = 1
    : CntClr.d = 1
  Transitions:           Type: dependent
    Case: default
          Goto: Scan
State: Scan
  Equations:
    : RAMOE = 1
    : RAMWE = !Clk
    : CntClr.d = !SyncLevel
  Transitions:           Type: dependent
    Case: VTC
          Goto: Record
    Case: default
          Goto: Scan
State: Record
  Equations:
    : RAMOE = 1
    : RAMWE = !Clk
    : CntClr.d = ATC
  Transitions:           Type: dependent
    Case: CntClr.q
          Goto: PlayBack
    Case: default
          Goto: Record
State: PlayBack
  Equations:
    : RAMOE = 0
    : RAMWE = 1
    : CntClr.d = ATC
  Transitions:           Type: dependent
    Case: default
          Goto: PlayBack
```

22-Jun-89 08:42 AM control.doc DASH-GATES Rev 4.11b

Form type: simulation Form name: control

	Clk	Freeze	VTC	ATC	AD	statereg	CntClr	RAMOE	RAMWE
1	~c	0	0	0	0	Live	1	1	1
2	~c	0	0	0	0	Live	1	1	1
3	~c	1	0	0	0	Scan	1	1	1
4	~c	1	0	0	0	Scan	0	1	1
5	~c	1	0	0	0	Scan	0	1	1
6	~c	1	0	0	0	Scan	0	1	1
7	~c	1	0	0	100	Scan	1	1	1
8	~c	1	0	0	0	Scan	0	1	1
9	~c	1	0	0	0	Scan	0	1	1
10	~c	1	1	0	0	Record	0	1	1
11	~c	1	0	0	100	Record	0	1	1
12	0	1	0	0	0	Record	0	1	1
13	1	1	0	0	200	Record	0	1	0
14	0	1	0	0	0	Record	0	1	1
15	~c	1	0	0	50	Record	0	1	1
16	~c	1	0	1	0	Record	1	1	1
17	~c	1	0	0	0	PlayBack	0	0	1
18	~c	1	0	0	0	PlayBack	0	0	1
19	~c	1	0	1	0	PlayBack	1	0	1
20	~c	1	0	0	0	PlayBack	0	0	1
21	~c	1	0	0	0	PlayBack	0	0	1
22	~c	0	0	0	0	Live	1	1	1
23	~c	0	0	0	0	Live	1	1	1
24	~c	0	0	0	0	Live	1	1	1
25	~c	1	0	0	0	Scan	1	1	1
26	~c	1	0	0	0	Scan	0	1	1
27	~c	1	0	0	0	Scan	0	1	1

Input signals	width	orientation
Clk	3	h
Freeze	6	h
VTC	3	h
ATC	3	h
AD	3	h

Output signals	width	orientation
statereg	8	h
CntClr	6	h
RAMOE	3	v
RAMWE	3	v

Run Mode: until-error Equations: reduced

Index

INDEX OF REFERENCED DEVICES

GENERAL INDEX

A

B

E

F

G

H

N

O

P

Q

R

S

T

U

V

W

X

Y

Z